MATERIALS SCIENCE AND TECHNOLOGIES

DOPING

PROPERTIES, MECHANISMS AND APPLICATIONS

MATERIALS SCIENCE AND TECHNOLOGIES

Additional books in this series can be found on Nova's website under the Series tab.

Additional e-books in this series can be found on Nova's website under the e-book tab.

PHYSICS RESEARCH AND TECHNOLOGY

Additional books in this series can be found on Nova's website under the Series tab.

Additional e-books in this series can be found on Nova's website under the e-book tab.

MATERIALS SCIENCE AND TECHNOLOGIES

DOPING

PROPERTIES, MECHANISMS AND APPLICATIONS

LIXIN YU
EDITOR

nova publishers
New York

For permission to use material from this book please contact us:
Telephone 631-231-7269; Fax 631-231-8175
Web Site: http://www.novapublishers.com

Library of Congress Cataloging-in-Publication Data

Doping : properties, mechanisms and applications / editor, Lixin Yu.
pages cm
Includes bibliographical references and index.
ISBN 978-1-62618-097-0 (hardcover)
1. Semiconductor doping. I. Yu, Lixin, 1972-
TK7871.85.D665 2013
621.3815'2--dc23
2013004719

Published by Nova Science Publishers, Inc. † New York

CONTENTS

PREFACE

As a powerful and efficiently method for altering and controlling the properties and performance of materials, doping can enable properties that are innately present in a materials. This book presents the overall survey of recent advances of organic electric materials and magnetism materials in doped materials and review recent results doped naonocrystals and bulk materials.

Doped bulk materials have been widely applied in many fields, such as optical display, communication, constructive materials, laser crystals, biological systems, etc. Doped bulk materials have present excellent performance and been applied. This book presents latest results regarding the growth, characterization and magnetic properties of bulk samples (composite fibers) of the optimally doped $La_{1-x}Ce_xMnO_3$ manganites by using the Laser Heated Pedestal Growth technique etc. As for inorganic semiconductors, doping has proven to be one of the most efficient techniques to adjust the electronic properties of the organic semiconductors. Doping of organic materials often causes significant changes in these materials and it is of both fundamental and practical interest. This book presents the works about alkali metal doping-induced electronic structure modification in several molecular organic semiconductors and two-stage model to describe the evolution of the energy levels for the doped-organic molecules. In addition, organic electronic materials have received widespread attention due to their potential application in micro- and opto-electronic devices, such as organic light emitting diodes (OLEDs), solar cells, thin film transistors, etc. Among them OLEDs is the most important device for its potential application in display technology. Tris (8-hydroxyquinolinato) aluminium (Alq_3) is presented in this book because it is an important organic material for the green light emitter OLED device. The efficient preparation technique for doping also introduced, such as ion implantation. Doped Nanomaterials, will present unique physical and chemical characteristics in contrast with micrometer materials due to size and quantum confinement effect. The doped nano materials have many potential applications in display, optical communications, sensors, biological probe and label, etc to the demands for further miniaturization of components and systems. This book also presents the recent advances in the doped transition metal oxide nanomaterials on the energy application. Rare earth-doped insulator nanomaterials represent a new type of high-efficiency luminescent materials. Because of their large surface to volume rations, insulator nanomaterials exhibit some novel properties that make them suitable for numerous useful applications, such as phosphors, biological labeling. In this book, rare earth doped nanomaterials in the renew resource application are also reviewed.

Chapter 1 - The development of electronic devices based on molecular organic semiconductors, such as organic light emitting diodes (OLEDs), organic thin film transistors (OTFTs), and organic photovoltaic cells (OPVs), has been intensively explored in the last two decades. To employ successfully organic molecules as the semiconducting materials in the devices, it is essential to be able to modify the electronic structures of these molecules in a controlled way. As for inorganic semiconductors, doping has proven to be one of the most efficient techniques to adjust the electronic properties of the organic semiconductors. Doping of organic materials often causes significant changes in these materials and it is of both fundamental and practical interest. However, due to the localization of the electronic wavefunctions and the resulting strong correlation effects between the electrons and the nuclei, doping theories developed in inorganic semiconductors may not be applied without careful justification. It therefore remains a fundamental and important issue as how the energy level shift depends on the doping density in organic semiconductors. In particular, it is important to investigate a guest-host system whose properties are otherwise well studied, in order to reveal fundamental patterns and rules governing the electronic structure of doping in organic semiconductors. In this context, alkali metals are good candidates for prototypical studies due to their simple electronic structures and low electron affinity. In this chapter, the authors will first review the works about alkali metal doping-induced electronic structure modification in several molecular organic semiconductors. Based on experimental studies, the authors proposed a two-stage model to describe the evolution of the energy levels for the doped-organic molecules. Density functional theory (DFT) calculation will be employed to explain the experimental results and the proposed model. Further, the reversal of the energy level shift induced by metal-deposition or oxygen exposure for the doped-organic molecules will be discussed. Finally, as a particularly interesting case, the authors will discuss the role of thin LiF insertion layer in metal/ tris-(8-hydroxyquinoline) aluminum (Alq) interface, which has been suggested to be a critical factor that determines the device performance.

Chapter 2 - In this Chapter the authors review some of the authors' latest results regarding the growth, characterization and magnetic properties of bulk samples (composite fibers) of the optimally doped $La_{1-x}Ce_xMnO_3$ manganites by using the Laser Heated Pedestal Growth technique. It has been widely recognized that, in comparison with the hole doped manganite (like, e.g., $La_{1-x}Ca_xMnO_3$), its electron doped counterpart $La_{1-x}Ce_xMnO_3$ is a metastable phase. That is why, before optimizing the growth conditions, the authors were able to grow polycrystalline multiphase fibers. Their composition and granularity were verified through Scanning Electron Microcopy with Energy Dispersive X-ray analysis. The magnetic behavior of the fibers was probed by AC susceptibility and magnetization measurements as a function of temperature. Several ferromagnetic transitions were identified. In addition to a weak transition at T_{C0}=45K (due to MnO_2 impurities), three more transitions with Curie temperatures T_{C1}=126K, T_{C2}=180K, and $T_{C,max}$=300K were observed and linked to fiber regions with cerium concentration of x=0.063, x=0.13, and x=0.3,respectively.

Chapter 3 - The transition metal oxide (TMO) nano-structures with the insulators have created new avenues for the applied research in future energy devices. In particular, manganese oxide (MnO_x) has been considered as the most promising TMO for next generation energy applications like supercapacitors and batteries due to its low cost, environmental friendliness and natural abundance along with its wide structural diversity combined with its unique chemical and physical properties. Furthermore, this attractive functional metal oxide exhibits different polymorphs depending on the arrangement of the basic MnO_6 structure. Nevertheless, the electrochemical performance of MnO_x depended on their morphology, porosity, specific surface area, electrical conductivity and ionic transport within the pore.

Herein, manganese oxide (MnO_2) nanostructures were prepared with chemical techniques using the precursors containing K^+ and NH_4^+. Nanorods with an α - MnO_2 structure and a diameter ranging from 25 to 40 nm, along with tipped needles with a β- MnO_2 structure and a diameter of 100 nm were obtained. Addition of NH_4^+ has increased the diameter of the nanorods obtained. The 25 nm diameter α MnO_2 nanorods showed the best catalytic activity for dissociation of HO_2^- formed during oxygen reduction in a KOH solution. The MnO_2 nanostructures preferably followed two-electron oxygen reduction mechanism in a LiOH solution. Potassium used as stabilizing agent enhanced the electrocatalytic properties of the manganese oxide for the oxygen reduction reaction in aqueous as well as in non-aqueous medium. The supercapacitive behavior shown by the NH_4^+ added MnO_2 is higher. For charge storage application NH_4^+ added MnO_2 has shown greater efficiency while K^+ added MnO_2 had possessed higher efficiency for catalytic application.

Chapter 4 - In this section for ion implantation as one of the most famous doping techniques, materials in both theoretical and experimental aspects are provided. For the former aspects, basic equipment anatomy by the major parts of the ion implanter, tools related with ion implantation other than the implanter, and the applications of ion implantation are handled. For the latter, experimental sources are provided and the authors are made to have closer look into the physics of the ion implantation and criteria for process control. The experimental resources have been obtained from the periodic semiconductor process educations held for graduate students and field engineers at the Inter-university Semiconductor Research Center (ISRC) of Seoul National University, Seoul, Korea which has been conducted more than 10 years and still going on. The ion implanter and the other process tools installed in the ISRC have been used throughout the chapter.

Chapter 5 - Stable social development requires novel approaches for energy production, distribution and storage combined with reasonable restrictions of the environmental impact. The fuel cell-based technologies, as well as the separation of gases from mixtures, particularly implemented into innovative power plant concepts, are therefore in the scientific and political focus.

At the very core of these technologies are selected materials from a number of classes ranging from purely organic-based matrices to the precious metals and alloys that all ought to meet well defined operation-relevant conditions and efficiencies.

Depending on the targeted application, each material foreseen as a base for component development has to be evaluated individually and its properties have to be modified in terms of corresponding application requirements. In order to elaborate tailored ceramic materials for gas separation and ion/electron transport at the relevant operating conditions and stability ranges, improved electrical or ionic conductivities and permeation rates are required. That can be achieved by doping and substitution which are actors on a nano-scale that usually lead to macroscopic impacts.

This chapter is dedicated to the fascinating world of tailoring ceramic materials for energy and environmental applications. Selected approaches to tune ceramics will be discussed to illustrate the versatile effects that compositional variation can have on the macroscopic properties, e.g. the conductivity of protons, oxygen ions and/or electronic carriers, stability, etc. The present chapter will therefore consider the structural features of selected material classes, as well as the principles of transport in bulk and microporous solids. It will furthermore illustrate and discuss the effects of selected additives and substituents on sinterability, electrical/electrochemical properties and stability of selected ceramic materials for energy and environmental applications. The material variety will cover ceramic materials with different crystal structures like fluorites, perovskites, pyrochlores, fergusonites, as well as selected zeolite structures.

Chapter 6 - External strain has been proposed to be an effective parameter to control vacancy, interstitial dopants and substitutional dopants in various semiconductor systems. In this paper, some recent progresses in tuning vacancy formation and doping by external strain are reviewed. Vacancies or dopants formation in semiconductors induce volume changes. Vacancies usually induce volume contraction in covalent semiconductors. Interstitial dopants usually induce volume expansion. Substitutional dopants induce volume change based on their atomic size and electronic environment. The strain tuned vacancy or dopant formation is based on the volume change. If the external strain is in the same direction as the volume change, the formation energies of the vacancy or dopant are reduced. Otherwise, the formation energy is increased. This effect can also be used to tune doping sites in a host. The external strain can be a very useful approach to tune the semiconductor defects and dopants and control their device properties.

Chapter 7 - Organic electronic materials have received widespread attention due to their potential application in micro- and opto-electronic devices, such as organic light emitting diodes (OLEDs), solar cells, thin film transistors, etc. Among them OLEDs is the most important device for its potential application in display technology. Tris (8-hydroxyquinolinato) aluminium (Alq_3) is an important organic material for the green light emitter OLED device. The light emission colours can be tuned in Alq_3 material for full-colour display by organic fluorescent dyes doping technique. A small amount of dopant can dramatically change the emission wavelength and emission intensity. However, the carrier mobility, the transient photoluminescence (PL), and electroluminescence (EL) of excitons are affected by the interaction between the dopants-host molecules in the film due to the appearance of local states in the vicinity of the dopants. The exciton recombination rate, diffusion coefficient and the diffusion length were estimated from the PL decay lifetime measurements and the exciton kinetic equation in doped films. The EL onset was delayed before achieving the saturation in doped devices. Carrier drift mobility in organic doped films is affected by the interaction between the dopants and surrounding host molecules. The temperature and electric field dependence charge carrier mobility was also studied in pure and doped Alq_3 films using Time-of-flight measurements technique. The co-dopants into Alq_3 improved the carrier mobility compared to the single dye-dope organic Alq_3 films. The carrier mobility in the doped Alq_3 films does not follow the Poole-Frenkel (PF) model closely. Instead, the mobility is in agreement with the PF model at two different ranges of electric fields separated by a critical field. The transient current-voltage measurements are modelled by considering the de-trapping of carriers from a discrete trap level to a Gaussian distribution model of carriers transport states. The trap depths in pure and doped films are also estimated.

Chapter 8 - Deals with rare earth elements and their doping features. A particular attention is paid to nano-scale effects, mainly Urbach tailing features in the particular case of Ytterbium along with the supply to renewable energy-related nano-devices materials.

In: Doping
Editor: Lixin Yu

ISBN: 978-1-62618-097-0

Chapter 1

DOPING OF ORGANIC ELECTRONIC MATERIALS

Ashraf Uddin[*]
School of Photovoltaic and Renewable Energy Engineering
University of New South Wales (UNSW), Sydney, Australia

ABSTRACT

Organic electronic materials have received widespread attention due to their potential application in micro- and opto-electronic devices, such as organic light emitting diodes (OLEDs), solar cells, thin film transistors, etc. Among them OLEDs is the most important device for its potential application in display technology. Tris (8-hydroxyquinolinato) aluminium (Alq_3) is an important organic material for the green light emitter OLED device. The light emission colours can be tuned in Alq_3 material for full-colour display by organic fluorescent dyes doping technique. A small amount of dopant can dramatically change the emission wavelength and emission intensity. However, the carrier mobility, the transient photoluminescence (PL), and electroluminescence (EL) of excitons are affected by the interaction between the dopants-host molecules in the film due to the appearance of local states in the vicinity of the dopants. The exciton recombination rate, diffusion coefficient and the diffusion length were estimated from the PL decay lifetime measurements and the exciton kinetic equation in doped films. The EL onset was delayed before achieving the saturation in doped devices. Carrier drift mobility in organic doped films is affected by the interaction between the dopants and surrounding host molecules. The temperature and electric field dependence charge carrier mobility was also studied in pure and doped Alq_3 films using Time-of-flight measurements technique. The co-dopants into Alq_3 improved the carrier mobility compared to the single dye-dope organic Alq_3 films. The carrier mobility in the doped Alq_3 films does not follow the Poole-Frenkel (PF) model closely. Instead, the mobility is in agreement with the PF model at two different ranges of electric fields separated by a critical field. The transient current-voltage measurements are modelled by considering the de-trapping of carriers from a discrete trap level to a Gaussian

[*] Corresponding author: Ashraf Uddin. School of Photovoltaic and Renewable Energy Engineering, University of New South Wales (UNSW), Sydney 2052, Australia. Email: a. uddin@unsw.edu.au.

distribution model of carriers transport states. The trap depths in pure and doped films are also estimated.

Keywords: Organic semiconductor, Doping, OLED, Alq_3, carrier mobility

1. Introduction

Organic electronic materials have attracted much interest due to their potential application in micro- and opto-electronic devices, such as organic light emitting diodes (OLEDs), solar cells, thin film transistors, etc. [1-4]. Among them OLEDs is the most important device for its potential application in display technology. Tris (8-hydroxyquinolinato) aluminum (Alq_3) is an important organic material for the green light emitter OLED device [5, 6]. The light emission colors can be tuned in Alq_3 material by organic fluorescent dyes doping technique [6]. Generally, the active layer Alq_3 is doped with a suitable dye to produce full-color OLEDs in order to increase the spectral region of the emission, to provide good charge transporter and reduce unwanted self-absorption effects. In this doping system, the exciton form in the host material is transfered nonradiatively to the guest materials through dipole–dipole coupling of donor (Alq_3) and acceptor (dopant) molecules. The Alq_3 appears to be one of the most suitable, efficient emitter material with a photoluminescence of 25–30%, and as a host for some laser dye–organic materials, such as red light emitter 4-(dicyanomethylene)-2-methyl-6-(pdimethylaminostyryl)-4H-pyran (DCM) and yellow light emitter (5, 6, 11, 12-tetraphenylnaphthacene (rubrene) and 3-(2'-benzothiazolyl)-7-diethylaminocoumarin (coumarin-6). A small amount of dopant can dramatically change the emission wavelength and emission intensity [7-9]. The lowest unoccupied molecular orbital (LUMO) and the highest occupied molecular orbital (HOMO) of Alq_3 is at ~ 3.0 and ~ 5.7 eV, respectively with respect to the vacuum level. The LUMO and HOMO of rubrene (Rb) is at 3.15 and 5.36 eV and that of coumarin-6 (C-6) is at 2.7 and 5.4 eV, and DCM is at 3.2 and 5.4 eV, respectively [10].

The transient electroluminescence (EL) of OLEDs provides knowledge of charge carrier injection (electron from cathode and holes from anode), transport to the recombination zone, formation of excitons and recombination to emit photons. There is no detailed information on transient EL decay in doped tris-(8-hydroxyquinoline) aluminum (Alq_3). The doping affects the EL response time delay and carrier mobility in films due to the interaction between the guest and host molecules. The induced local states in positional and energetic disorders in the film directly influence the drift mobility of the carrier. If the dopant has polarity properties it induces the formation of deeper localized states [11]. These traps strongly affect the charge transport property at equilibrium conditions and most carriers occupy those deep states. The redistribution of the local field inside the doped sample affects the EL onset. The onset is the time lag between the applied pulse bias and the first appearance of EL.

The response time of OLEDs is one of the most essential criteria for the commercial application in pulse light sources, high resolution display, and optical communications in high frequency area. The reason for the EL delay can be charge-carrier injection, charge-carrier transport, buildup of space charges and formation of an excited state and the radiative decay of the excited state. The best way to determine some of these delay processes is by time-resolved EL measurement [12-16]. Transient EL measurements have the advantage to provide information directly on the dynamics of the characteristics and the carrier mobility from the emitting device [16-19]. The transient EL of OLED has been the subject of intense technological as well as fundamental research because the response time of these devices provides an essential criterion for their possible applications in pulse light sources, high resolution displays and high frequency optical communication. The delay time in an OLED starts when the leading front of the charge carriers (electrons) reach the internal organic-organic interface and ends when they recombine with the injected opposite charge carriers (holes) waiting at the interface [20]. The delay time can be divided into two components: the charge injection time and the charge transport delay time. The charge transport delay time depends on the carrier mobility.

The performance of OLEDs and organic photovoltaic (OPV) mainly depends on the diffusion of exciton. The accurate determination of the exciton diffusion length, diffusivity and life time is very important in optimizing the design of OLEDs and organic solar cells. Various theoretical models with different levels of the geometric and electronic structures details have been employed to describe exciton transport in organic materials [21]. Recently, it has been shown that the presence of traps limits the exciton diffusion in conjugate polymers [22]. The doping of Alq_3 affects the exciton diffusion length and life time's due to the interaction between the guest and host molecules. The induced local states in positional and energetic disorders in the film directly influence the movement of excitons. These traps strongly affect the charge transport property at equilibrium conditions and most carriers will occupy those deep states.

The interaction between randomly located and oriented dipoles is one of the major causes of energetic disorder in the doped film. When the structure of the dye-dope molecules is different from the guest molecules, an additional local modification of the polarization energy is appeared due to a static distortion of the structure. The polar dopants enhance the dipolar disorder, and decrease the carrier mobility in the film [11]. The localization of charge carrier due to the positional and energetic disorders [23, 24] is directly influenced the transport of carriers. The presence of positional disorder gives the energetic disorder due to the potential energy of interaction between the distance interacting molecules.

The mechanisms of electroluminescence from the organic layer have several steps such as charge injection, transport, capture and radiative recombination of positive and negative charges [15, 25]. Generally, the light emission is generated in pure Alq_3 layer close to the anode or hole transporting layer. At high injection level the accumulation of charge carriers as space charge near the interface can significantly modify the electrical potential inside the device. The energy level offset in the interface region is the main reason of buildup of interfacial space charges. The significant differences of electron and hole mobilities in the Alq_3 layer can be another source of interfacial space charge formation. The information of the charge and field distribution inside the device and its variation with applied bias is very important for the physical understanding of device operation, device optimization for efficiency, response and stability.

For conventional OLED devices, the traps at interface between hole transport and organic active layer has induced the recombination of electron-hole pairs and showed the delay of response time in OLEDs devices, which is proven from transient electroluminescence (EL) measurement [10, 26]. This delayed is due to the injected electrons from cathode hopping across the trapping sites in organic active layer giving rise to field-dependent mobilities and transit time spread due to disorder structure. The transport state is further influenced and complicated if the amorphous solid films doped with guest molecules [27].

The differential capacitance measurement has been used in poly p-phenylenevinylene (PPV) and showed that both the transport and trap state distributions within the polymer can be determined [28]. The dielectric constant of all Alq_3 films is increased with reducing frequency and it is found as doping concentration dependent.

The slope of Mott-Schottky plot indicates that traps act as electron acceptors in all films. The density of trap state is found as 4×10^{17} cm^{-3} in pure Alq_3 film. The Gaussian distribution width is found as 0.16 ± 0.02 eV in pure Alq_3 film and in doped films it is 0.30 ± 0.04 eV. The estimated trap depth in pure Alq_3 film is found as 0.65 ± 0.05 eV and in doped Alq_3 films it is 0.71 ± 0.15 eV for all dopants.

2. Emission Properties of Dopants Coumarine-6 and Rubrene in Alq_3 Films

The molecular structures of green light emitter host materials Alq_3, and three dopants DCM, rubrene (Rb), coumarin-6 (C-6) are shown in Figure 1. One of an ideal schematic OLED device structure is shown in Figure 2.

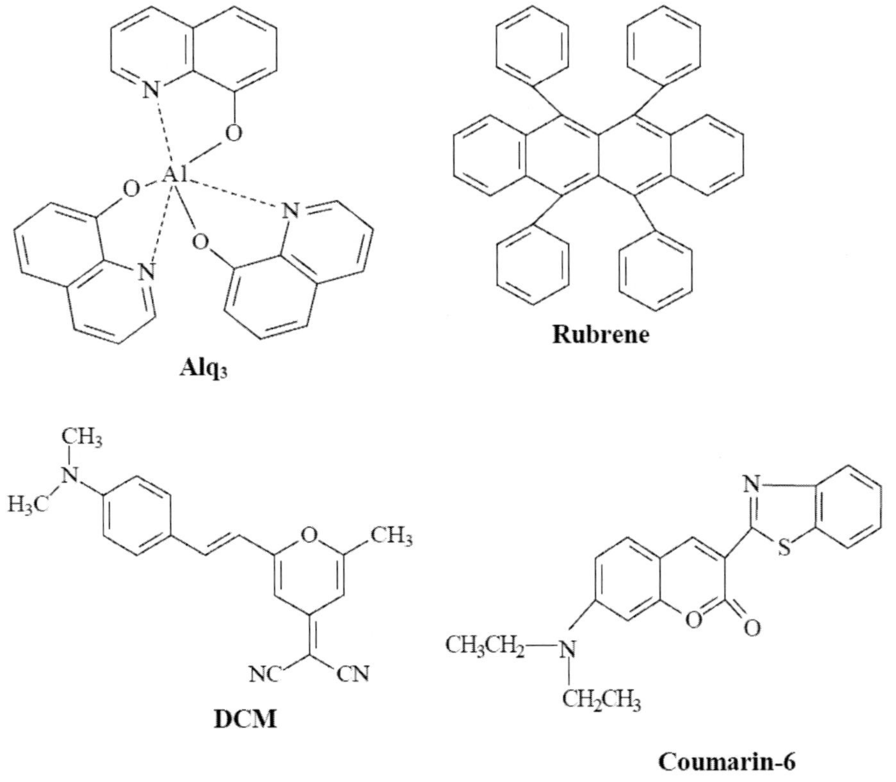

Figure 1. The molecular structure of organic host materials Alq_3, and dopants rubrene, DCM and coumarin 6.

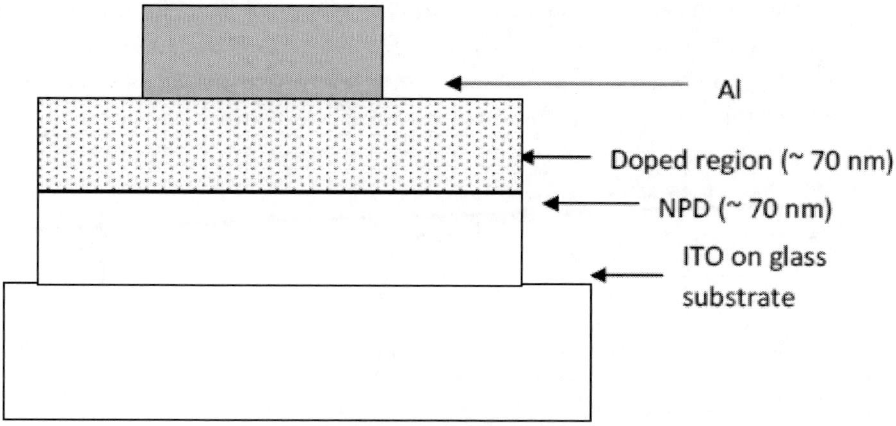

Figure 2. The structure of double-layer OLED device. The ITO and Al layers are used as anode and cathode, respectively.

The device is composed of a hole transport layer, N,N'-Di-[(1-naphthalenyl)-N,N'-diphenyl]-1,1'-biphenyl)-4,4'-diamine (NPD) and emitting layer Alq_3 doped with C-6 and Rb. The device structure is {ITO/NPD (70 nm)/Alq_3:dopant (70 nm)/Al} as shown in Figure 2.

The absorption spectra of C-6 and Rb are overlapped with PL spectrum of Alq_3 as shown in Figure 3(a). This indicates that the energy transfer from Alq_3 to C-6 and Rb can be effectively occurred. The laser excitation light wavelength 390 nm was used for the PL measurements. The main absorption peak in the C-6 film is appeared at ~ 420 nm region and two weak absorption peaks are also appeared at ~ 452 and ~ 498 nm regions.

The PL peak is appeared at ~ 520 nm from pure Alq_3 film. Figure 3(b) shows the EL and PL spectra of Alq_3:C-6 film with doping concentration 1.0 wt. % and the PL spectrum of pure C-6 film.

The PL peak position of pure C-6 film is at ~ 570 nm wavelength. The EL and PL spectra of Alq_3:C-6 film are almost superpose on each other at ~ 515 nm wavelength region. The EL spectrum is a bit wider than the PL spectrum. The PL emission peaks at around 570 nm from pure C-6 is disappeared from the Alq_3:C-6 film. This indicate that the complete energy transfer from the Alq_3 to the dopant C-6 is occurred at the doping concentration of 1 wt.%. Figure 3(c) shows the emission spectra of Alq_3:Rb film with doping concentration of 1 wt. % and PL spectrum of pure Rb. The PL peak of pure Rb is appeared at ~ 556 nm with a shoulder peak at ~ 595 nm wavelength regions. Both of the PL and EL peaks of Alq3:Rb appeared at ~ 550 nm with a weak Alq_3 emission shoulder at ~ 520 nm region. The PL and EL spectra are almost superposed on each other. The appearance of Alq_3 peak in Alq_3:Rb film indicate that the complete energy transfer did not take place from the Alq_3 to Rb molecules at the doping concentration of 1 wt. %.

The current-voltage characteristics of co-doping OLEDs are shown in Figure 4. The bias voltage of the co-doped OLED devices is improved compared to single doped Alq_3:C-6 device for the same current density.

For example, the bias voltage is reduced to ~ 7 V in the co-doped device compared to ~ 14 V in single doped film for the current density of $1x10^{-2}$ mA/cm^2. The co-doped devices show the better performance with high current density at low voltage.

The luminescence efficiency of OLEDs from Alq_3:(C-6:Rb) films for different co-doping concentration at room temperature are shown in Figure 5. The maximum luminescence efficiency ~ 6.5 cd/A is observed for the Rb doping concentration of ~ 0.7 wt.% which is nearly double compared to pure Alq_3 film luminescence efficiency.

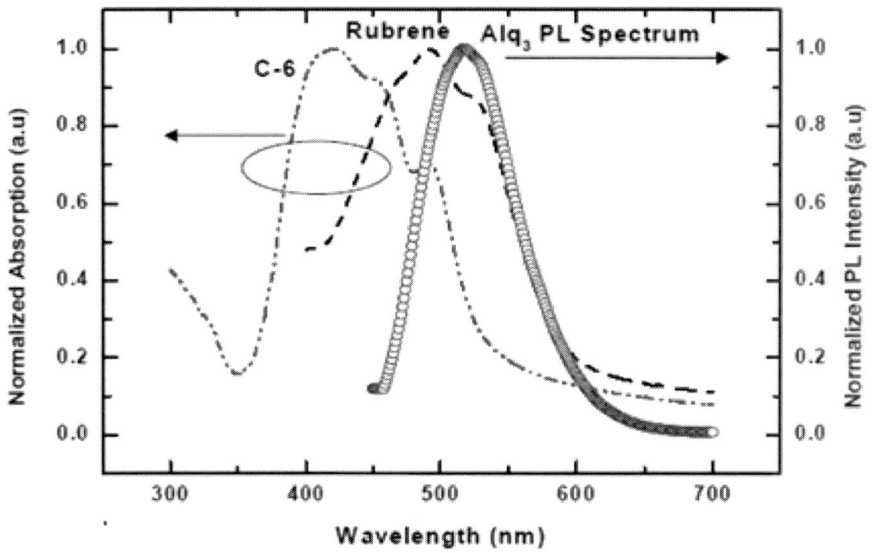

(a)

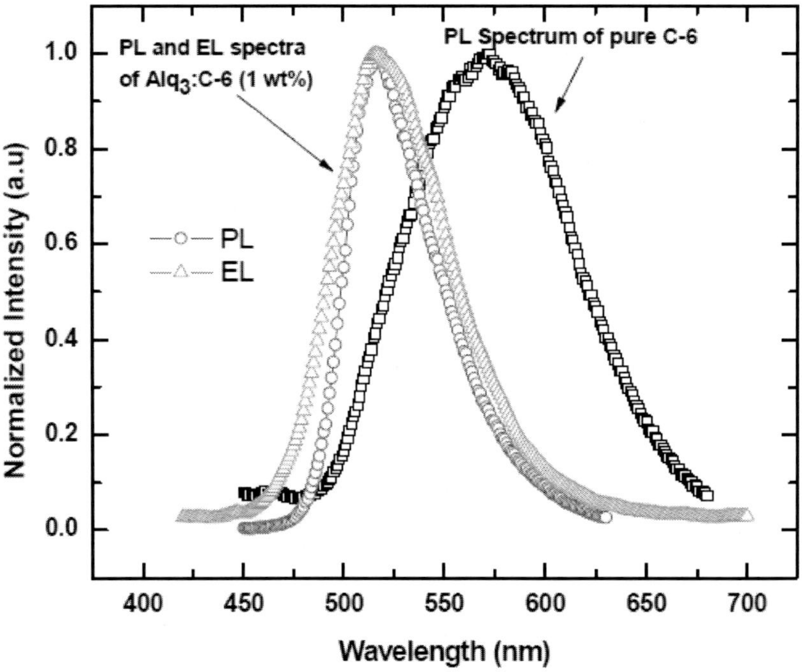

(b)

Figire 3. (Continued).

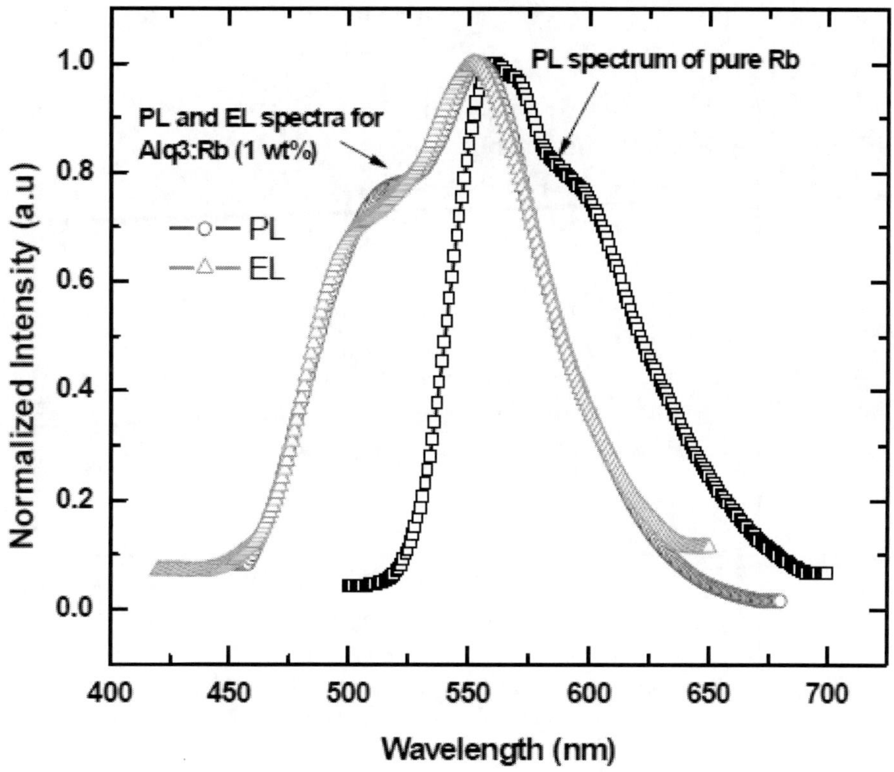

(c)

Figure 3. (a) The absorbance spectra of pure coumarin 6 (C-6) and rubrene (Rb) overlapping with the PL spectrum of pure Alq_3. (b) The PL spectra of pure C-6 and Alq_3:C-6(1 wt.%) films and the EL spectrum of OLED device of Alq_3:C-6(1 wt.%) film. (c) The PL spectra of pure Rb and Alq_3:Rb(1 wt.%) films and the EL spectrum of OLED device of Alq_3:Rb(1 wt.%) film at room temperature.

A small amount of Rb doping up to ~ 0.7 wt.% in Alq_3 film is sharply increased the luminescence efficiency. After the doping concentration of 0.7 wt.% the luminescence efficiency of OLEDs of Alq_3:Rb film is reduced with doping concentration. The induced defects related with Rb molecules may be responsible to reduce the luminescence efficiency in Alq_3:Rb films after the doping concentration of 0.7 wt.%. The maximum brightness of ~ 3000 cd/m^2 of OLEDs at current density of 45 mA/cm^2 by co-doping of C-6:Rb (1:2) in Alq_3 film are shown in Figure 5. The maximum luminescence efficiency is ~ 7 cd/A in C-6:Rb (1:2) co-doped device which is higher than OLED in Alq_3:Rb film. The luminance efficiency is increased rapidly as the current density increased and then declined gradually after a certain current density as shown in Figure 5. The additional Rb doping in Alq_3:C-6 film shows the improvement of devices performance.

The schematic energy band diagram of OLED device is shown in Figure 6. The possible emission mechanism in Alq_3:{C-6:Rb} films is the formation of exciton by the injected electrons and holes from the cathode and anode respectively in the host molecules of Alq_3. The LUMO and HOMO of Rb exist between the HOMO and LUMO of Alq_3. The exciton can directly transfer to Rb; or it is transferred to Rb molecules via intermediate C-6 molecules and recombines to produce the emission.

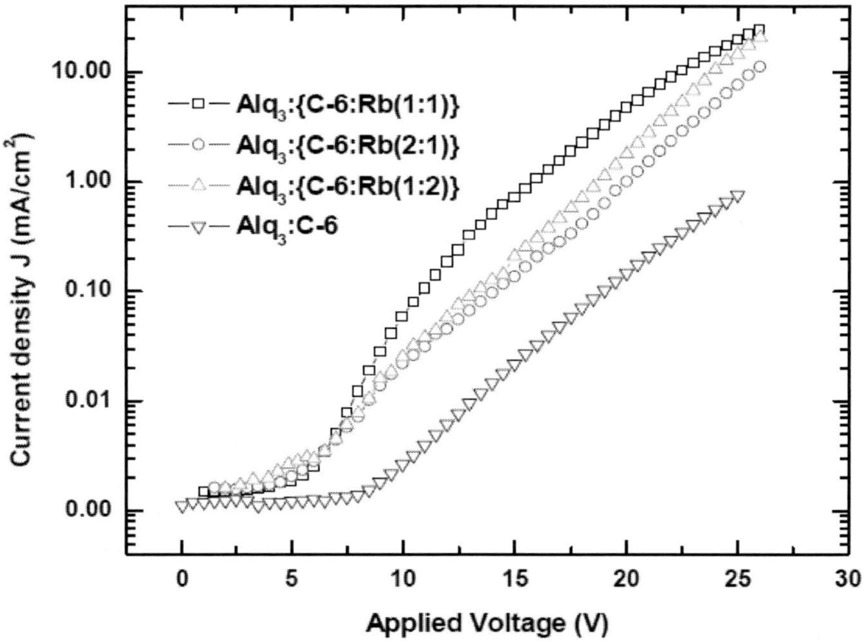

Figure 4. The current-voltage characteristics of OLED devices for different dopants. The co-doping of Alq_3:[C-6:Rb] films are reduced the bias voltage for the same amount of device current density.

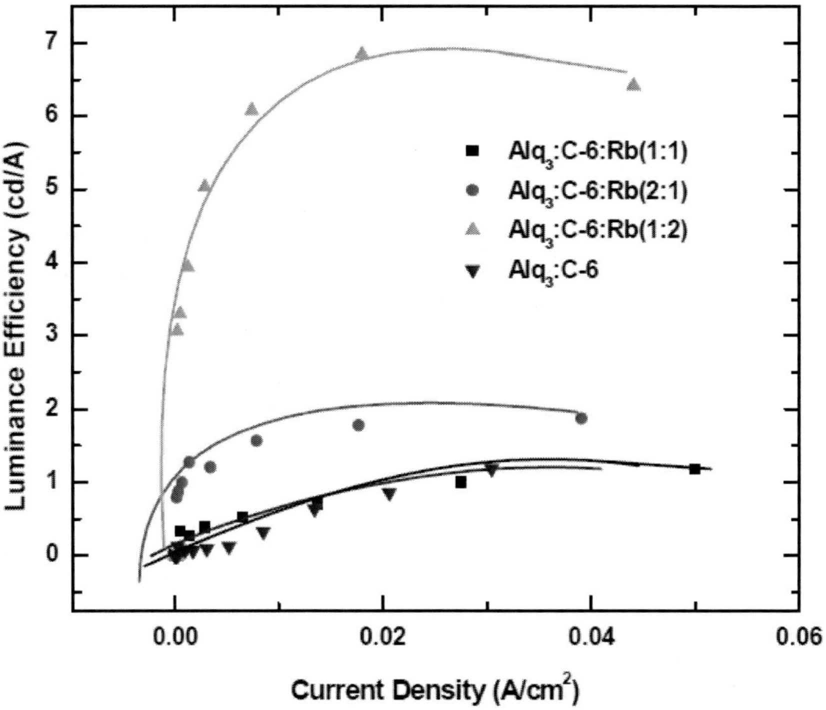

Figure 5. The plot of luminescence efficiency versus current density of OLEDs of single and co-doped Alq_3 films. The maximum efficiency of ~ 7 cd/A is observed for the co-doping film of Alq_3:{C-6:Rb (1:2)}.

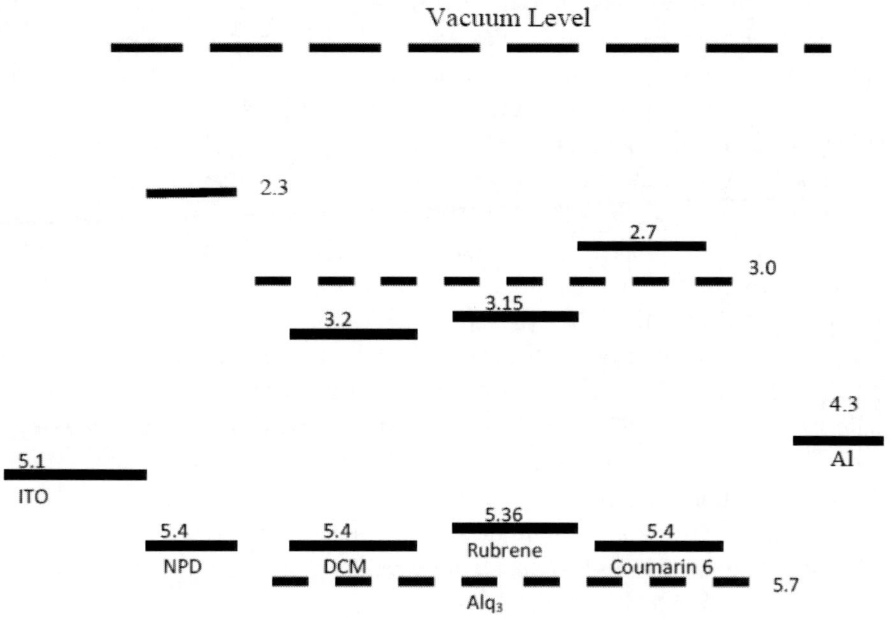

Figure 6. The schematic energy level diagram of the OLED devices with NPD and Alq_3 doped with C-6, Rb and DCM. The vacuum level is the reference energy.

This process may be active in all co-doped devices. Injected electrons in Alq_3 molecules may directly transfer to Rb molecules and injected holes in Alq_3 layer may transfer to Rb via C-6 molecules, and then recombination take place with electrons at Rb molecules to produce light in co-doped films. It is also possible to form exciton between the co-dopant molecules such as electron in Rb molecules and holes in C-6 molecules and then recombine to produce light as the HOMO level of Rb (- 5.36 eV) and C-6 (-5.4 eV) are nearly in the same level. Exciton concentration quenching is a serious problem in the Alq_3 film. A suitable dopant concentration (~ 1.0 wt.%) of co-doping of C-6 and Rb reduce the exciton quenching effect. The co-doped device improved current efficiency without increasing drive voltage is shown in Figure 4, so the power efficiency is also improved as shown in Figure 5. From above results we can conclude that improved EL efficiency based on co-doping method is mainly owed to the successful suppression of the exciton concentration quenching. When the two dopants are co-doped in Alq_3, the energies transferred from Alq_3 and adjacent C-6 to Rb by Förster energy transfer mechanism are consistent with the dominant emission of Rb in the EL spectra (Figure 3) of the mixture [29]. Thus one can conclude that the competition process of energy transfer might be utilized the energy of host more adequately, therefore the efficiency is improved greatly. The improvement could be attributed to the effective utilization of host energy by both energy transfer and trapping in the electroluminescence process and the depression of concentration quenching between the dopants molecules.

2.1. Emission Properties of Dopant DCM in Alq_3 Films

The current (J) - voltage (V) characteristics were measured in Alq_3:DCM films for different DCM concentrations at room temperature (RT) as shown in Figure 7. The sample structure is shown in inset. The current density is considerably increased with DCM concentration up to 1.5 wt.% at a given bias voltage V. The current is sharply reduced after the DCM concentration above 1.5 wt.%. The change of bias voltage with DCM concentration in Alq_3 films is shown in Figure 8 for a current density $J = 0.01\ Acm^{-2}$.

The DCM is capable of trapping both electron and holes. At the higher dopant levels the J – V curves is noticeably shifted to the higher voltage, indicating that the dopant can influence the carrier transport properties in Alq_3:DCM films. The voltage shift is as much as 12 V at concentration above 24 wt.%. The voltage required to drive the doped or undoped OLEDs is quite low 7 – 9 V to achieve the enough emission. The average electric field on the structure is as high as $1.2\ x10^6$ V/cm for 7 V bias across the organic layer, which is typically 60 nm thick. This high field is apparently needed for the injection of electrons from the Al cathode.

The Al cathode has the highest work function differences with Alq_3 compare to other cathode materials such as Mg, Ca, etc. Al is more stable and less reactive with oxygen. The hole injection from the ITO anode is believed to be ohmic.

The carrier injection efficiency of Al cathode at a constant electric field should be exponentially dependent on the magnitude of the injection barrier. The injection barrier between Alq_3 and Al cathode can be considerably reduced by inserting an intermediate layer such as LiF, NaF, NaCl, etc. [4].

It is believed that an intermediate states participating in the injection process, so that injection does not occur in a single hop.

The interface dipoles at the contact may significantly alter the effective injection barrier for each cathode material [30].

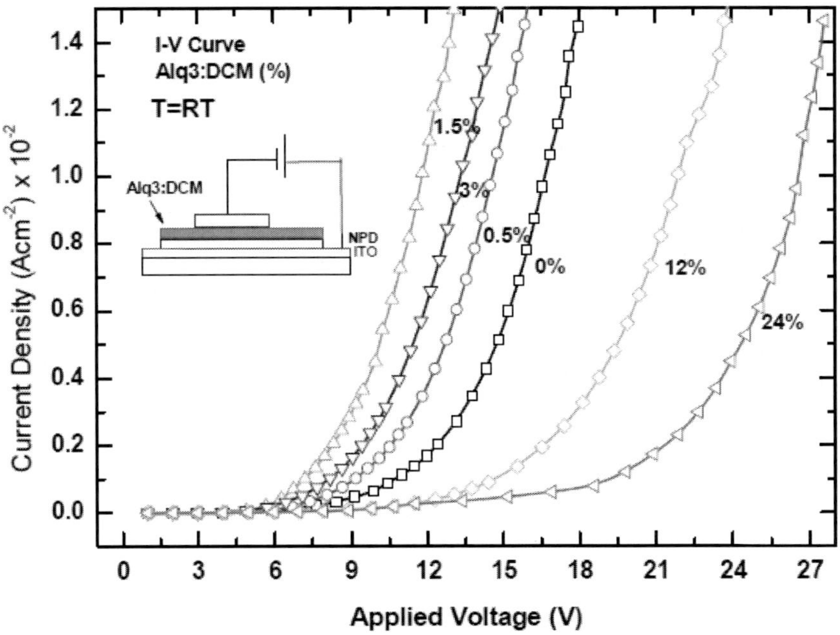

Figure 7. The plot of current density versus bias voltage characteristics at RT for different concentration of DCM in Alq_3:DCM layers. The structure of sample is shown in inset. The current density is sharply increased with DCM concentration up to 1.5 wt.% at the same bias voltage.

In addition, the organic material may possess extrinsic trap states due to impurities or damage introduced in the film deposition process and cathode metal diffusion into the organic [31].

One can assume that charges from the metal are readily injected into a broad distribution of interfacial states from which they must hop into a narrower, higher energy distribution of transport states. Although transport is dominated in this case by the effects of disorder, electric field screening by the metal contact also reduces the LUMO and HOMO splitting energy at the interface. The carrier injection properties of Al cathode can be described by charge hopping mechanism [4]. The problem of carrier injection at the cathode has been treated in terms of the orbital energies of the electrode and the electron transporting material.

Figure 9 shows the EL peak energy shifting of Alq_3:DCM samples for different DCM concentrations. The EL spectra were taken from OLEDs of Alq_3:DCM films with their intensity normalization to a referential Alq_3 film prepared on the same substrate. The inset in Figure 9 shows the EL spectra for different DCM concentrations.

The peak energy is shifted from the green to the red region with DCM concentrations. The EL peak energy is sharply shifted to the lower energy side for DCM concentration upto 1.5 wt.%. The peak intensity is also increased about 2.5 times for 0.5 wt.% DCM concentrations compare to pure Alq_3 intensity at the same current density.

The EL efficiency is estimated at a current density of 0.015 A/cm^2 for different DCM concentration in Alq_3:DCM samples from room temperature to 130 °C. The highest EL efficiency is found for DCM doping concentration around 0.5 wt.% as shown in Figure 10.

The overall EL efficiency is reduced with increasing temperature. It is estimated that if more than one DCM molecule is present in about 40 Alq_3 molecules on average exciton quench is likely to occur [9].

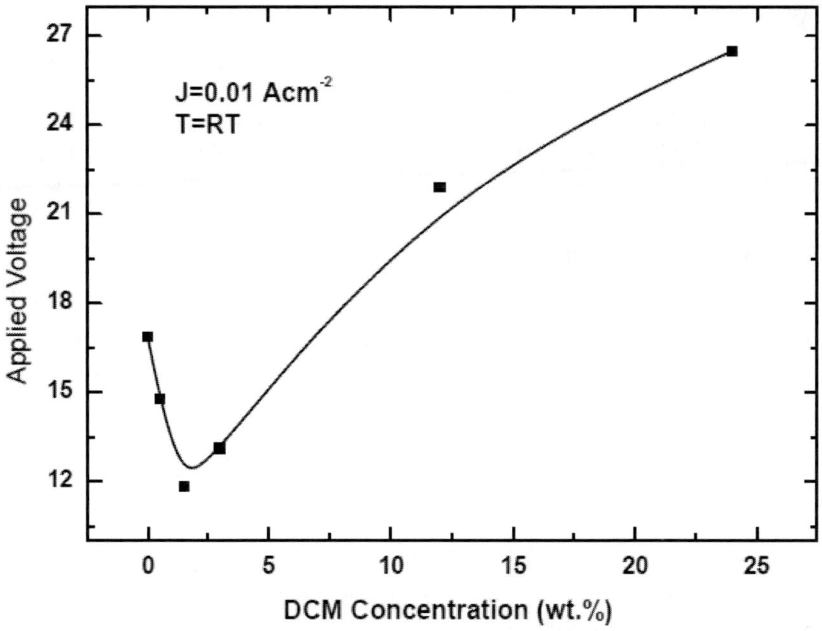

Figure 8. The change of bias voltage with DCM concentration in Alq_3:DCM films for a current density J = 0.01 A/cm^2. The minimum bias voltage 12.0 V is observed for the DCM concentration of 1.5 wt.%.

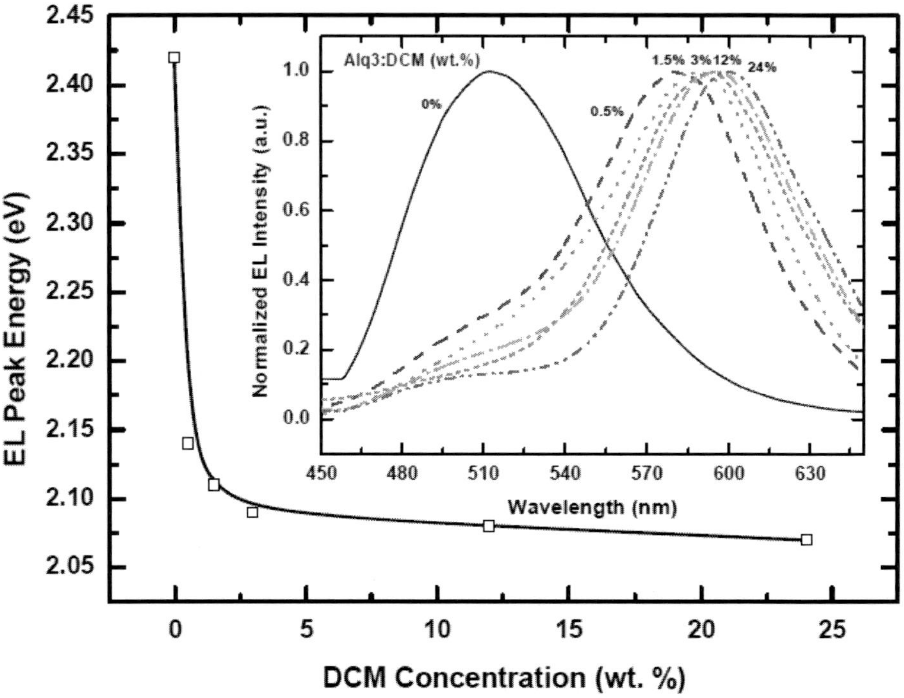

Figure 9. The plot of EL peak energy as a function of DCM concentration in Alq_3:DCM films. The EL spectra are shown in inset with normalized EL intensity.

The lifetime of DCM-originated excitons is much shorter than that of Alq_3-originated. The Alq_3-originated excitons reach the DCM molecule will transfer energy to it.

When the DCM concentration is increased the possibility for DCM molecules to get together also increase, which leads to enhance the emission at longer wavelengths. If too many molecules get together the emission would be smeared due to concentration quench. The lowest doping concentration ~ 0.5 wt.% is needed for the efficient EL devices.

Figure 11 shows the EL light intensity for injection current density for a number of doped and an undoped OLED samples. The light output is directly proportional to the injection current.

For the doped samples the EL intensity is higher than the undoped sample for the entire current range. The highest light output is observed for the DCM doping concentration 0.5 wt.% in Alq_3. The EL intensity is reduced with increasing temperature from RT to 130 °C.

3. Doping Effects on Initial Rise of Electroluminescence

The experimental set-up is shown in Figure 12. The schematic bilayer OLED structure grown on glass substrates covered with a 100 nm thick patterned ITO anode having a surface resistance of 30 V/square is shown inset in Figure 12. The observed green emission in undoped Alq_3 based OLED is due to the recombination of accumulated holes and electrons at the interfacial layer of NPD/Alq_3.

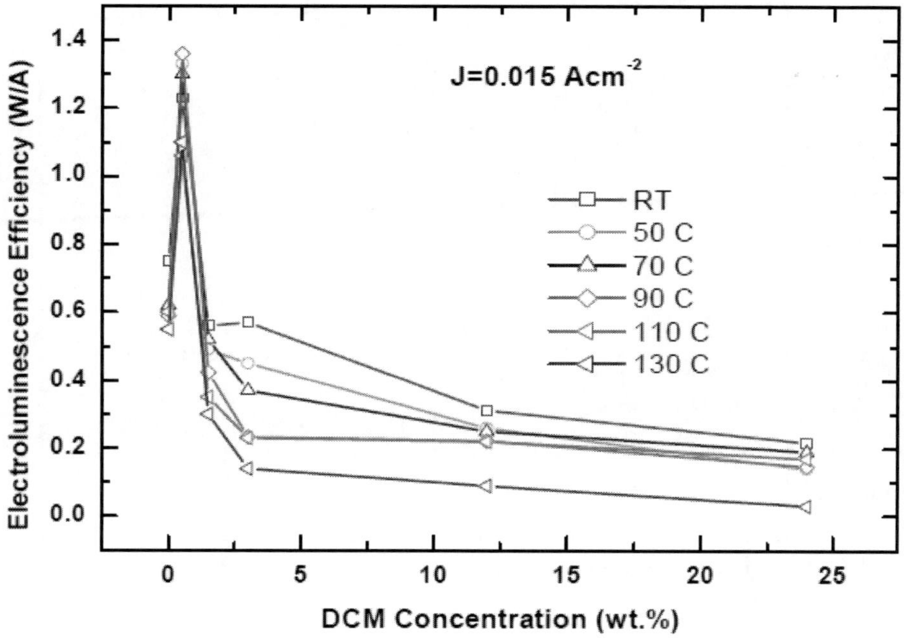

Figure 10. The plot of EL efficiency as a function of DCM concentration in Alq_3:DCM films at different temperatures. The maximum EL efficiency is observed for DCM concentration of 0.5 wt.%.

Figure 13(a-c) shows the transient behavior of the EL in pure Alq_3 and for each dopant for different concentrations as a parameter. A voltage pulse of 11 V and the length 5 - 6 μs with the repetition rate 1 kHz is applied. Figure 13(a) shows that the EL onset has a delay time of only 0.7 μs, in pure Alq_3. This delay time is comparable with the best reported values [32]. In Figure 13(a) the transient EL is shown for different DCM doping concentration in Alq_3.

The delay time, t_d, is increased to 1.9 μs at 24 wt.%. After this the signal is increased and tends to saturate. The EL saturation or rise time, t_r, can be defined by the intercept of their tangents (not shown in Figure 13).

The saturation time t_r is the time until the positive and negative charge distributions are interpenetrated. Actually, the rise time is governed by the charge carrier motion into the bulk of the sample. In a bi-layer OLED structure the situation is more complicated due to the possible charge carrier accumulation at the interfaces.

The t_d is the time taken by the electrons to reach the internal NPD/Alq_3 interface and recombine with the injected opposite charge (holes) carriers waiting at the interface. Whereas the t_r reflects the buildup of the carrier density in the NPD/Alq_3 interface recombination zone. The decay pattern of the EL at the end of the applied voltage pulse reflects the depletion of the charge carrier reservoir established during the proceeding on-phase. It is believed that the EL onset time t_d is controlled by the growth of interfacial charge densities and the concomitant redistribution of the electric field inside the sample rather than the charge carrier transport [33]. Figure 13(b) and (c) show the transient EL in OLEDs of Alq_3 for C-6 and Rb dopants with different concentrations, from 0.5 - 24 wt.%, respectively. The EL onset time, t_d, and the rise time, t_r, are summarized in Figures 14(a) and (b) as a function of the doping concentration.

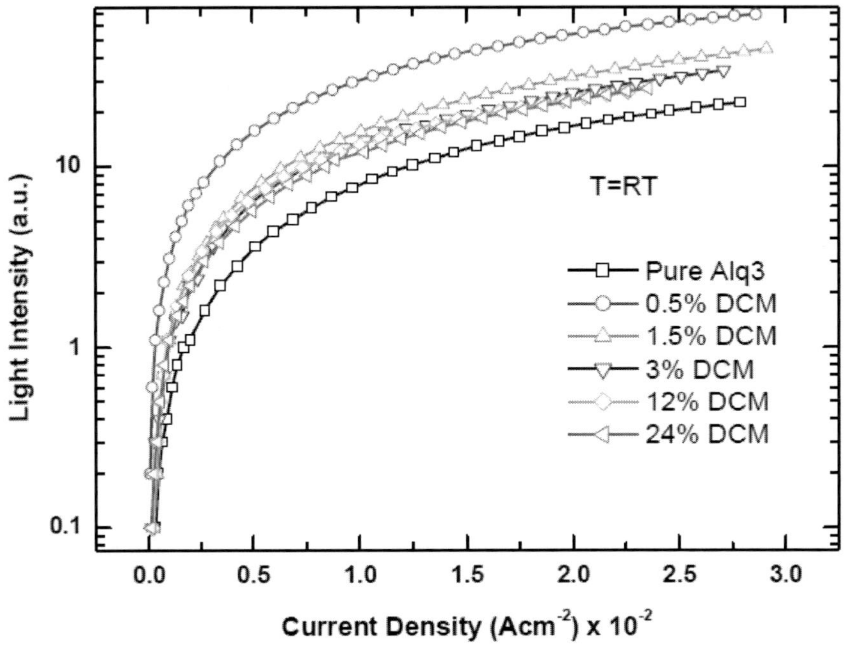

Figure 11. The plot of EL intensity as a function of current density for different DCM concentration in Alq_3:DCM films at room temperature.

The t_d is raised sharply at the lower doping concentration up to one or a few wt.%. After that the onset time is increased nearly linearly with the DCM doping concentration. The largest measured EL onset (delay) time, $t_{d,}$ is for the Alq_3:DCM layer and increased from 0.7 to 1.9 μs with the doping concentration. The corresponding time dalay in the Alq_3:C-6 and Alq_3:Rb films are increased only to 1.5 μs and 1.0 μs, respectively for the highest doping. Especially Alq_3:Rb film shows a low onset time. The stronger polar nature of the DCM molecule compared to other two dopants may be responsible for the higher t_d in the Alq_3:DCM films [34].

Also the EL (switch-on) time, t_r, is increased with doping concentration for all dopants, as shown in Figure 14(b). The time constant is found to be about the double as compared to t_d but otherwise similar dependence as for the t_d. At the lower concentration region t_r is increased sharply in the Alq_3:DCM films with doping concentration up to ~1 wt.%. The value of t_r is notably smaller for the Alq_3:C-6 and Alq_3:Rb films for the same doping concentration. The increase of t_d and t_r with doping means the OLEDs become slower in response with increasing doping concentration.

The polarity of dopant molecules may play an important role to increase the t_d and t_r in the doped samples. The dipole moments of DCM is the highest, ~ 6 D (in the ground state) and ~ 26 D (in the excitated state) [35] compared to C-6 of 7 D (ground state) and ~ 9 D (excitated state) [36].

However, the dopant Rb has non-polar property due to the symmetry of the molecular structure.

The transient EL behavior, especially at low electric fields, is related to the interfacial charging. The delay time t_d is governed by the accumulation of space charge, both holes and electrons, at the NPD/Alq_3 interface.

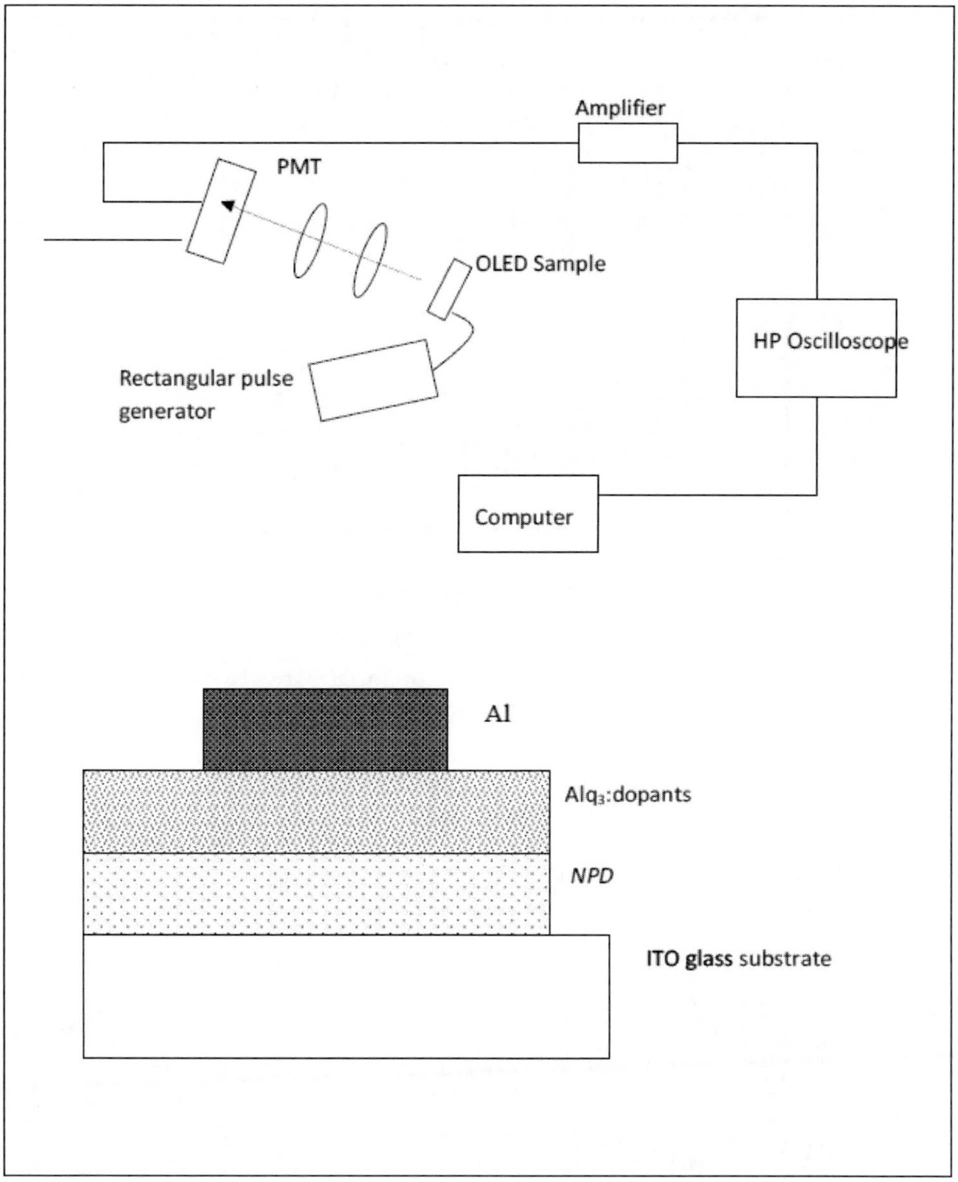

Figure 12. The inset is the experimental setup for the transient EL measurements. The schematic structure of the bi-layer OLED device is shown on ITO coated glass substrate.

At higher electric fields the charge (hole) accumulation at this interface may play a minor role, as it is controlled by the highly field dependent mobility electron.

The EL delay time in the doped devices is longer than for the undoped Alq_3 at a constant electric field (11 V bias). The DCM doped sample shows the longest delay time t_d, for the stronger dipole nature of DCM followed by C-6 and Rb doping.

For lightly doped devices (0.5 to 1 wt.%) the t_d is influenced and delayed about 40% (DCM), 21% (C-6) and 15% (Rb) in the Alq_3 films, as compared to the pure Alq_3 film. This indicated that the charge carrier transport is influenced by the dopants. The injected electron is believed to transport by hopping between localized states.

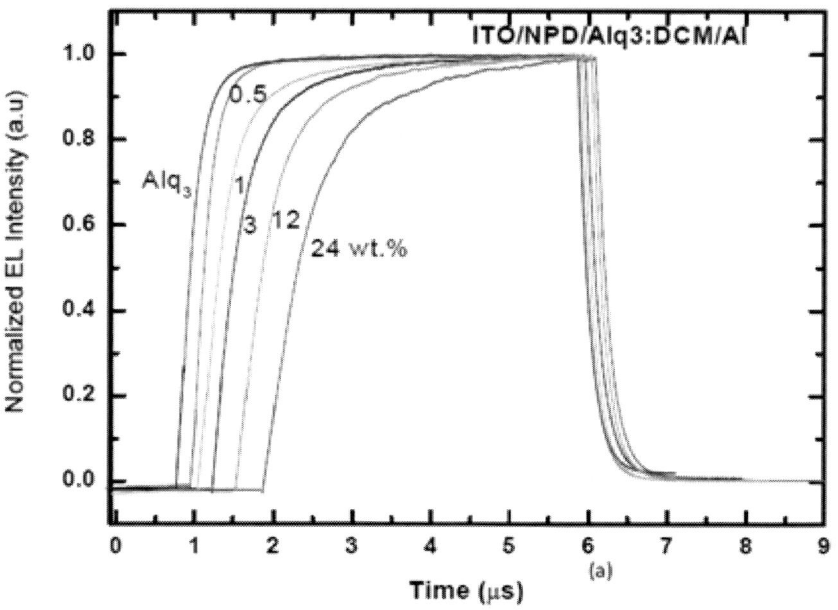

(a)

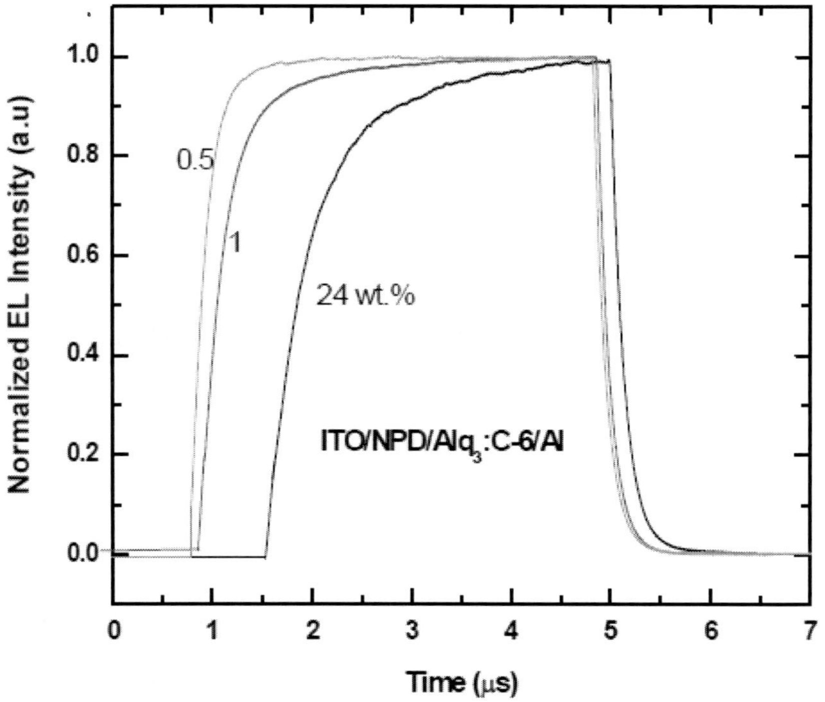

(b)

Figure 13. (Continued).

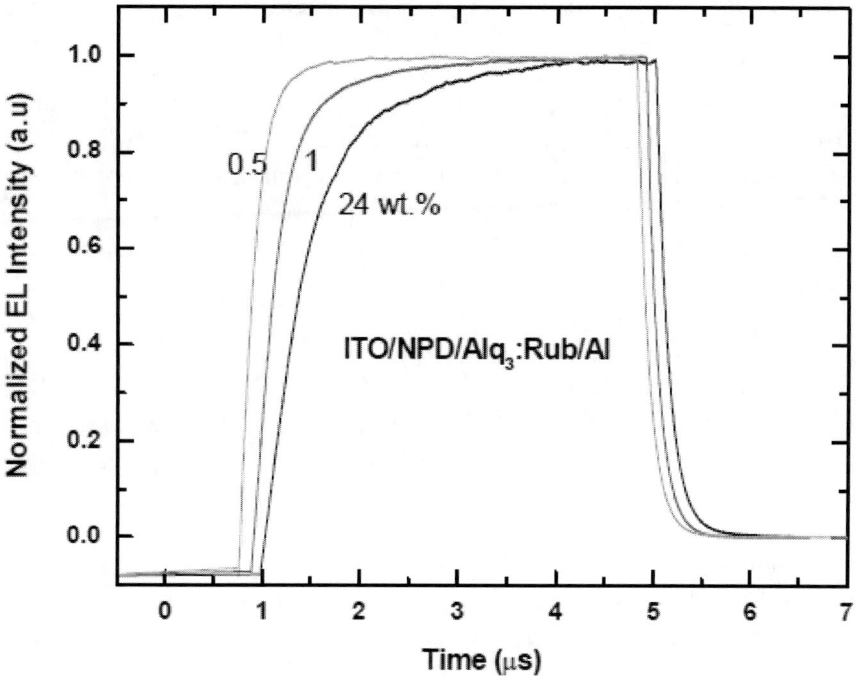

(c)

Figure 13. The transient EL with a rectangular voltage pulse of 11 V with pulse length 5 - 6 μs and a repetition rate of 1 kHz for different doped OLEDs. (a) The transient EL in pure and DCM doped devices for different doping concentrations with voltage pulse length 6 μs. (b) The transient EL in C-6 doped devices for three different doping concentrations with voltage pulse length 5 μs and (c) the transient EL in Rb doped devices for three different doping concentrations with voltage pulse length 5 μs.

The free electrons injected from the Al cathode in the doped Alq_3 film may hop between Alq_3 LUMO and the dopant states. Both have a Gaussian distributed localized states in the film. The energy level diagram of a device sample, under flat band condition, is shown in Figure 6.

The three doping cases are illustrated. The lowest unoccupied molecular orbital (LUMO) and the highest occupied molecular orbital (HOMO) energy levels of DCM are located at 3.2 and 5.4 eV respectively, which are well coupled within the LUMO and HOMO of Alq_3.

The HOMO of Rb and C-6 are located at 5.36, 5.4 eV and LUMO at 3.15, 2.7 eV respectively. The transient EL behavior of the bi-layer ITO/NPD/Alq_3/Al OLED clearly shows the existence of energy barriers for both electrons and holes at the NPD/Alq_3 interface. At the NPD/Alq_3 interface region the electrons and holes can be transferred to the LUMO and HOMO of dopants molecules before they recombine [6].

From the transient EL measurements in pure and doped Alq_3 devices it is possible to estimate the carrier mobility, μ, based on the equation (1) [32]:

$$\mu = \frac{L}{t_d E}, \qquad (1)$$

where L is the active material thickness and E is the electric field $E = \frac{(V - V_{bi})}{d}$, with d as the device thickness.

The symbol, V, is the voltage applied across the device, and V_{bi} is the built-in voltage calculated from the difference between the work functions of ITO and Al ($V_{bi} \approx 0.7$ eV). The delay time t_d is used because it is determined the buildup of space charges at the NPD/Alq_3 interface inside the devices, and the charge-carrier transport behavior. The electric field E was < 10^6 V/cm. The carrier mobility is obtained from Equation (1) is shown in Figure 15(a) as a function of doping concentration for three dopants.

The mobility is in the range of 0.5 – 1.2×10^{-5} $cm^2V^{-1}s^{-1}$ and is decreased with doping concentration. In a separate experiment the charge carrier mobility is measured by the time-of-flight (TOF) method in the doped, ~ 1.0 μm thick, Alq_3 films as shown in Figure 15(b). The details of the TOF measurements have been described elsewhere [34]. The estimated carrier mobility from the transient EL measurements is lower than the values obtained through the TOF method but the trend with doping concentration is same in both measurements.

The lower mobility in the transient EL measurements may be due to the accumulation of space charges at the NPD/Alq_3 interfacial barrier as well as the dopants trapped the charges in Alq_3 film and near the NPD/Alq_3 interfac results the redistribution of electric field inside the OLED affects the motion of carriers.

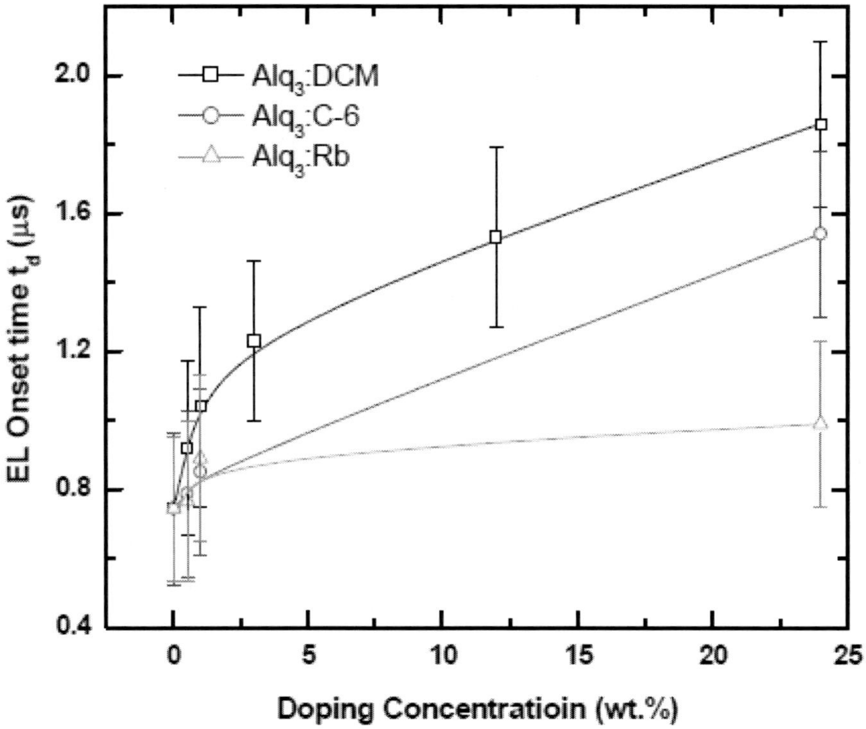

(a)

Figure 14. (Coninued).

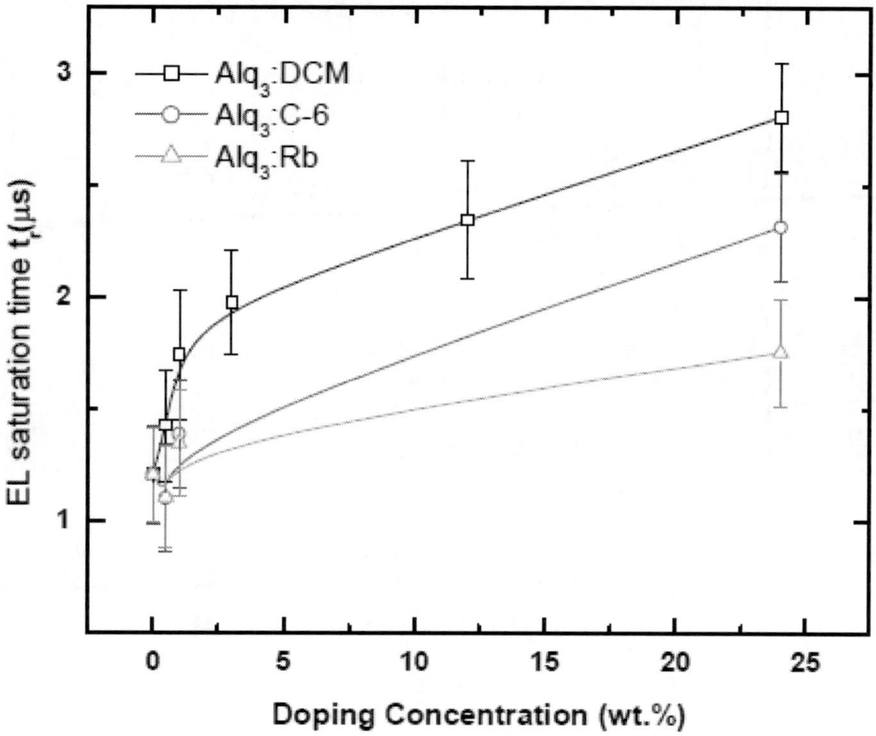

b)

Figure 14. (a) The plot of transient EL onset time t_d from OLEDs as a function of the doping concentration (wt.%) in Alq_3 films with three different dopants DCM, C-6 and Rb. The value of t_d is increased with doping concentrations. (b) The plot of the EL saturation time, t_r, with doping concentration for three dopants. Also t_r is increased with doping concentration for all three dopants.

The carrier mobility in Alq_3:Rb film is higher than in the Alq_3:DCM and Alq_3:C-6 films due to the less polar effects. The polar dopants DCM and C-6 in Alq_3 films increase the dipolar disorder $\sigma_{dipolar}$, and consequently decrease the carrier mobility. The non-polar Rb induces the disorder of density-of-state due to van der Waals component σ_{vdW} with Alq_3 molecules, but this van der Waals effect is weak compared to $\sigma_{dipolar}$ [37].

The polarity of the DCM molecules in the Alq_3 films induced deep localized states [38] and these traps strongly affect the charge mobility at equilibrium conditions [34].

4. Doping Effects on Exciton Behaviours

Figure 16 shows the normalized PL spectra from Alq_3:DCM films at room temperature for different doping concentrations. The emission peaks of these films red-shifted with doping concentrations as shown in Figure 16.

The PL peak position is sharply shifted with DCM concentration up to 1 wt.% in Alq_3 film. Details explanation about the PL peak shifting with doping concentration in Alq_3 film is available elsewhere [6].

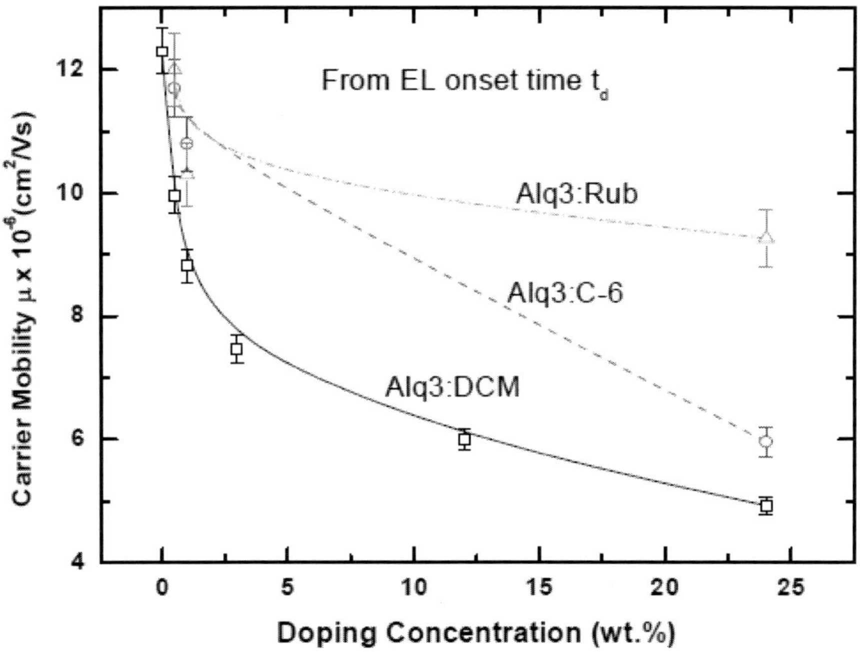

a)

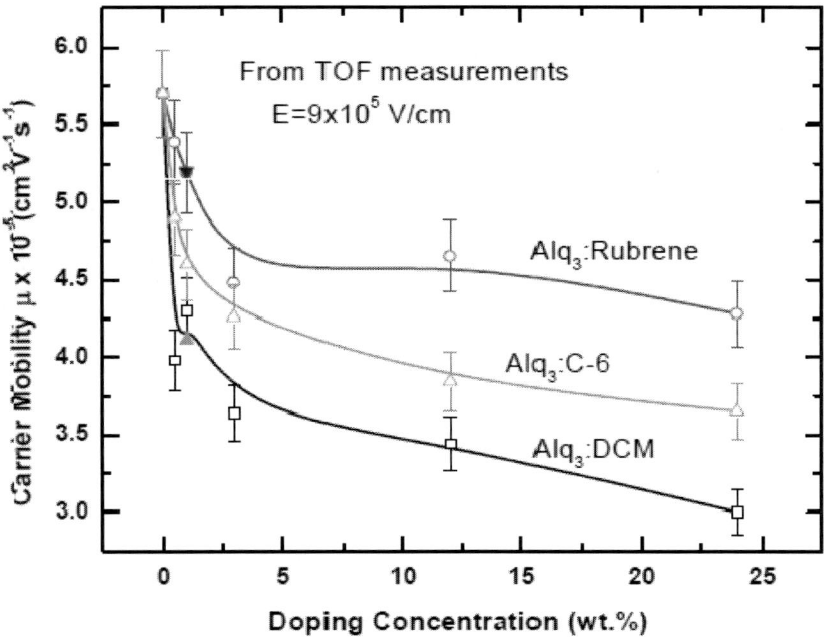

b)

Figure 15. (a) The plot of the estimated carrier mobility from the measured EL onset time t_d in the doped films as a function of the doping concentration. The applied electric field E was less than 10^6 V/cm. The maximum carrier mobility reduction is the factor of two or less. (b) The carrier mobility is estimated from the TOF measurements. The carrier mobility from TOF measurements is nearly one order of magnitude higher than the mobility estimated from the EL onset time t_d.

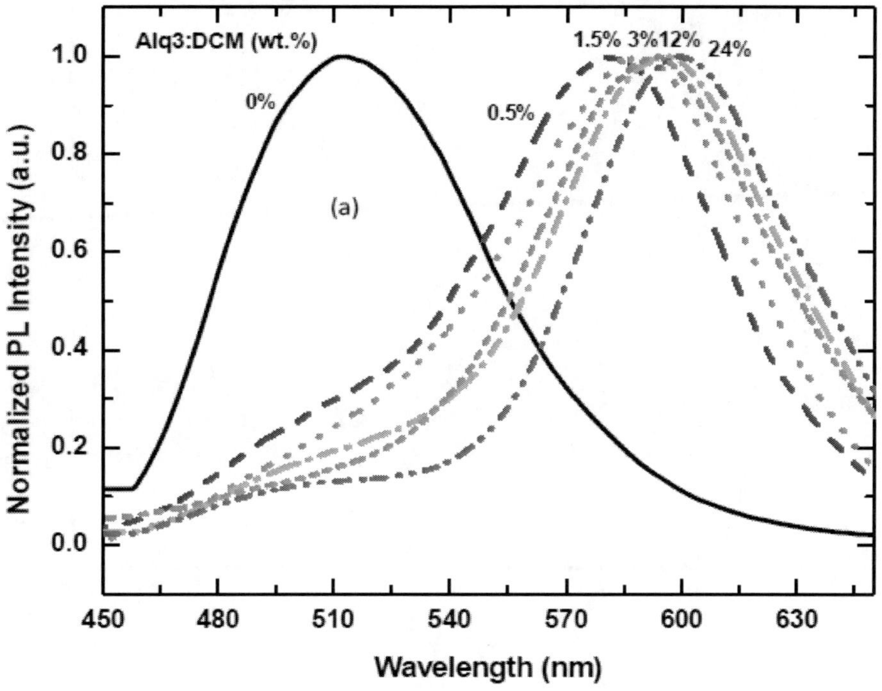

Figure 16. The normalized PL spectra versus the emission wavelength of Alq_3:DCM films for different doping concentrations. The red-shifted emission wavelength is observed in both doped films.

The PL decay was measured in doped films at room temperature for different doping concentrations as shown in Figure 17. The PL decay time in pure Alq_3 and Alq_3:DCM films for 3 different doping concentrations are compared to the Nd:YAG laser decay in Figure 17. The transient PL decay is faster with higher DCM concentrations. Similar phenomenon is also observed in Rb doped films.

The simulated fast initial rise time and emission lifetime of transient PL are estimated from the empirical signal intensity (*I*) equation (2):

$$I = A(1 - \exp(-\frac{t}{t_1}))\exp(-\frac{t}{t_2}) \tag{2}$$

where A is a constant, t is the signal scanning time, t_1 and t_2 are the fast initial rise time and emission lifetime respectively. The fast initial rise time t_1 is attributed to energy transfer from Alq_3 molecules to dopant molecules through Förster resonance energy transfer [39].

The inset in Figure 18 shows an example of the simulated dotted line which was fitted to the experimental data for Alq_3:DCM (24 wt.%) film. The rise time t_1 is sharply decreased from ~$1x10^{-9}$ s to ~6×10^{-10} s as shown in Figure 18 when the Alq_3 films are lightly (≤ 1 wt.%) doped.

This indicates the faster energy transfer takes place from Alq_3 to dopant molecules even at the lower doping concentrations. The rise time t_1 is slightly increased with increasing doping concentrations in both dopant films. The polar nature of dopant molecules may be responsible for the increase of time t_1 for higher doping concentration in Alq_3 film.

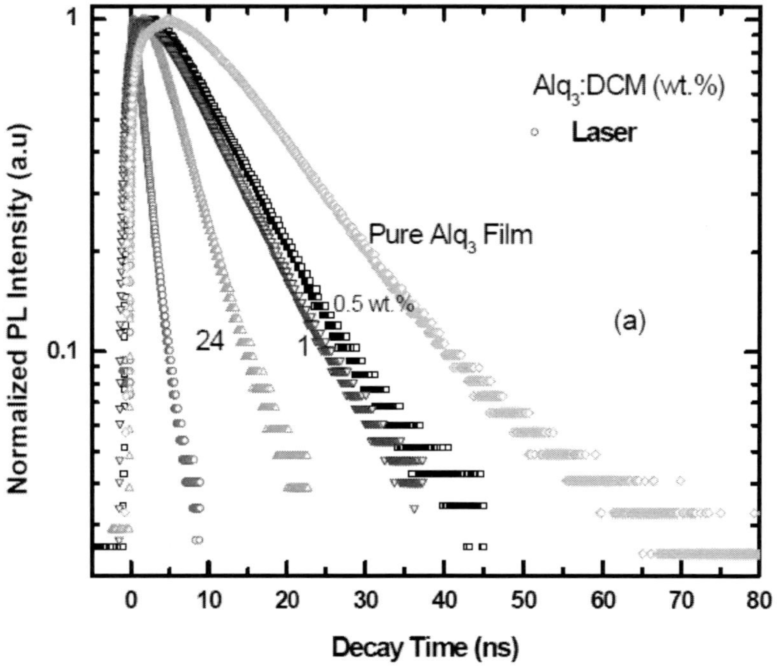

Figure 17. The normalized PL decay curve in Alq_3:DCM films for different doping concentrations. The PL decay in pure Alq_3 is compared with three different concentrations in Alq_3:DCM films.

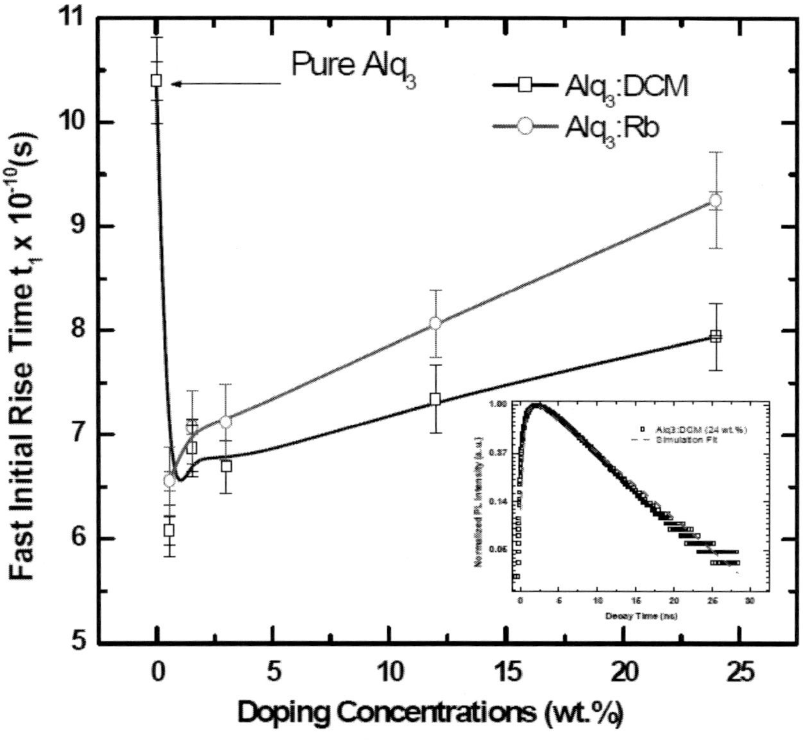

(a)

Figure 18. (Continued).

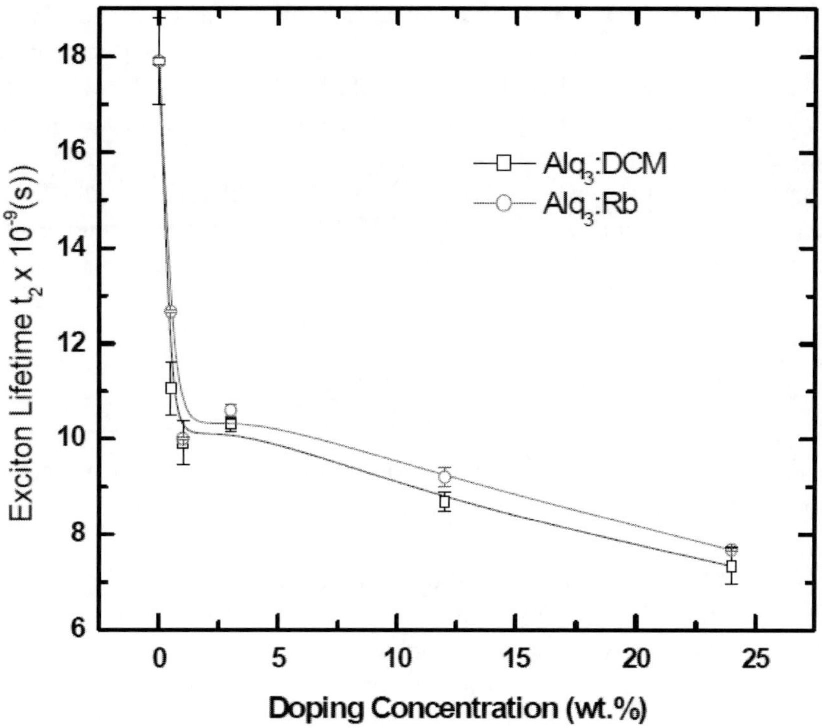

(b)

Figure 18. The plot of fast initial rise time t_1 and the exciton lifetime t_2 versus the dopants concentration in figures (a) and (b) in Alq_3:DCM and Alq_3:Rb films, respectively. The inset in figure (a) shows the simulated dotted line fitted to the experimental data.

The exciton lifetime in pure Alq_3 is found as ~ 20 ns. The lifetime t_2 is reduced to ~ 8 ns with doping concentration for both dopants in Alq_3 film. The sharp reduction of exciton lifetime is believed to be due to the induced defects in doped films. The estimated recombination rate constant γ_{ss} of singlet excitons in Alq_3:DCM and Alq_3:Rb films by fitting PL data to the exciton kinetic equation describes the concentration of singlet molecular excitons [40]. The recombination rate of excitons is increased with doping concentration for both dopants. The estimated values of γ_{ss} are in the rage of $2 – 5 \times 10^{-12}$ cm^3s^{-1}.

The dopant molecules act as a recombination center in the film. However, the recombination rate is increased slightly after the doping concentration ≥ 1 wt.%.

The exciton diffusion coefficient D_s and diffusion length L_D are also estimated in doped and undoped films. The estimated diffusion coefficient D_s of singlet excitons is from the equation in reference [40].

The estimated D_s for different doping concentrations are recorded in Table I. The value of D_s in pure Alq_3 is found as $(2.7 \pm 0.6) \times 10^{-7}$ cm^2s^{-1}. The diffusion coefficient D_s is increased to $0.5 – 1.4 \times 10^{-6}$ cm^2s^{-1} in both doped films, which is nearly one order higher than in pure Alq_3 film. The reported values of exciton diffusion coefficient in Alq_3 films are varied from $\sim 10^{-4}$ to $\sim 10^{-8}$ cm^2s^{-1} [40, 41]. The diffusion coefficient estimated from the electroluminescence measurements is affected by several factors such as high electric field, quenching by charge carriers, electrodes, etc. which is not comparable with the PL measurement values. Again the values of diffusion coefficient can be different in PL measurements for thin (~ 70 nm) and thick (~ 1 μm) films. The exciton quenching rate in the surface of the thin film is much higher than that in thick film as the surface concentration of exciton in thin film is much higher than that in thick film. The surface quenching rate is proportional to the exciton concentration at the surface. The efficient energy transfer from host (Alq_3) to guest (dopant) molecules of smaller energy gap is favorable for short lifetime which behaves as exciton traps and results in a short exciton diffusion length. The increase of diffusion coefficient in the doped films may be due to the increase of traping of exciton in dopant molecules. The exciton diffusion length L_D of singlet excitons is estimated from the equation (3) [40],

$$L_D \cong \sqrt{Zt_2D_s} \qquad (3)$$

where Z is equal to 6, 4 or 2 for three-, two- or one-dimensional diffusion, respectively. The t_2 is the exciton lifetime which we have measured. The L_D value is estimated and is shown in Figure 19 by assuming the exciton diffusion is three-dimensional both in pure and doped films. The exciton diffusion length in doped films is reduced to 18 - 35 Å from ~ 57 Å in pure Alq_3 film. The exciton diffusion coefficient D_s is increased with doping concentration in Alq_3 films as the dopant molecules act as a recombination centre. The capture rate of exciton in dopant molecules is accelerated with doping concentration which we can realize by the increase of D_s values. This assumption is also supported by the shifting of PL spectra from green to red colors with doping concentration as shown in Figure 16.

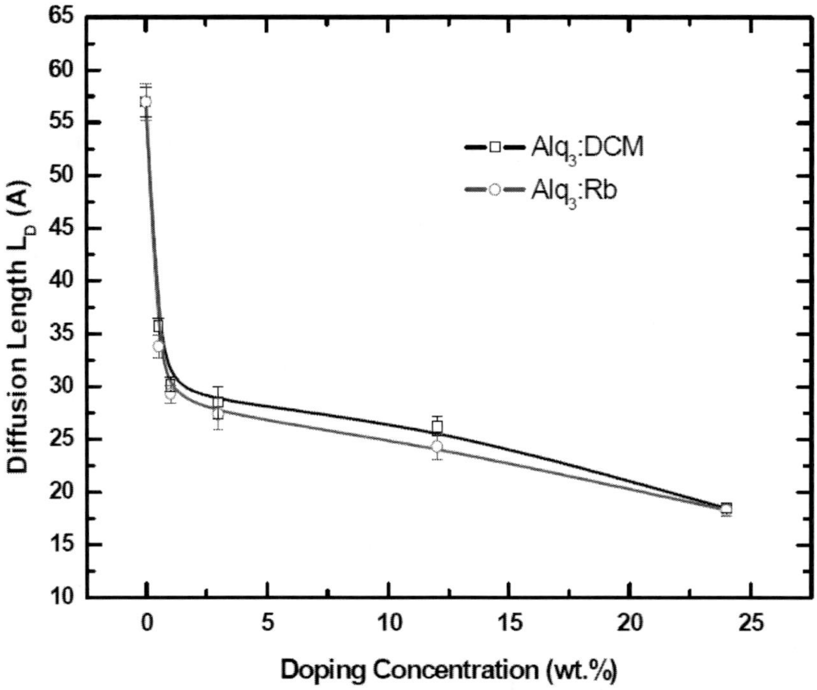

Figure 19. The plot of diffusion length L_D of singlet exciton with doping concentration in Alq_3 films. The diffusion length is sharply reduced with doping concentration up to 1 wt.%.

The lifetime of exciton in dopant molecules is controlled by induced defects in the films which are supported by the increase of exciton recombination rate.

5. Doping Effects on Charge Carrier Mobilities

Figure 20 shows the transient photocurrent in pure Alq_3 and Alq_3:DCM films for different DCM concentrations. The TOF transient in both films can be regarded as dispersive transport. The inset in Figure 20(a) shows the two lines crossed at a point which indicated flight time of carriers [38, 42].

Carrier's average mobility model [43, 44] is used to simulate TOF photocurrent instead of using the well-known mobility relation: $\mu = \frac{d^2}{(V * \tau)}$, where V is the applied voltage, τ is the flight time and d is the sample thickness. The applied model for this study is modified from Reference [43], by assuming the carriers hopping velocity v is a Gaussian function across the Alq_3 bulk film. The photocurrent is depended on the carriers average drift velocity and density.

The modified photocurrent is defined as:

$$i(t) = \frac{q_o}{d}\exp(\frac{-t}{\tau'})[\int_0^t \exp(\frac{t'}{\tau'})[\int_0^{d/t} vf(v,t)dv]dt'/\tau'] \quad , \qquad (4)$$

and the charge velocity distribution

$$f(v,t) = \frac{1}{\sqrt{2\pi\alpha^2/t}}\exp(\frac{-(v-v_o)^2}{2\alpha^2/t}) \qquad (5)$$

where α^2/t and v_o are the variance and the mean for Gaussian function distribution, respectively, t is the photocurrent measured time, q_o is the total amount of charge reached the electrode and τ' is the carrier drift time.

By fitting the simulated photocurrent to the experimental data four constant parameters, τ', α, v_o and q_o are estimated.

Table 1. The estimated exciton diffusion coefficient D_s in pure Alq_3 and in doped films for two dopants DCM and Rb with different doping concentrations

Doping conc. Wt. %	Alq_3:DCM D_s (cm^2s^{-1}) x 10^{-7}	Alq_3:Rb D_s (cm^2s^{-1}) x 10^{-7}
0	2.7 ± 0.6	
0.25	4.6 ± 0.5	4.2 ± 0.8
0.5	5.1 ± 1.0	6.2 ± 1.7
1.0	7.6 ± 0.8	7.8 ± 1.3
24.0	12.7 ± 0.6	14.3 ± 1.5

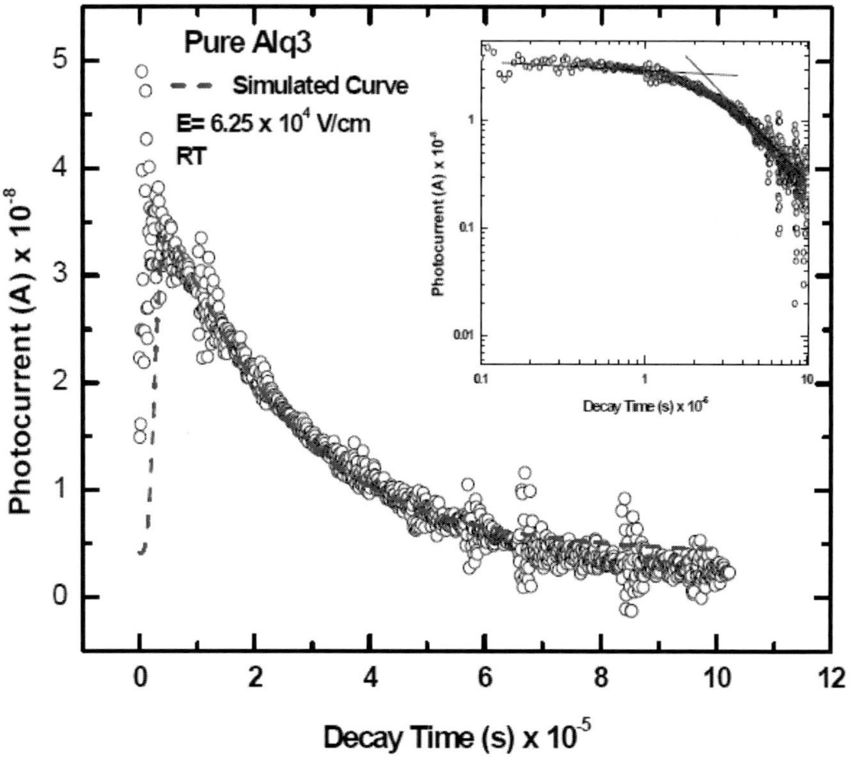

(a)

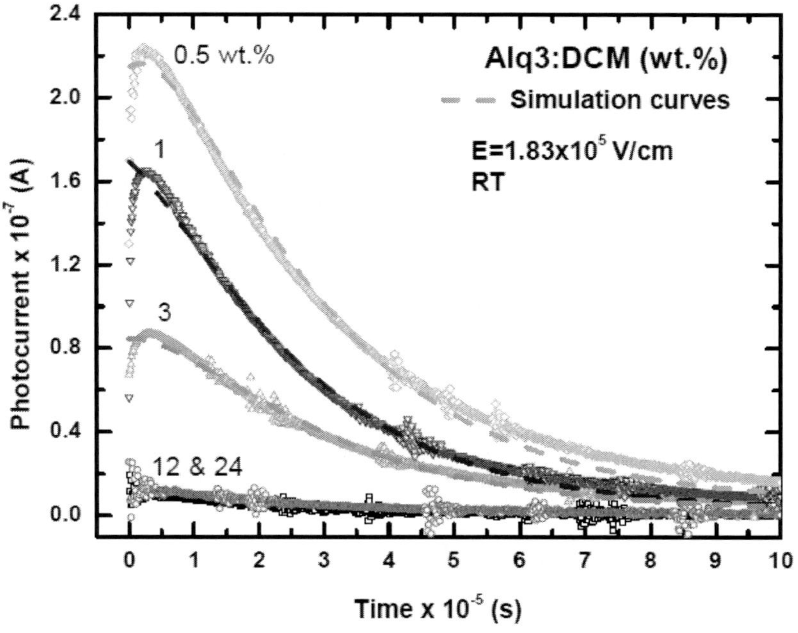

(b)

Figure 20. The typical transient photocurrent curves, (a) in pure Alq_3 film, (b) in Alq_3:DCM films for various concentrations. The inset in (a) shows the two lines crossed at the point indicated as flight time τ. The dotted lines in (a) and (b) are the simulated curves fitted with experimental data at RT.

The average velocity $\langle v \rangle$ is defined as [44]

$$\langle v \rangle = \frac{\int_0^\infty dt \int_0^\infty v f(v,t) dv}{\int_0^\infty dt \int_0^\infty f(v,t) dv} \tag{6}$$

The integrand in the numerator is the carrier hopping velocity v weighted by the Gaussian function distribution $f(v,t)$ within the time dt. The denominator is the probability of finding a carrier with any hopping velocity v and it is the total number of carriers with hopping velocity in the system. The average mobility $\langle \mu \rangle$ is estimated from the relation:

$$\langle \mu \rangle = \langle v \rangle / E, \tag{7}$$

where E is the applied electric field. Figure 20(a) and (b) show the fitted simulation photocurrent (doted lines) from equation (4) to the experimental curves in pure Alq_3 and Alq_3:DCM films with various DCM doping concentration respectively. The electron mobility in each doped film is estimated. The measured photocurrents indicate that the heavily doped (12 and 24 wt.%) films has the lower conductivity than lightly doped film. This may be due to the increase of trapping of carriers in heavily doped bulk films.

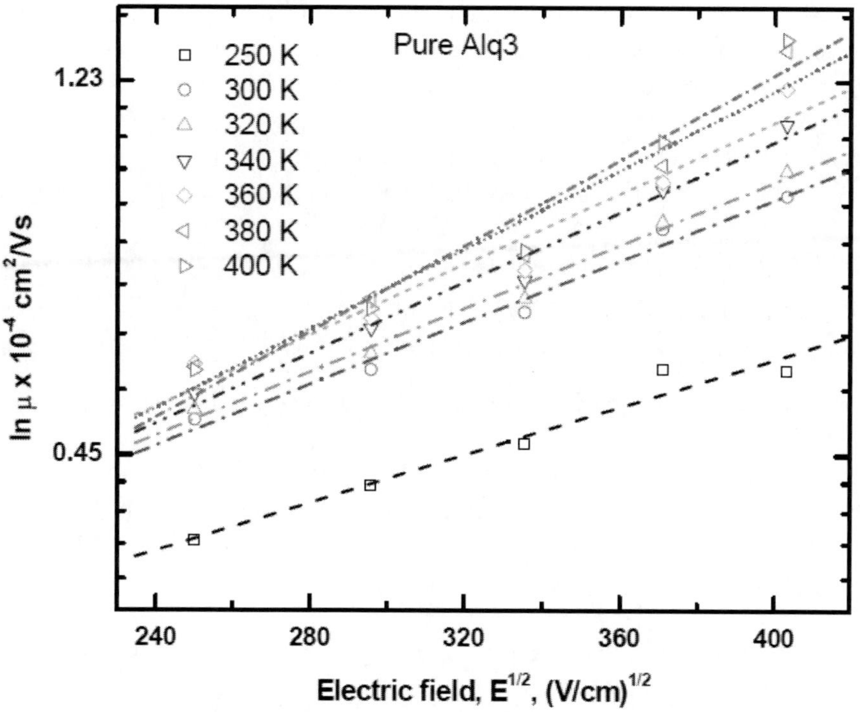

(a)

Figure 21. (Continued).

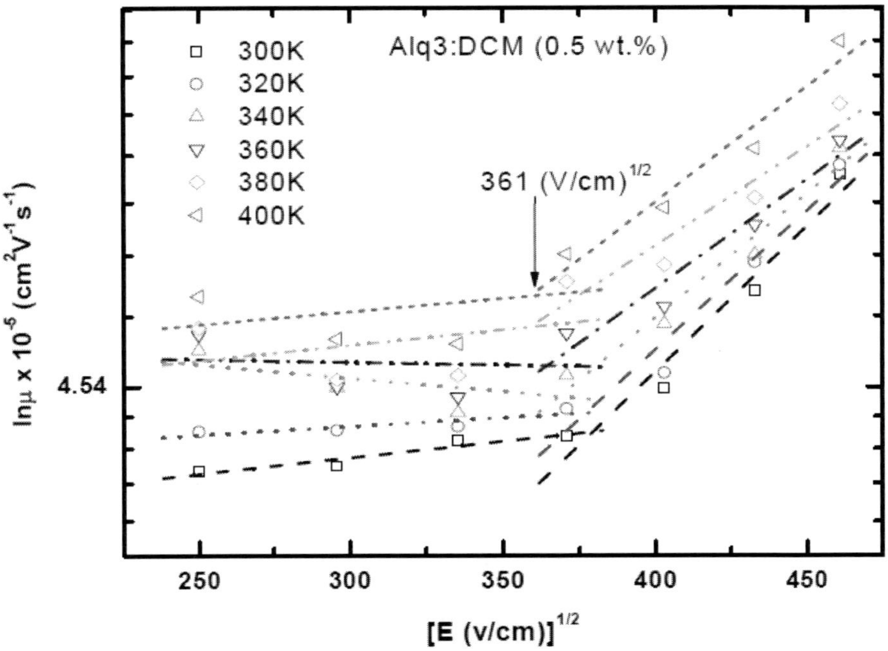

(b)

Figure 21. (Continued).

The summarized field dependence electron mobility at various temperatures in pure Alq_3 and Alq_3:DCM films for various DCM doping concentrations as shown in Figure 21 (a) - (c). The carrier mobility in both the pure and doped films is increased with temperature.

This is in contrast to the carrier transport in inorganic semiconductors such as Si where the carrier mobility decreases with temperature due to the increase of phonons scattering. The mechanism of carrier transport in organic films is by phonon assisted hopping. The local potential barrier of carrier trapping is modified with temperature due to the increased lattice vibrations. The carriers can hop more easily out of the barrier to another site with temperature. Figure 21(a) shows the field dependence electron mobility in pure Alq_3, which is closely followed the Poole-Frenkel (PF) model of mobility $\mu \propto \exp(\beta\sqrt{E})$, where β is a constant at a fixed temperature [45, 46]. The field dependence electron mobility is separated into two discrete regions of square root field ($\sqrt{E}$), which is indicated by an arrow in each curve as shown in Figure 21(b) and (c).

This discrete region phenomenon is observed even in lightly doped 0.5 wt.% Alq_3:DCM film. The carrier mobility in the doped film is only in agreement with the PF model at two different ranges of square root fields. The electron mobility characteristic in doped films into two discrete regions can be defined as $\mu_{1,2} \propto \exp(\beta_{1,2}\sqrt{E_{1,2}})$.

Here, it is indicated β_1 and β_2 for before and after the critical point of $\sqrt{E_c}$. The electron mobility in doped films is increased faster after $\sqrt{E_c}$. The observed value of β_2 is greater than β_1 . This phenomenon is not reported elsewhere.

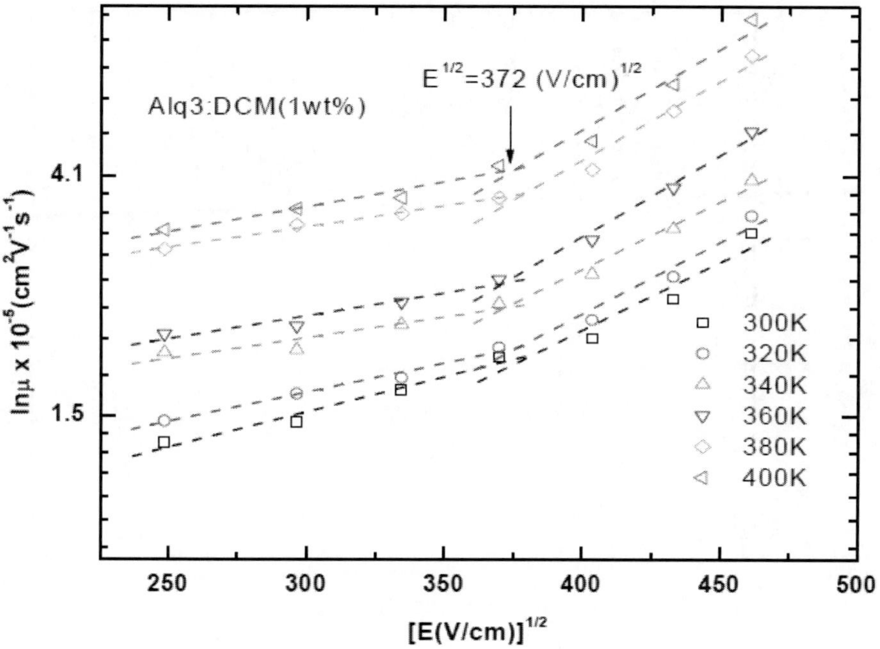

(c)

Figure 21. The plot of electron mobility versus at various temperatures (a) in pure Alq_3 film, (b)-(c) in Alq_3:DCM films for various concentrations. The arrow marks at each figure show the critical electric field of electron mobility. The $\sqrt{E_c}$ is increased as the DCM concentration is increased in Alq_3.

The values of $\sqrt{E_c}$ are shifted to the higher field with DCM concentration in Alq_3:DCM films and is saturated at DCM concentration ≥ 3 wt.%. The results are summarized in Figure 22. The values of are changed from 360 to 405 $(V/cm^2)^{1/2}$ with DCM doping concentration up to 24 wt.%. The saturated values of is attributed to the maximum polarity effect of DCM on Alq_3 molecules when the biasing field is parallel to the orientation of the dipole in films [11]. The electric field and temperature dependence carrier mobility in Alq_3:DCM bulk films can be described by the empirical equation [23, 24]:

$$\mu(E,T) = \mu_\infty \exp[-(\frac{2\sigma}{3kT})^2]\exp[\beta\sqrt{E}] , \tag{8}$$

where T is the absolute temperature, k is the Boltzmann constant, μ_∞ is the high temperature limit of the mobility, and $\beta = C[(\frac{\sigma}{kT})^2 - \Sigma^2]$; where C is a temperature and field independent constant related to the inter-site distance, σ is the energetic disorder of the film, Σ is a temperature independent parameter related to positional disorder. The electron mobility in doped-Alq_3 films depend exponentially on $\sqrt{E}$ at a fixed temperature according to Equation (8). The constant parameters value is obtained by fitting the simulated curve from Equation (8) with experimental results.

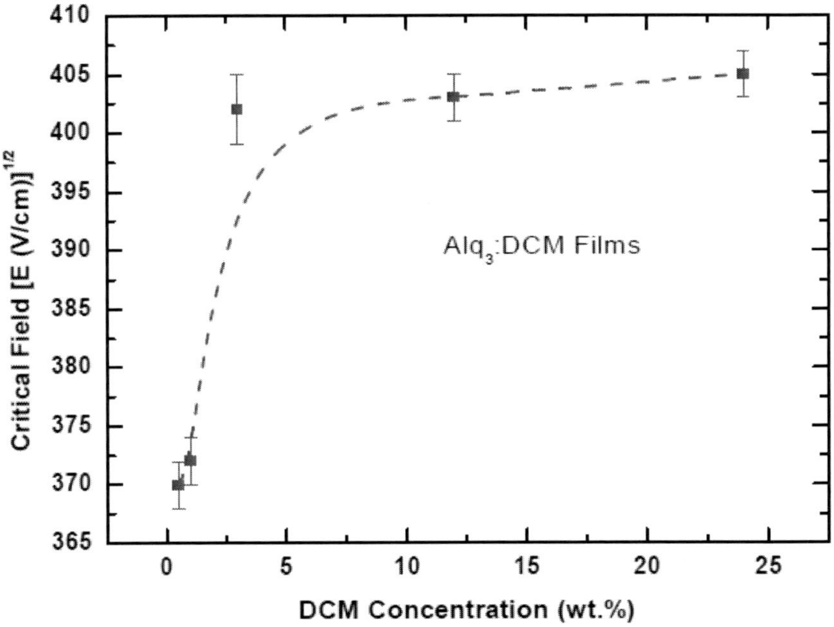

Figure 22. The plot of critical electric field $\sqrt{E_c}$ versus DCM concentrations in Alq_3 bulk films. The $\sqrt{E_c}$ is saturated at the higher concentrations of DCM ≥12 wt.% in Alq_3.

The parameters value is also depended on the nature of the charge-transporting material and the polarity of the organic material. The charge carrier transport in disorder molecular solids is changed with field and temperature due to the local (random) modifications of the polarization energy.

For this model one should be assumed (i) the distributions of intersite distances and site energies are Gaussian, (ii) the process is incoherent. The condition (i) defines the shape of the density-of-states (DOS) function [24],

$$N(\xi) = \frac{N_{tot}}{(2\pi\sigma^2)^{1/2}} \exp\left(\frac{-(\xi-\xi_m)^2}{2\sigma^2}\right), \tag{9}$$

where N_{tot} is the total DOS, ξ_m is the energy of the maximum distribution and σ is the energetic disorder. The width of the DOS is assumed as a superposition of van deer Waals σ_{vdW} and dipolar σ_{dip} components. Then

$$\sigma^2 = \sigma_{vdW}^2 + \sigma_{dip}^2 \tag{10}$$

with the dipolar component related to the permanent dipole moment of the species (*m*) and its concentration, where

$$\sigma_{dip} = \frac{Kx^b m}{\varepsilon a_0^2}, \tag{11}$$

where K and b are constants, ε is the relative electric permittivity of the medium and a_0 is the inter-site distance and x is the mole fraction.

One can apply this approach to calculate the electron mobility in Alq_3:DCM bulk films and estimate the energetic and positional disorders in films.

Figure 23(a) shows the zero-field mobility $\ln\mu(E=0,T)$ vs $(1/T^2)$ plot for pure Alq_3 bulk film as derived from Equation (8). The linear curve is extrapolated for each temperature and reveale the linear dependence as predicted in Equation (8).

One can plot the zero-field mobility $\mu(E=0,T)$ vs $(1/T^2)$ curve for each DCM doping concentration in Alq_3 film. From the slope of $\mu(E=0,T)$ vs. $(1/T^2)$ plot, the energetic disorder σ is estimated for each DCM concentration in Alq_3 film.

The results are summarized in Figure 23(b). The energetic disorder σ is estimated as σ_1 and σ_2 according for two discrete linear lines separated by $\sqrt{E_c}$ in each doped film. The energetic disorder σ_1 ($> \sigma_2$) is higher in each doped film.

The energetic disorder σ_2 is considerably reduced at $\sqrt{E} > \sqrt{E_c}$ due to the lower influenced of DCM molecules polarity and van der Waals interaction in the films. The higher DCM concentration in Alq_3:DCM films show higher energetic disorder compared to lightly doped films. The energetic disorder σ_1 is increased from 0.013 eV (in pure Alq_3 film) to 0.09 eV with DCM doping concentration up to 24 wt.%. Similar observations are also made in Reference [11].

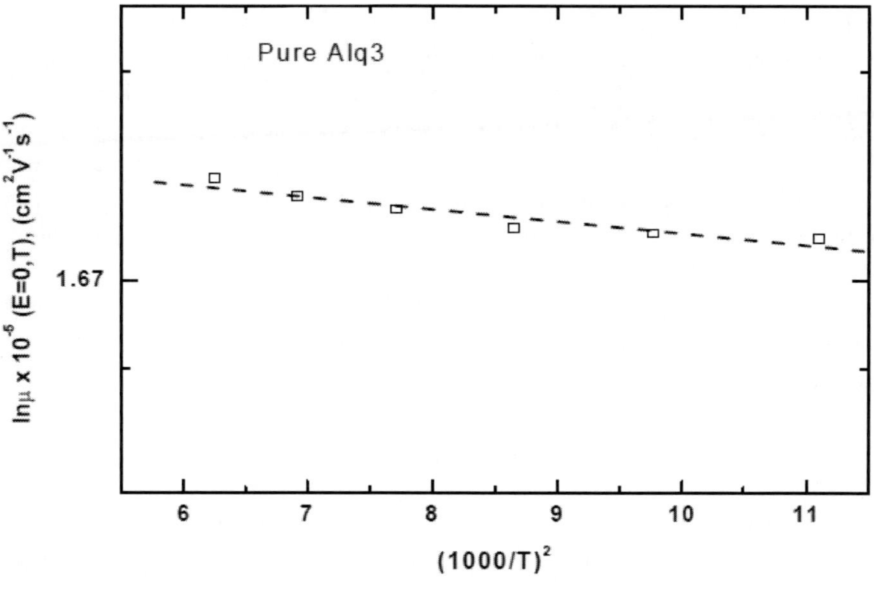

(a)

Figure 23. (Continued).

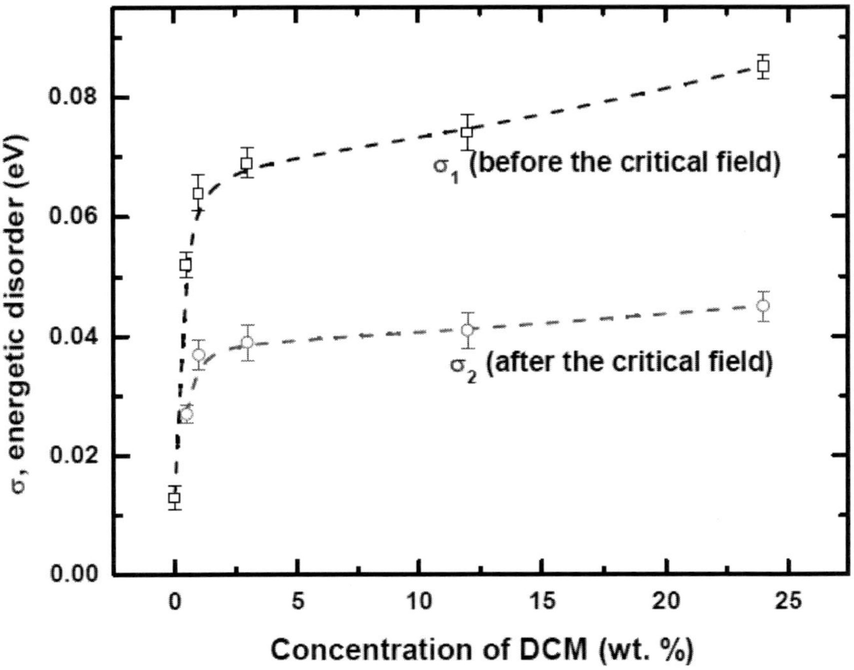

(b)

Figure 23. (a) The zero field mobility vs $1/T^2$ in pure Alq_3 film. The energetic disorder is estimated from this zero field mobility plot. (b) The plot of energetic disorders before and after the $\sqrt{E_c}$ in Alq_3:DCM for various concentrations.

The energetic disorder has lower influence on charge carrier mobility at the higher electric field in amorphous film. The DCM molecule has a tendency to aggregate the molecular structure asymmetry due to the strong dipolar interactions [47]. At highly doped films (such as 12 and 24 wt.%) the DCM molecules may be arranged themselves in an anti-parallel manner due to the strong attraction between molecules of reduce distance.

The DCM molecules have also strong interaction with surrounding Alq_3 molecules due to the phenoxide donor and a pyridyl acceptor in the quinolate ligands of Alq_3 as shown in Figure 1. These interactions rise the positional and energetic disorders in the doped films. The dipole orientation dominates the geometric disorder than the fluctuation in intermolecular separation.

At the higher field, Alq_3 and DCM molecules may be polarized and well aligned with each other's and reduce the effect of dipolar contribution in the DOS function. The energetic disorder σ will mainly depend on van der Waals component.

That's why the higher electron mobility ($\beta_2 > \beta_1$) when the bias field is greater than the critical field $\sqrt{E_c}$.

Figure 24(a) shows the plot of β as a function of for 1 wt.% DCM doped in Alq_3 before and after the critical field. The β has the linear relationship with $(\frac{\sigma}{kT})^2$. The constant C is estimated from the slope of the linear curve. The linear line is extrapolated to obtain the positional disorder, $\sum$.

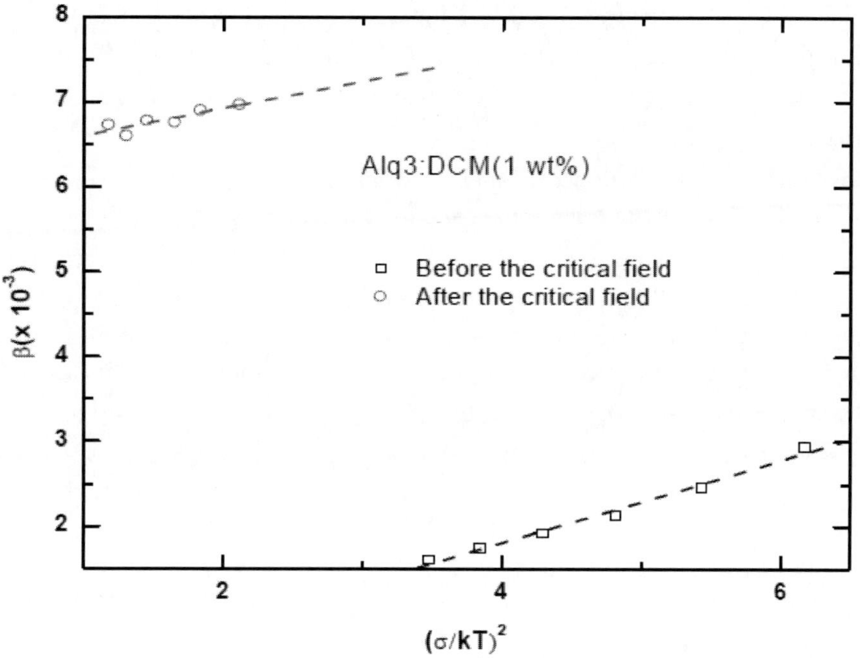

(a)

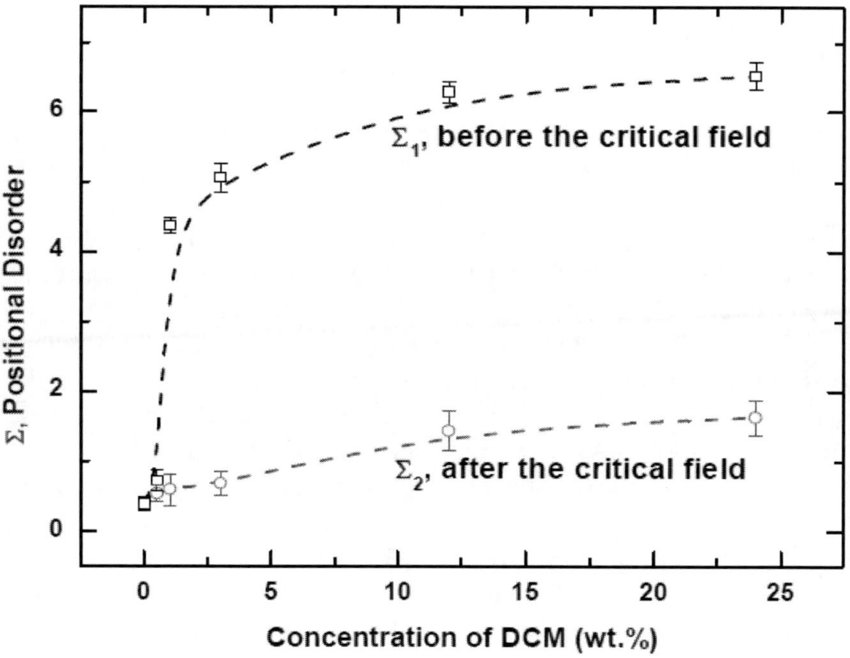

(b)

Figure 24. (a) The example of the plot of corresponding β vs. $(\sigma/kT)^2$ in Alq_3:DCM (1 wt.%) film. The constant C and position disorder $\sum$ are estimated from these plots. (b) The plot of positional disorders before and after the critical field $\sqrt{E_c}$ in Alq_3:DCM films for various concentrations.

The plot of β as a function of curves for each DCM doping film is to estimate the constant C and positional disorder, $\sum$. Figure 24(b) shows the results of positional disorder for different DCM concentrations. The is assigned to and for two regions of field. The values of $\sum_1$ are found higher than $\sum_2$.

The positional disorder in doped films is related with the geometric disorders associated with randomness both in dipole orientations and intermolecular separations.

The $\sum_1$ values are changed from 0.3 to 6.5 with DCM doping concentration up to 24 wt.% in Alq_3. The Coulombic interaction between the randomly distributed polar DCM and Alq_3 molecules is responsible for the increased positional disorder with doping concentration. This also leads to broaden the effective density of states distribution. The positional disorder $\sum_2$ values are changed from 0.3 to 1.9 with DCM doping concentration up to 24 wt.%. The positional disorders are saturated at the higher doping concentration of DCM. The increase of positional disorder gives the energetic disorder due to the positional molecular dependence potential energy of interaction.

This has been proven from the above result of energetic disorder σ ($\sigma_{\text{van der}}$ Waals and $\sigma_{\text{dipole-dipole}}$), which is estimated higher as (i) the doping concentration of DCM increased, and (ii) the bias field lower than the critical field $\sqrt{E_c}$. It is recognized that the dipole-dipole interaction between randomly located and oriented dipoles is one of the major causes of positional and energetic disorders in doped films. The random distribution of DCM dopant molecules Coulombically interact with carriers localized in intrinsic hopping sites. The DCM molecule has higher dipolar interactions due to consist of donor-like dimethylamine, and acceptor-like dicyanomethylene. The Alq_3 molecule also shows the dipole characteristic where a phenoxide donor and a pyridyl acceptor formed the quinolate ligands. The higher densities of randomly distributed DCM molecules interact either via attractive or repulsive among dimethylamine or dicyanomethylene with phenoxide or pyridyl of Alq_3 molecules. The degree of interaction strongly depends on the density and the polarity of DCM molecules. The intermolecular interaction forces stretch molecular bonds and shift charge distribution on molecules, which effectively broadens the DOS distribution. In addition to the broad of DOS distribution, disordered DCM doped films probably have deeper localized states originated from chemical and structural defects. These deeper localized states are normally located well below the intrinsic DOS distribution, which is often referred as deep traps. These traps are strongly affected the charge transport characteristics in doped Alq_3 film.

The energy of the localized states is affected by the nearest dopant molecules. The total density of deep localized states is increased with increasing dopant concentration. The increase of DCM concentration in Alq_3 bulk films induce the additional deep traps and broaden the deep tail of the DOS. The change of DOS distribution by doping is affected the electron carrier hopping mobility in Alq_3:DCM films.

6. Doping Effects on Trap Centers

Figure 25 shows the differential capacitance as a function of bias voltage at different frequencies from 40 kHz to 1 MHz in pure Alq_3 and Alq_3:DCM films of doping concentration

1 wt.%. This single layer device on ITO makes it simple to understand the carrier transport as well as trapping nature in pure and doped Alq_3 films. The bias dependent measurements show that from near zero to positive bias the capacitance is nearly frequency independent. The capacitance is increased with increasing negative biases on Al electrode. The highest capacitance increment is observed at 40 kHz measurements in all samples. The capacitance increment is reduced with increasing frequency up to 600 kHz after that the capacitance is again increased with increasing frequency. The carrier displacement is a factor to influence the capacitance under the bias field. Carriers are hopping across the localized states in the film in Gaussian form on the application of an electric field.

The transient C-V measurements from a reverse to forward bias voltage sweep can be able to fill all bulk traps within the Alq_3 film or electrode/organic interfaces. The parallel plate capacitance equation with Alq_3 film between two electrodes can be written as

$$C = \frac{\varepsilon_r \varepsilon_0 A}{d}, \tag{12}$$

where d is the Alq_3 film thickness, A is the cross-sectional area (cm^2) of Al film and ε_0 is the permittivity of free space. ε_r is the dielectric constant of Alq_3 material or its relative permittivity.

The dielectric constant ε_r is directly proportional to the capacitance. The capacitance of the sample is increased due to the polarization of Alq_3 film in negative bias regions. The positive and negative charges are displaced in the film with respect to their equilibrium positions which is indicated in Equation(12).

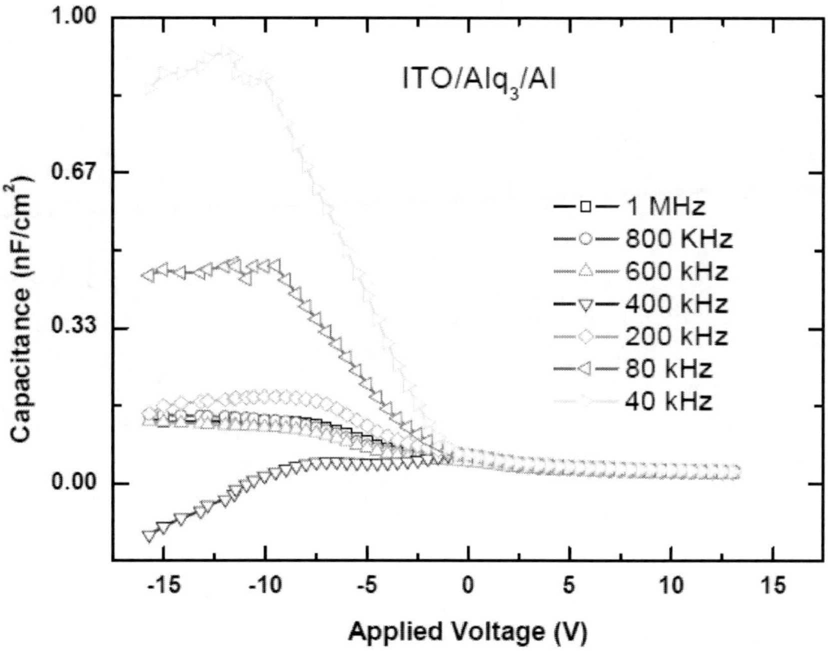

(a)

Figure 25.(Continued).

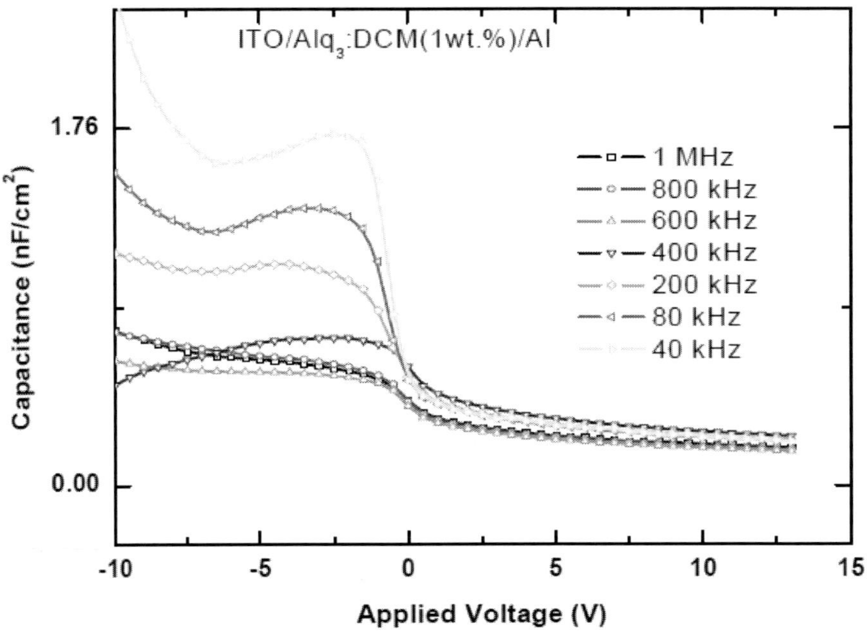

(b)

Figure 25. The C-V measurements of organic film sandwiched between two ITO and Al electrodes at room temperature at different frequencies. As an example (a) For pure Alq_3 film, (b) for Alq_3:DCM film. The doping concentration was 1 wt.%.

By neglecting the depletion layer effects with two electrodes, the dielectric constant ε_r of pure and doped Alq_3 films are calculated from Equation (12) for parallel plate configuration as

$$\varepsilon_r = \frac{dC}{A\varepsilon_0} \tag{13}$$

The dielectric constant ε_r versus frequency measurements are summarized in Figure 26 for all samples. The dielectric constant of all films is increased as the frequency reduced. The dielectric constant of all doped Alq_3 films are also frequency and doping concentration dependent. At low frequency, the dielectric constant was increased with applied bias due to the increased of charge separation. The reduction of dielectric constant with increasing frequency is due to the incomplete dielectric relaxation of carriers. The real part of dielectric constant ε_r in a single relaxation model in a solid is given by [48],

$$\varepsilon_r = \varepsilon_U + \frac{\varepsilon_R - \varepsilon_U}{1+\varpi^2\tau_E} \tag{14}$$

where, ε_U is the unrelaxed dielectric constant at higher frequencies; ε_R is the relaxed dielectric constant at low frequencies, ϖ is the frequency and τ_E is the dielectric relaxation time. The ε_r decreases with increasing frequency ϖ from a value of ε_R to ε_U.

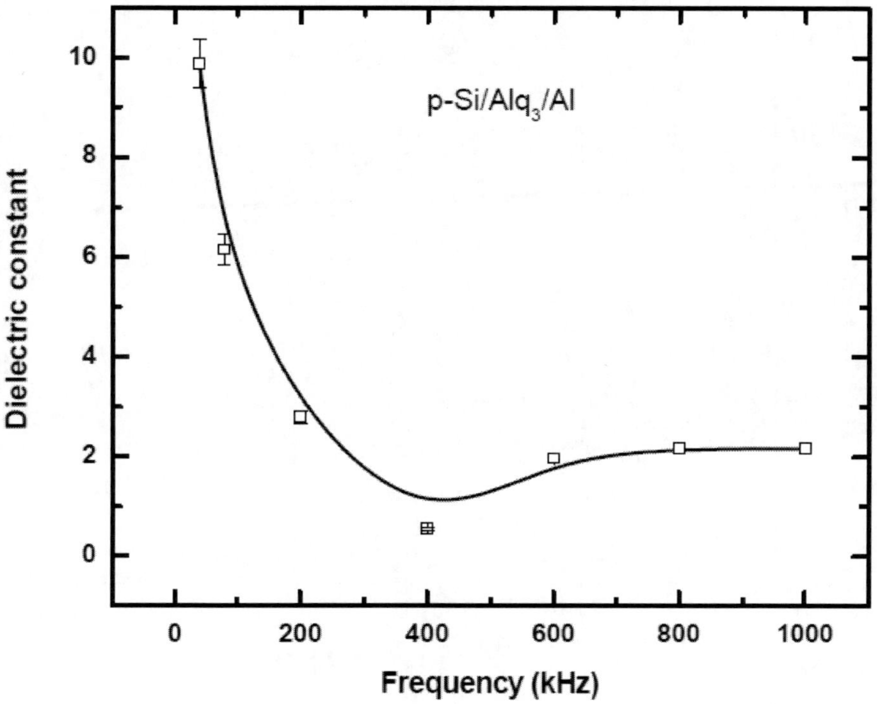

(a)

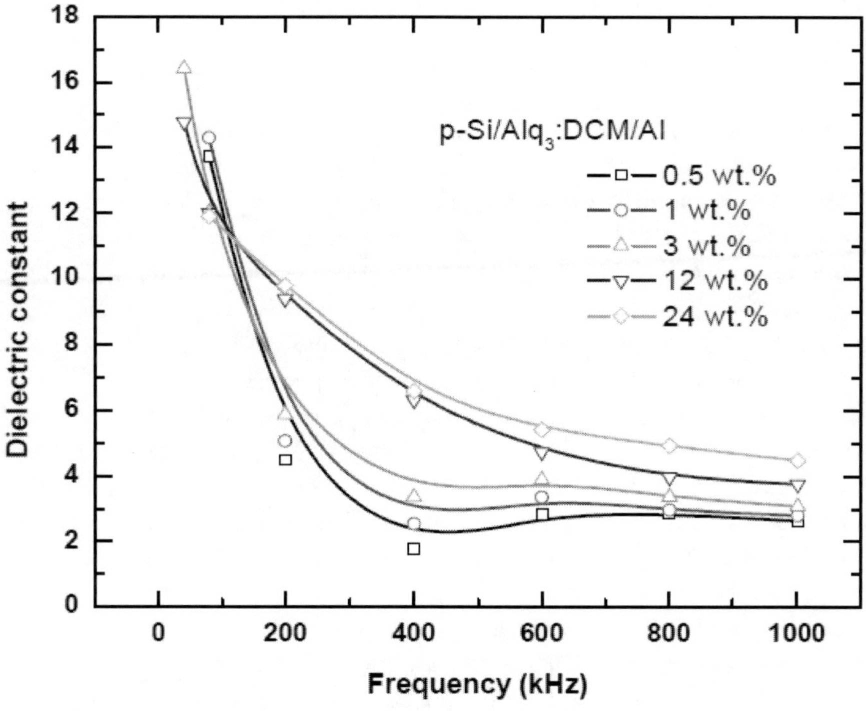

(b)

Figure 26. (Continued).

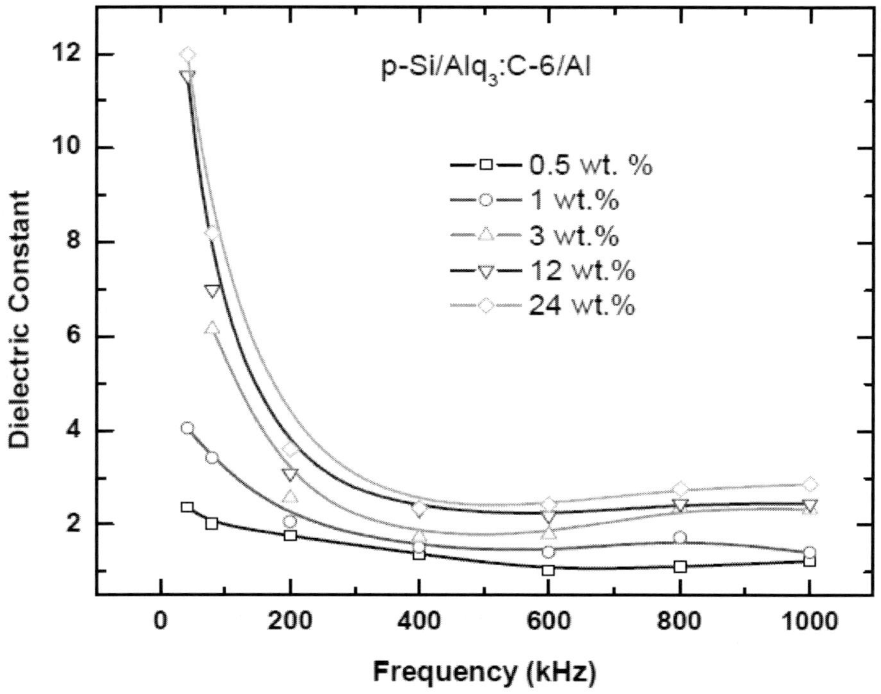

(c)

Figure 26. The dielectric constant of Alq_3 films as a function of frequency for different doping concentration. (a) For pure Alq_3 film, (b) for Alq_3:DCM film, and (c) for Alq_3:C-6 film. The higher doping concentration gives higher dielectric constant for all dopants. The dielectric constant is increased with reducing frequency in all films.

The increase of ε_r with decreasing frequency is the indication of interfacial polarizations [49-51]. The interfacial polarization increases the capacitance of the sample [49].

The strong dipole moment of dopant molecules such as DCM and C-6 are induced stronger polarization compared to dopant Rb molecules in Alq_3 film. The DCM molecules have arylidene laser dye with a dimethylamino [-$N(CH_3)_2$] electron donor group linked to dicyanovinyl [=$C(CN)_2$] electron acceptor groups by an effective π-conjugated pathway of alternating double and single bonds links the two units have strong interaction with phenoxide donor and pyridyl acceptor in the quinolate ligands of Alq_3 molecules [38].

The C-6 has benzothiazolyl and diethylamino groups, which have strong electron affinities to form localized states to trap carriers. Thus, as the concentration of these dopants were increased in Alq_3 films, the stronger interaction with Alq_3 molecules induced polarization in doped film for higher capacitance [38]. However, Rubrene ($C_{42}H_{28}$) is a molecule comprised of tetracene with four additional phenyl rings attached at the center positions, which is a good charge transporter and non-polar structure compare to DCM and C-6. Rubrene doped in Alq_3 film can trap carriers, and acts as recombination center for hole and electron. When carriers are impeded in their motion, space charges and macroscopic field distortions occur. When the carriers are hopping under the influence of an electric field, it is possible to localize or trap in doped films as well as in the interface of electrode and doped films which is indicated in our observation of higher dielectric constant values in Alq_3:DCM and Alq_3:C-6 films [Figure 26(b) and (c)]. The trapping of charge is one of the reasons for the observed large increase of dielectric constant at low frequencies. Dipoles between the trapped carriers increase the polarization vector. The dipole orientation can dominate the geometric disorder than the fluctuation in intermolecular separation. The interactions of polar dopant molecules broad the total density of states, induced the local electric fields and increased the intermolecular interaction forces in the films. The polarity of DCM and C-6 molecules in Alq_3 films may induce the deeper localized states and normally located below the intrinsic density-of-state (DOS) distribution of Alq_3 [24]. These traps can strongly affect the charge transport property at equilibrium conditions and most carriers can occupy these deep states. The Coulombic interaction between the randomly distributed polar DCM molecules and Alq_3 molecules can increase the positional disorder and leads to broaden the effective density of states distribution. Another reason for the increase of dielectric constant ε_r at low frequencies is the increase of polarization arising from the rotation of dopant molecules in doped Alq_3 films under applied field. The polarization is also enhanced by high density trapped charges in doped films. The DCM and C-6 possess stronger dipole moments where they have stronger interaction with surrounding Alq_3 molecules due to their strong dipole characteristic.

Figure 27 shows the $\frac{1}{C^2}$ versus V plot from capacitance-voltage characteristics measurements of pure and doped Alq_3 films at a frequency of 40 kHz. The bias voltage V is swept from negative to positive values on Al electrode.

All observed transients are non-exponential in disordered organic films in which transport states, traps or both, are distributed in Gaussian form. The $\frac{1}{C^2}$ versus V plots from all doped Alq_3 films are shown in Figure 27(b) – (c) for different doping concentrations. The curves Figure 27(b) and (c) shows two slopes as a function of bias voltage V.

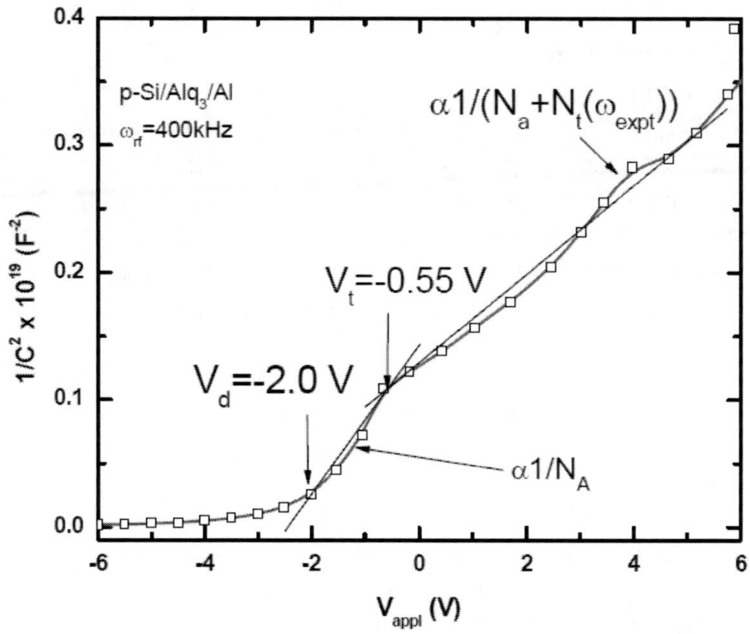

(a)

Figure 27. (Continued).

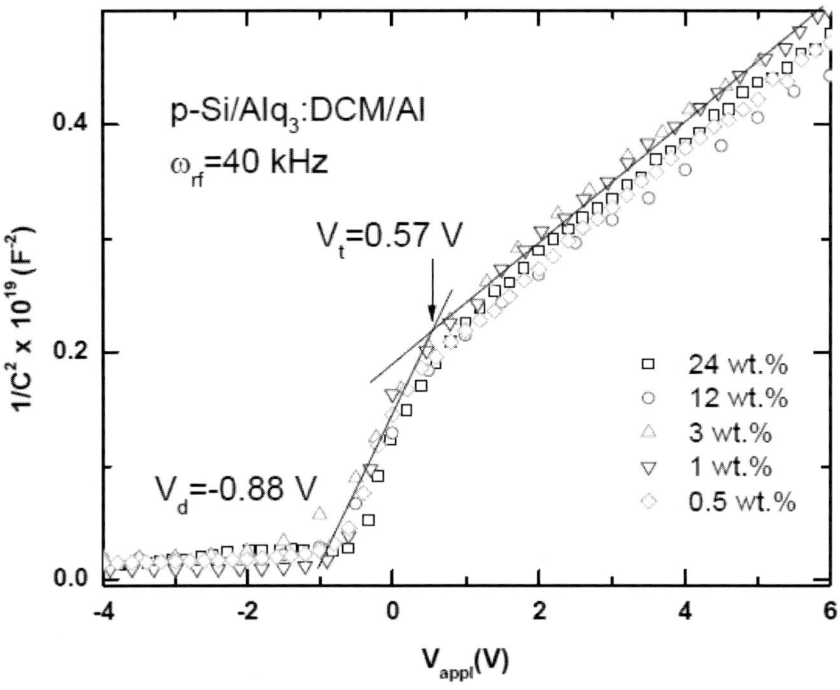

(b)

Figure 27. (Continued).

The curve behavior shows the majority carriers' traps in organic layer sandwiched between two electrodes. Electrons injected into Alq_3 layer encountered the interface barrier formed between electrodes. Normally, the transport current is limited in the Alq_3 film by the electron injection from the Al cathode [4].

The Al cathode forms a Schottky contact with Alq_3 layer for electron with a barrier height of ~1.3 eV [52]. This interface barrier height limits the electron injection into organic film. These injected electrons can be captured or emitted from the interface traps.

The positive slope indicates that the traps are acted as electron acceptors in the film. For the reverse to forward bias, the slope changes slightly in Alq_3:Rb films which indicates the presence of a single abundant deep level. The electron trap concentration N_t can be estimated from the slope of $\frac{1}{C^2}$ versus V plot in each doped Alq_3 film. The slope changes in presence of deep trap levels with bias voltages. The positive bias voltages put the Fermi level E_F below the deep level somewhere in the depletion region. The larger gradient of slope at positive bias arises from the density of acceptor traps N_t in all doped Alq_3 films and is proportional to $1/N_t$. This slope is increased significantly for polar dopants in Alq_3:DCM and Alq_3:C-6 films. The Alq_3:Rb films has lower slope compare to Alq_3:DCM and Alq_3:C-6 films.

At the slope changing bias V_t with reducing positive bias, the traps pulled below the Fermi level at the interface and become permanently filled. They can no longer respond in the experiment. If the trap state lies at interfaces of Al/Alq_3 and/or Alq_3/silicon, the value of V_t can give the trap depth from $E_t = q(V_t - V_d) + E_F$; where V_d is the diffusion potential which is estimated from the plot as shown in Figure 27.

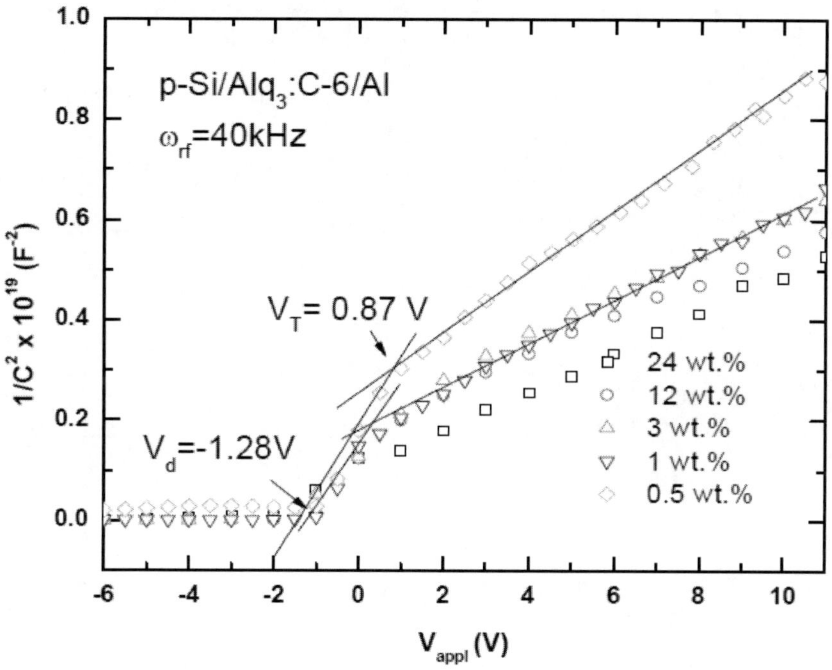

(c)

Figure 27. Typical $\frac{1}{C^2}$ versus V plots for different doping concentration at room temperature. (a) For pure Alq_3 film (b) for Alq_3:DCM films, (c) for Alq_3:C-6 films at 40 kHz. Each of the curves from figure (a)-(c) shows two distinct slopes as a function of bias. This behavior is the characteristic of a Schottky diode with majority carrier traps. The indication of V_d and V_t values represent the diffusion potential and slope changing bias, respectively.

The estimated trap depth E_t is increased from 0.65 ± 0.05 eV to 0.71 ± 0.15 eV as the dopant concentration was increased from 0.5 to 24 wt% in Alq_3:DCM films. The trap depth E_t in all doped films are shown in Figure 28.

The estimated E_t value in pure Alq_3 film is higher than the reported values of 0.2-0.4 eV from thermally stimulated current measurements [53, 54]. No comparison can be made for the Alq_3:DCM, Alq_3:C-6 and Alq_3:Rb films as for the first time it is estimated. The higher values of trap depth in pure Alq_3 film may be due to the oxidation.

Figure 29 shows the slightly change of capacitance with time for each sample on a log-linear plot. The data are recorded at room temperature for an 8.0 V applied bias on Al cathode after being stored samples in nitrogen atmosphere for 4 hours at 0 V bias. The increase of differential capacitance shows the amount of negative charge decreases in the depletion region. The trap carriers are escaped even after 4 hours which indicates some traps are very deep within the energy gap.

The polarity of DCM and C-6 molecules in Alq_3:DCM and Alq_3:C-6 films induce the formation of deeper localized states and normally located below the intrinsic DOS distribution of Alq_3. The trapping states are in the forbidden gap and strongly affect the charge transport property at equilibrium conditions. The acceptor type traps can capture and release electrons with depending trap depth.

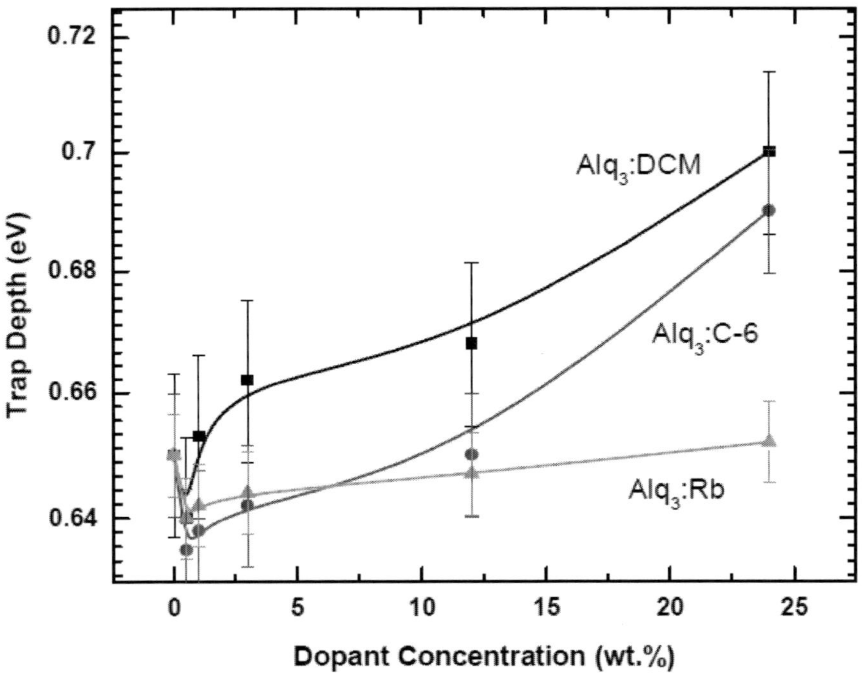

Figure 28. The plot of trap depth energy E_t versus the dopant concentration (wt.%) for different films. The trap energy E_t is raised from 0.65 ± 0.05 eV to 0.71 ± 0.15 eV as the dopant concentration is increased in Alq_3:DCM films. The change of E_t in Alq_3:Rb films with doping concentration is almost negligible.

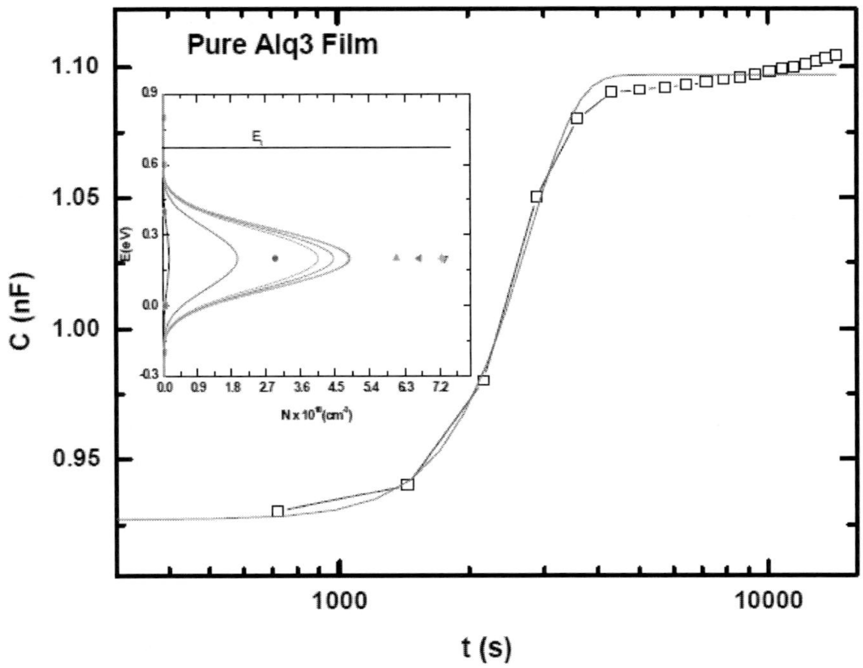

(a)

Figure 29. (Continued).

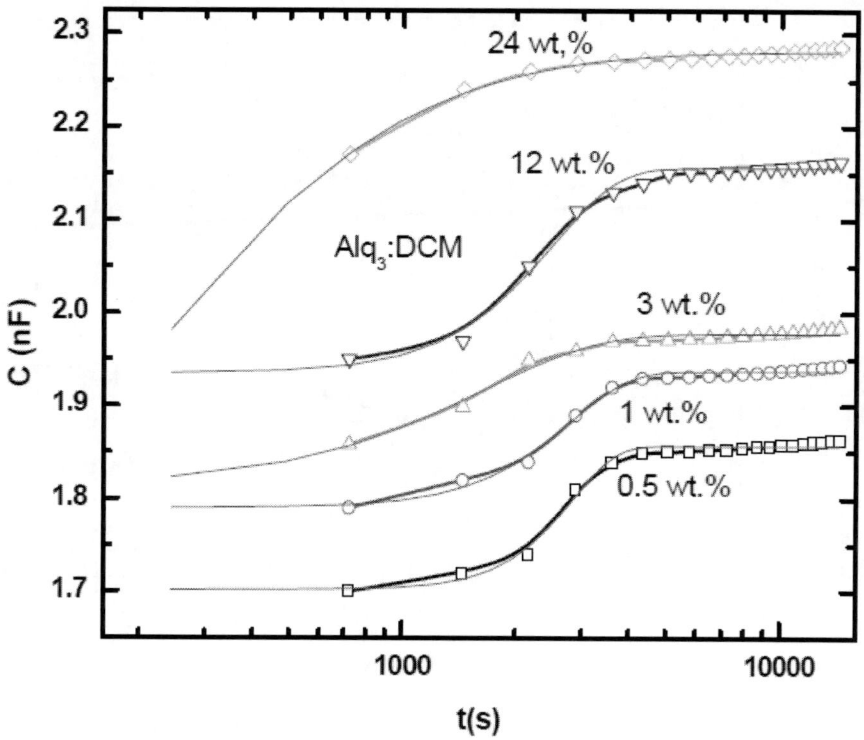

(b)

Figure 29. (Continued).

The simulated model from Ref.30 is well fitted to the experimental data at room temperature as shown in Figure 29. The carrier mobility in these doped films is determined by time-of-flight measurements in the order of 10^{-5} $cm^2V^{-1}s^{-1}$ at room temperature [38].

The average transit time of carriers across the depletion region is in the order of 10 μs. It is assumed that the re-trapping of carriers is negligible on the time scale of experiment. The decay function *f(t)* is calculated for a range of different E_t, and Gaussian width σ. The inset in Figure 29(a) shows the density of states N_{DOS} of different E_t and σ based on the Gaussian distribution model. The relative permittivity of each sample is taken from Figure 2 at 40 kHz. The N_{DOS} is taken as 5 x 10^{18} cm^{-3} for the simulation. The Gaussian width σ is found as 0.16 ± 0.02 eV in pure Alq_3 film, and increase to ~ 0.30 ± 0.04 eV in doped films. The estimated Gaussian width in pure Alq_3 film is in agreement with reported values [55]. The polarity of DCM and C-6 molecules broad the total density of states, induced the local electric fields and is increased the intermolecular interaction forces in the films to form the deeper localized states and normally located below the intrinsic density-of-states distribution of Alq_3.

For Alq_3:Rb films, the van der Waals component of Rb molecules is the only contributor to induce energetic disorder in the film. The geometric disorders are generally associated with randomness both in dipole orientations and in intermolecular separations. The fluctuation in intermolecular separation should dominate the geometric disorder in Rb molecules. The estimated values of the acceptor trap densities N_t are found as 4x10^{17} cm^{-3}, 1x10^{18} cm^{-3}, 9x10^{17} cm^{-3} and 5x10^{17} cm^{-3} in pure Alq_3, Alq_3:DCM, Alq_3:C-6 and Alq_3:Rb films, respectively.

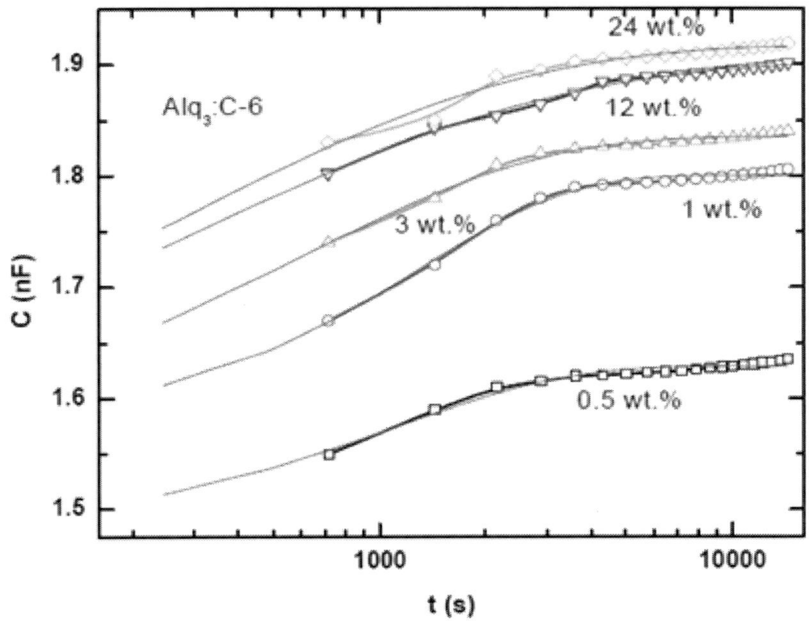

(c)

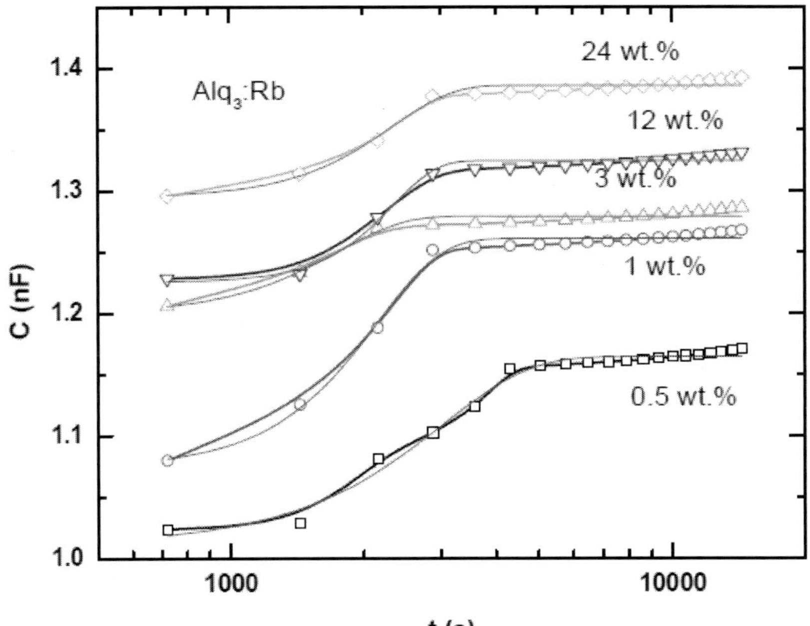

(d)

Figure 29. Log-linear plot of differential capacitance C with time t(s) for different films measured at room temperature. The simulated model from Ref. 30 is well fitted to experimental results (a) For pure Alq_3 film, (b) for Alq_3:DCM films, (c) for Alq_3:C-6 films and (d) for Alq_3:Rb films for different doping concentrations. The capacitances are increased considerably up to 4000 seconds. The capacitance continues to increase even after 4 hours. The inset in (a) shows the Gaussian distribution width of various σ, trap depth E_t and trap concentration N_t are varied to fit the experimental data.

Based on the simulation model, the N_t is found increasing with distance from the Al cathode and its value is approximately doubling at the distance nearly 60-75 nm. The dopant in Alq_3 film changes the dielectric behaviors of films with trap density N_t and trap depth E_t.

These trap states are also influenced the injected electrons from Al into the Alq_3 film. The injected electron are transported by hopping between these trap sites and giving rise to field-dependent mobility and transit time spread.

The initially injected charge carriers are trapped as the probability of trapping process increases with increasing transit time. The derived mobility of carriers becomes smaller in the low-field regime than the actual trap-free charge carrier mobility due to one or more additional stops at trapping sites. The measured activation energy of traps is expected to increase from the actual value due to the increasing contribution from de-trapping processes at decreasing electric field. The activation energy of de-trapping processes depends on trap depth E_t. The longer time emission of trapped carriers suggests that the traps are deep within the energy gap in all doped films.

CONCLUSION

The luminescence properties of dopants C-6, Rb and DCM in Alq_3 films are investigated at room temperature for different doping concentrations. The maximum luminescence efficiency of 6.5 cd/A is observed for the Rb doping concentration of ~ 0.7 wt.% in Alq_3 which is nearly double compared in pure Alq_3 device emission. The luminescence efficiency is improved to ~ 7 cd/A in the co-doping in Alq_3:{C-6:Rb(1:2)} film.

The bias voltage is reduced in co-doped films compared to pure Alq_3, Alq_3:Rb and Alq_3:C-6 devices for the same current density. The direct recombination of holes and electrons in the dopant molecules may be responsible for the improvement of the luminescence efficiency. The EL peak position shifted from green to red color with DCM doping concentration in Alq_3. The highest EL efficiency was found for DCM doping concentration of 0.5 wt.%. The EL intensity is increased about 2.5 times than the pure Alq_3. The EL onset delay time and the offset rise time are measured as a function of doping concentration from 0.5 to 24 wt.% in Alq_3 film. The EL onset (delay time) t_d is limited by the formation of NPD/Alq_3 interfacial space charge. The charge carrier transport behavior in Alq_3 determines the delay time t_d. The delay time t_d is increased from 0.78 to 1.86 μs with doping concentration. The EL saturation time t_r is also increased from 1.2 to 2.8 μs with doping concentration. The carrier mobility in pure and doped Alq_3 devices is estimated from the EL delay time t_d. The estimated carrier mobility is found as 0.5 – $1.2\text{x}10^{-5}$ $cm^2V^{-1}s^{-1}$ in doped films which is lower than the values obtained from the TOF method due to the formation of interfacial barriers and space charges.

The exciton recombination rate is found to increase with doping concentration for both dopants DCM and Rb in Alq_3 film. The exciton diffusion coefficient in pure Alq_3 film is found as (2.7 ± 0.6) x 10^{-7} cm^2s^{-1}. The estimated diffusion coefficient of excitons in doped films is 0.5 – 1.4 x 10^{-6} cm^2s^{-1}, which is nearly one order higher than in pure Alq_3 film. The diffusion length of exciton in doped films is reduced to 18 - 35 Å from ~ 57 Å in pure Alq_3 film. The exciton lifetime in pure Alq_3 is found as ~ 20 ns and it is reduced to ~8 ns with doping concentration for both dopants. The electron mobility is reduced with DCM doping concentration in Alq_3 films. The field dependence electron mobility is separated into two discrete regions of a critical electric-field $\sqrt{E_c}$. The values of $\sqrt{E_c}$ is shifted to the higher values with DCM concentration and saturated. The similar trend is also observed for the positional and energetic disorders in DCM doped films. The energetic disorder is increased from 0.013 eV to 0.09 eV with DCM concentration up to 24 wt.% in Alq_3. The positional disorder is also changed from 0.3 to 6.5 with DCM concentration. The results indicate strong Coulombic and dipole-dipole interactions between DCM and Alq_3 molecules. It is realized that the dipole-dipole interaction between randomly located and oriented dipoles is one of the major causes of energy disorder in organic materials.

The dielectric constant of Alq_3, Alq_3:DCM, Alq_3:C-6 and Alq_3:Rb films are increased with reducing frequency. The dielectric constant is depended on doping concentration and dopant dipole moments. The estimated trap energy E_t is found as 0.65 ± 0.05 eV and 0.71 ± 0.15 eV in pure and doped Alq_3 films, respectively. The estimated trap densities N_t are found as $4\text{x}10^{17}$ cm^{-3}, $1\text{x}10^{18}$ cm^{-3}, $9\text{x}10^{17}$ cm^{-3} and $5\text{x}10^{17}$ cm^{-3} in pure Alq_3, Alq_3:DCM, Alq_3:C-6 and Alq_3:Rb films, respectively. The density of states N_{DOS} = $5\text{x}10^{18}$ cm^{-3} and Gaussian distribution width $\sigma = 0.16 \pm 0.02$ eV in Alq_3 film are also estimated.

REFERENCES

[1] M. A. Baldo, D. F. O' Brien, M. E. Thompson and S. R. Forrest, "Excitonic singlet-triplet ratio in a semiconducting organic thin film", *Phys. Rev*. B 60 (1999) 14422.

[2] T. Li, W. Li, X. Li, B. Chu, Z. Su, L. Han, Y. Chen, Z. Hu and Z. Zhang, «Improved performances of red organic light-emitting devices by co-doping a rubrene derivative and DCJTB into tris-(8-hydroxyquinoline) aluminum host", *J. Lumin*. 130 (2010) 1676; J. Chen and D. Ma, «Improved color purity and efficiency by a coguest emitter system in doped red light-emitting devices", *J. Lumin*. 122-123 (2007) 636.

[3] C. W. Tang, S. A. VanSlykex and C. H. Chen, "Electroluminescence of doped organic thin films", *J. Appl. Phys*. 65 (1989) 3610.

[4] A. Uddin, C. B. Lee, X. Hu and T. K. S. Wong, "Interface injection-limited carrier-transport properties of Alq3", *Appl. Phys*. A 78 (2004) 401.

[5] B. J. Chen, X. W. Sun, T. K. S. Wong, X. Hu and A. Uddin, "Organic light-emitting devices with *in situ* postgrowth annealed organic layers", *Appl. Phys. Lett*. 87 (2005) 06 3505.

[6] A. Uddin, C. B. Lee, X. Hu, T. K. S. Wong and X. W. Sun, "Effect of doping on optical and transport properties of charge carriers in Alq3" *J. Crystal Growth* 288 (2006) 115.

[7] S. R. Forrest, "Ultrathin Organic Films Grown by Organic Molecular Beam Deposition and Related Techniques", *Chem. Rev*. 97 (1997) 1793.

[8] B. J. Jung, C. B. Yoon, H. K. Shim, L-M Do and T. Zyung, "Pure-Red Dye for Organic Electroluminescent Devices: Bis-Condensed DCM Derivatives", *Adv. Funct. Mater*. 11 (2001) 430.

[9] G. Y. Zhong, J. He, S. T. Zhang, Z. Xu, Z. H. Xiong, H. Z. Shi, X. M. Ding, W. Huang and X. Y. Hou, "*In situ* photoluminescence investigation of doped Alq", *Appl. Phys. Lett*. 80 (2002) 4846.

[10] A. Uddin, C. B. Lee and T. G. Andersson, "Initial rise of transient electroluminescence in doped Alq3 films", *Phys. Status Solidi* A 207 (2010) 2334.

[11] J. Sworakowski, "Effect of polar molecules on the transport and localization of charge carriers in molecular materials", *Brazil J. Phys*. 29, 318(1999).

[12] C. Hosokawa, H. Tokailin, H. Higashi, and T. Kusumoto, "Transient electroluminescence from hole transporting emitting layer in nanosecond region", *Appl. Phys. Lett*. 63, 1322 (1993)

[13] V. R. Nikitenko and H. von Seggern, "Nonequilibrium transport of charge carriers and transient electroluminescence in organic light-emitting diodes" *J. Appl. Phys*. 102, 103 708 (2007).

[14] A. J. Pal, R. Osterbacka, K. M. Kallman, and H. Stubb, "Transient electroluminescence: Mobility and response time in quinquethiophene Langmuir–Blodgett films", *Appl. Phys. Lett*. 71, 228 (1997)

[15] S. W. Feng, M. C. Shih, C. J. Huang and C. T. Chung, "Impacts of dopant concentration on the carrier transport and recombination dynamics in organic light emitting diodes", *Thin Solid Films* 517, 2719 (2009).

[16] J. Kalinowski, N. Camaioni, P. Di Marco, V. Fattori and A. Martelli, "Kinetics of charge carrier recombination in organic light-emitting diodes", *Appl. Phys. Lett*. 72, 513 (1998)

[17] D. Braun, D. Moses, C. Zhang and A. J. Heeger, "Transient electroluminescence from polymer light emitting diodes", *Synth. Met*. 55-57, 4145 (1993)

[18] S. Kirstein, G. Cohen, D. Davidov, U. Scherf, M. Klapper, K. Chmil, K. Müllen' "Transient and DC electroluminescence of some new conjugated polymers", *Synth. Met*. 69, 415 (1995).

[19] P. W. M. Blom and M. C. J. M. Vissenberg, "Charge transport in poly(*p*-phenylene vinylene) light-emitting diodes", *Mat. Science and Eng*. 27, 53(2000)

[20] Y. Ohmori, N. Tada, Y. Kurosaka and K. Yoshino, "R-G-B Emitting organic EL device and its pulse response characteristics", *Synth. Met*. 102, 1099 (1999)

[21] S. Athanasopoulos, E. V. Emelianova, A. B. Walker and D. Beljonne, "Exciton diffusion in energetically disordered organic materials", *Phys. Rev*. B 80 (2009) 19520 9.

[22] S. Athanasopoulos, E. Hennebicq, D. Beljonne and A. B. Walker, "Trap Limited Exciton Transport in Conjugated Polymers", *J. Phys. Chem*. C112 (2008) 11532.

[23] V. I. Arkhipov, E. V. Emelianova, P. Heremans and G. J. Adriaenssens, "Equilibrium hopping conductivity in disordered materials", *Journal of Optoelectronics and Advanced Materials* 4, 425 (2002).

[24] E. V. Emelianova and G. J. Adriaenssens, "Stochastic approach to hopping transport in disordered materials", *Journal of Optoelectronics and Advanced Materials* 6, 1105 (2004).

[25] A. Uddin and C. B. Lee, "Exciton behaviours in doped tris(8-hydroxyquinoline) aluminum (Alq3) Films", *Phys. Status Solidi* C 8, No. 1, 80–83 (2011).

[26] S. Nowy, W. Ren, A. Elschner, W. Lovenich and W. Brutting, "Impedance spectroscopy as a probe for the degradation of organic light-emitting diodes", *J. Appl. Phys*. 107 54501 (2010).

[27] H. L. Kwok, "Modeling negative capacitance effect in organic polymers", *Solid-state Electronics* 47, 1089 (2003).

[28] A. J. Campbell, D. D. C. Bradley, E. Werner and W. Brutting, "Transient capacitance measurements of the transport and trap states distributions in a conjugated polymer", *Organic Electronics* 1, 21 (2000).
[29] Z. H. Kafafi, H. Murata, L. C. Picciolo, H. Mattoussi, C. D. Merritt, Y. Lizumi, and J. Kido, *Pure Appl. Chem.* 71 (1999) 2085.
[30] A. J. Campbell, M. S. Weaver, D. G. Lidzey and D. D. C. Bradley, "Bulk limited conduction in electroluminescent polymer devices", *J. Appl. Phys.* 84 (1998) 6737.
[31] G. Parthasarathy, P. E. Burrows, V. Khalfin, V. G. Kozlov and S. R. Forrest, "A metal-free cathode for organic semiconductor devices", *Appl. Phys. Lett.* 72 (1998) 2138.
[32] S. Barth, P. Muller, H. Riel, P. F. Seidler, W. Rieb, H. Vestweber and H. Bassler, "Electron mobility in tris(8-hydroxy-quinoline)aluminum thin films determined via transient electroluminescence from single- and multilayer organic light-emitting diodes", *J. Appl. Phys.* 89 (7), 3711 (2001).
[33] V. R. Nikitenko, Y. H. Tak and Bassler, "Rise time of electroluminescence from bilayer light emitting diodes", *J. Appl. Phys.* 84, 2334(1998).
[34] C. B. Lee, A. Uddin and C. C. Teo, "Effect of polar dopant on energetic and positional disorders in tris(8-hydroxyquinolinato) aluminum *(*Alq3*)", Appl. Phys.* A 83, 115 (2006).
[35] S. K. Pal, D. Sukul, D. Mandal, S. Sen and K. Bhattacharyya, "Solvation dynamics of DCM in micelles", *Chem. Phys. Lett.* 327, 91(2000).
[36] U. S. Raikar, C. G. Renuka, Y. F. Nadaf, B. G. Mulimani, and A. M. Karguppikar, *J. Fluorese* 16, 847(2006).
[37] H. H. Fong, S. K. So, W. Y. Sham, C. F. Lo, Y. S. Wu and C. H. Chen, "Effects of tertiary butyl substitution on the charge transporting properties of rubrene-based films", *Chemical Physics* 298, 119(2004).
[38] C. B. Lee, A. Uddin and T. G. Andersson, "Investigation of charge carrier mobility in 5,6,11,12-tetraphenylnapthacene (rubrene) and coumarin 6 doped Alq3 films", *Solid State Comm.* 142, 206(2007).
[39] C. Wu, Y. Zheng, C. Szymanski and J. McNeill, "Energy Transfer in a Nanoscale Multichromophoric System: Fluorescent Dye-Doped Conjugated Polymer Nanoparticles", *J. Phys. Chem.* C 112 (2008) 1772.
[40] R. Priestley, A. D. Walser and R. Dorsinville, "Temperature dependence of transient photoluminescence in tris(8-hydroxyquinoline) aluminum (Alq_3)", *Opties Communications* 158 (1998) 93.
[41] M. A. Baldo and S. R. Forrest, "Transient analysis of organic electrophosphorescence: I. Transient analysis of triplet energy transfer", *Phys. Rev.* B 62 (2000) 10958.
[42] G. G. Malliaras, Y. Shen, D. H. Dunlap, H. Murata and Z. H. Kafafi, "Nondispersive electron transport in Alq_3 ", *Appl. Phys. Lett.* 79, 2582 (2001).
[43] J. C. Scott, L.Th. Pautmeier and L.B. Schein, "Mean mobilities of charge carriers in disordered media", *Phys. Rev.* B 46, 8603 (1992).
[44] R. M. Eisberg *"Fundamentals of Modern Physics"* John Wiley and Sons, 1961.
[45] P. E. Burrows, Z. Shen, V. Bulovic, S. R. Forrest and M.E. Thompson, "Relationship between EL and current transport in OLED", *J. Appl. Phys.* 79, 7991 (1996).
[46] S. Naka, H. Okada, H. Onnagawa, Y. Yamguchi and T. Tsutsui, "Carrier transport properties of organic materials for EL device operation", *Synth. Met.* 111, 331 (2000).

[47] G. Y. Zhong, Z. Xu, J. He, S. T. Zhang, Y. Q. Zhan, X. J. Wang, Z. H. Xiong, H. Z. Shi, X. M. Ding, W. Huang and X. Y. Hou, "Aggregation and permeation of 4-(dicyanomethylene)-2-methyl-6-(p-dimethylaminostyryl)-4H-pyran molecules in Alq", *Appl. Phys. Lett.* 81, 1122 (2002).

[48] M. Ershov, H. C. Li, L. Liu, M. Buchanan, Z. R. Wasilewsky and A. K. Jonscher, "Negative Capacitance Effect in Semiconductor Devices", IEEE Trans. *Electron Devices* 45, 2196 (1998).

[49] S. Jayalekshmi and M. G. K. Pillai, "Preparation and characterization of a new high quality polymer thin film from citral", *Thin Solid Films* 122, 197 (1984).

[50] A. Von Hippel, E. P. Gross, J. G. Jelatis and M. Geller, "Photocurrent, Space-Charge Buildup, and Field Emission in Alkali Halide Crystals", *Phy. Rev.* 91, 568 (1953).

[51] H. Birey, "Dielectric properties of aluminum oxide films", *J. Appl. Phys.* 49, 2898 (1978).

[52] M. Stoßel, J. Staudigel, F. Steuber , J. Blassing , J. Simmerer , A. Winnacker, H. Neuner, D. Metzdorf, H.-H. Johannes and W. Kowalsky, "Electron injection and transport in 8-hydroxyquinoline aluminium", *Synthetic Metals* 111–112, 19 (2000).

[53] J. Staudigel, M. Stoßel, F. Steuber, J. Blassing, and J. Simmerer, "Activation energies in organic light emitting diodes comprising ohmic contacts both for electron and hole injection", *Synthetic Metals* 111–112, 69 (2000).

[54] Y. S. Jeong, D. Troadec, A. Maliton, B. Ratier, R. Antony and G. Veriot, "Dielectric studies of Alq_3 and transport mechanisms", *Synthetic Metals* 127, 195 (2000).

[55] J. Steiger, R. Schmechel and H. V. Seggern, „Energetic trap distributions in organic semiconductors", *Synthetic Metals* 129, 1 (2002).

In: Doping
Editor: Lixin Yu

ISBN: 978-1-62618-097-0

Chapter 2

DOPED MANGANESE OXIDE NANOSTRUCTURES FOR ENERGY APPLICATION

R. S. Kalubarme,[1,*] Chan-Jin Park[1] and P. M. Shirage[2,†]

[1]Department of Material Science and Engineering, Chonnam National University, Buk-gu, Gwangju, Republic of Korea
[2]Department of Condensed Matter Physics and Materials Science, Tata Institute of Fundamental Research (TIFR), Colaba, Mumbai

ABSTRACT

The transition metal oxide (TMO) nano-structures with the insulators have created new avenues for the applied research in future energy devices. In particular, manganese oxide (MnO_x) has been considered as the most promising TMO for next generation energy applications like supercapacitors and batteries due to its low cost, environmental friendliness and natural abundance along with its wide structural diversity combined with its unique chemical and physical properties. Furthermore, this attractive functional metal oxide exhibits different polymorphs depending on the arrangement of the basic MnO_6 structure. Nevertheless, the electrochemical performance of MnO_x depended on their morphology, porosity, specific surface area, electrical conductivity and ionic transport within the pore.

Herein, manganese oxide (MnO_2) nanostructures were prepared with chemical techniques using the precursors containing K^+ and NH_4^+. Nanorods with an α - MnO_2 structure and a diameter ranging from 25 to 40 nm, along with tipped needles with a β-

* R. S. Kalubarme: Department of Material Science and Engineering, Chonnam National University, 77 Yongbong-ro, Buk-gu, Gwangju - 500 757 Republic of Korea. Tel: +82-62-530-0278, Fax:+82-62-530-1699; E-mail: ramkalubarme@yahoo.co.in.

† P. M. Shirage: Department of Condensed Matter Physics and Materials Science, Tata Institute of Fundamental Research (TIFR), Dr. Homi Bhabha Road, Colaba, Mumbai- 400 005. E-mail: p.shirage@tifr.res.in and paras.shirage@gmail.com.

MnO_2 structure and a diameter of 100 nm were obtained. Addition of NH_4^+ has increased the diameter of the nanorods obtained. The 25 nm diameter α MnO_2 nanorods showed the best catalytic activity for dissociation of HO_2^- formed during oxygen reduction in a KOH solution. The MnO_2 nanostructures preferably followed two-electron oxygen reduction mechanism in a LiOH solution. Potassium used as stabilizing agent enhanced the electrocatalytic properties of the manganese oxide for the oxygen reduction reaction in aqueous as well as in non-aqueous medium. The supercapacitive behavior shown by the NH_4^+ added MnO_2 is higher. For charge storage application NH_4^+ added MnO_2 has shown greater efficiency while K^+ added MnO_2 had possessed higher efficiency for catalytic application.

1. Introduction

The transition metal oxide (TMO) nano-structures with the insulators have created new avenues for the applied research in future energy devices. In particular, manganese oxide (MnO_x) has been considered as the most promising TMO for next generation energy applications like supercapacitors [1, 2] and batteries [3] due to its low cost, environmental friendliness and natural abundance along with its wide structural diversity combined with its unique chemical and physical properties. The various M^{n+}–MnO_2 (i.e. where M^{n+} denotes the counter cation) structures have been continuously studied for the last 3 decades. This attractive functional metal oxide exhibits the tunnel (1D), layer (2D) and spinel (3D) structures of MnO_2 depending on the arrangement of the basic MnO_6 structure. The one-dimensional MnO_2 structure with regular porous lattices have been classified by Turner and Buseck such as 1×1 pyrolusite (β-MnO_2), 2×2 hollandite (α-MnO_2), and 1×1, 1×2 intergrowth nsutite (γ-MnO_2). The layered structure of the δ type and a spinel structure of λ type are the examples of the polymorphs of MnO_2 structure. In addition, the size and morphology of MnO_x plays a key role in determining the properties for practical applications. In particular, nanostructured MnO_x are the promising candidate materials for the electrocatalysis [4] and energy devices like, supercapacitors [1, 2] and batteries [3]. Nevertheless, the electrochemical performance of MnO_x depended on their morphology, porosity, specific surface area, electrical conductivity and ionic transport within the pore.

In this study, MnO_2 nanostructures were synthesized using a chemical technique to obtain the different morphologies. The structural evolution of the MnO_2 as a result of precursor material and their concentration is thoroughly investigated. In addition, the effects of the crystal structure, size and shape of MnO_2 on their electrochemical performance are also analyzed.

2. Experimental

For the synthesis of MnO_2, the precursor used was $MnSO_4$. The oxidizers used for the hydrothermal technique contain the sources of K^+ and $NH4^+$ ions. The ratios of precursor to oxidizer used in the present investigation were 1:0.5, 1:1, and 1:2. Chemical with desired composition were dissolved in deionized water with vigorous stirring.

The resulting mixture was transferred to the Teflon lined stainless steel autoclave that was further placed in the oven maintained at 150 °C for 24 h. After the reaction, the autoclave was cooled to room temperature. The precipitates were washed with deionized water and absolute alcohol in centrifuge machine followed by overnight drying in oven.

The structure and morphology of the samples was examined by X-ray diffractometer (XRD) with Cu K_α radiation and scanning electron microscopy (SEM), respectively. X-ray photoelectron spectroscopy (XPS) was used for compositional analysis.

The electrodes for electrochemical and supercapacitor measurement were prepared by pasting a mixture of MnO_2, acetylene black and PTFE on Ni foam followed by drying in vacuum oven and roll pressing. A typical electrochemical cell for the measurement consisted of a MnO_2 based electrode as the working electrode, a Ni foam as the counter electrode, and Hg/HgO as a reference electrode. For oxygen reduction studies, the cyclic voltammograms of the electrodes were obtained over the potential range of 200 to -800 mV vs. Hg/HgO with a scan rate of 1 $mV{\cdot}s^{-1}$. The charge-discharge curves for nanostructured manganese oxide materials were recorded in 1M NaOH electrolyte and above mentioned working, counter and reference electrodes with applied current of 1mA during testing.

Air electrodes with a composition of 80 wt.% carbon, 10 wt.% MnO_2 catalyst and 10 wt.% poly vinylidene fluoride (PVDF) were prepared as mentoned above to form a cathode sheet and dried at 100 °C under vacuum for 4 h.

In a glove box under Ar atmosphere, Li/O_2 cells with an air window of 1.35 cm^2 were assembled by stacking a Li foil, a Celgard™ 5500 membrane, a carbon air cathode and an air window in sequence into a 2032 coin cell. An electrolyte prepared in an argon-filled glove-box by adding 1 M $LiSO_3CF_3$ in a propylene carbonate (PC)/Dimethyl Carbonate (DMC) by the ratio of 1:1 (by weight) was injected into the cell. The galvanostatic discharge tests were conducted using the cells at a current density of 0.2 mA/cm^2. Specific capacity of the cells was calculated with reference to the weight of carbon in the air cathode.

3. RESULTS AND DISCUSSION

MnO_2 can be obtained from the oxidation of a $MnSO_4$ precursor by $KMnO_4$ during hydrothermal treatment. The chemical reaction between Mn^{2+} and MnO_4^- ions to produce MnO_2 is as follows:

$$3Mn^{2+} + 2MnO_4^- + 2H_2O \rightarrow 5MnO_2 + 4H^+ \qquad (1)$$

Figure 1(a) shows XRD patterns of the dried precipitates obtained by the hydrothermal reaction using the precursor and oxidizer with a ratio of 1:0.5 (sample A_1), 1:1 (sample A_2), and 1:2 (sample A_3).

The XRD peaks corresponding to A_1 and A_2 were indexed to the tetragonal α-MnO_2 crystal structure [space group: I4/m (87)] with lattice constants of a=9.784 Å and c=2.863 Å, indicating that high purity and crystallinity. In contrast, the XRD pattern corresponding to A_3 showed peaks corresponding to the tetragonal β-MnO_2 crystal structure [space group: P42/mnm (136)] with lattice constants of a = 4.39 Å and c = 2.87 Å.

With the Mn^{2+} as the main Mn source, MnO_2 nanorods could be obtained by liquid-phase oxidation method. NH_4F was used as oxidizer reagent for $MnSO_4$, and the whole reaction is given in equation (2):

$$MnSO_4 + 2NH_4F + 2H_2O \rightarrow MnO_2 + (NH_4)_2SO_4 + 2HF + 2H^+ \quad (2)$$

The XRD pattern for the powder samples resulted from the reaction of precursor and oxidizer with the ratio 1:0.5 (B_1), 1:1 (B_2) and 1:2 (B_3) were shown in Figure 1(b). The x-ray plots for sample B_1 contains some impurity peaks along with the peaks corresponding to α-MnO_2 structure. With increase of the oxidizer in the hydrothermal reaction; the impurity peaks vanishes completely and phase pure α-MnO_2 powder is obtained for ratio 1:2. The crystallinity of the formed material is found to be increased with the increased amount of oxidizer. The lattice parameter calculated from the obtained peak is a = 9.801 Å and c = 2.874 Å, which is greater than the A_1 and A_2 powders.

Figure 2 shows TEM images of the synthesized MnO_2 powders. The length and diameter of the nanorods varied with the $MnSO_4$: $KMnO_4$ ratio. The α-MnO_2 nanorods have an average diameter of 40 nm and 25 nm for samples A_1 and A_2, respectively. In sample A_1, the nanorods were connected to each other at the ends or the middle of their bodies. This attached structure of nanorods reduces their effective area. In contrast, sample A_2 showed well grown nonorods without attachment between each other.

Sample A_3 showed the needle shaped nanostructures with a mean diameter of 100 nm with tips. The average length of the nanorods was 10 μm and 0.5μm for samples A_1 and A_2, respectively. The presence of K in the α-MnO_2 was verified by EDS, which is needed to stabilize the structure.

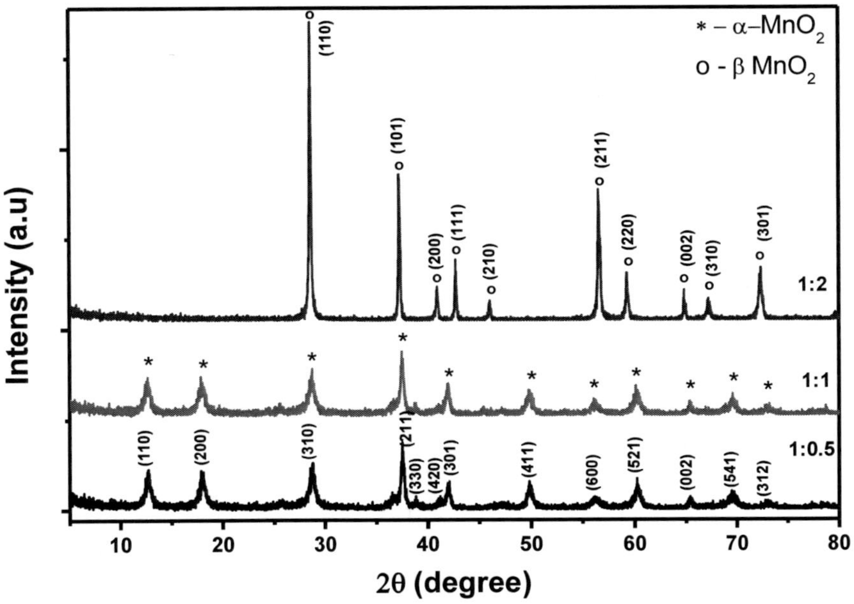

Figure 1. (a) XRD pattern of the product formed by a hydrothermal reaction of $MnSO_4$ and $KMnO_4$ with a ratio of 1:0.5, 1:1, and 1:2.

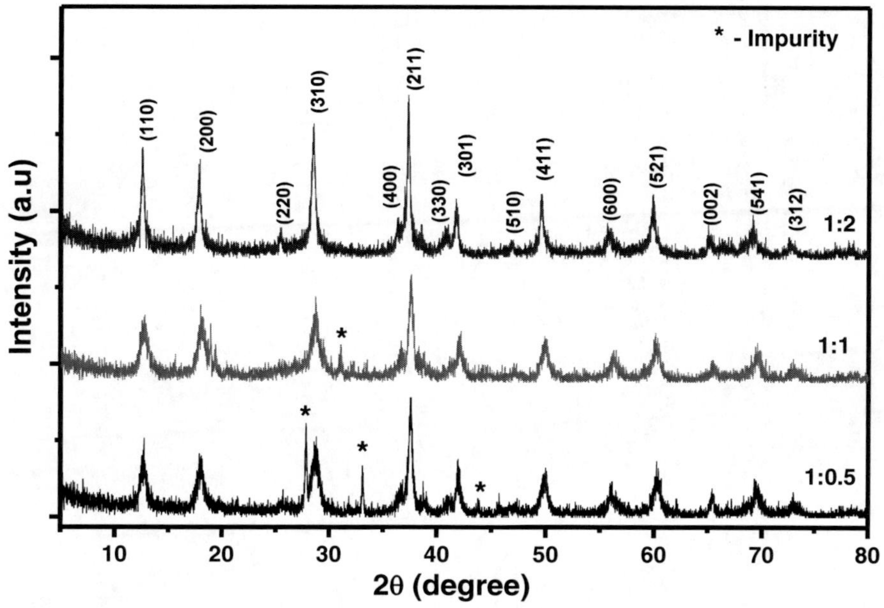

Figure 1. (b) XRD pattern of the product formed by a hydrothermal reaction of $MnSO_4$ and NH_4F with a ratio of 1:0.5, 1:1, and 1:2.

The phase transformation mechanism for MnO_2 was suggested by Wang et al. [5], which was further confirmed by Shen et al. [6] in an study using in-situ synchrotron powder XRD.

According to them, δ-MnO_2, a layered structure of MnO_2, is the primary product formed during the hydrothermal process, which further transforms to tunnel structures.

In addition, α-MnO_2 nanorods agglomerate and transform to β-MnO_2 nanostructures [7]. The formation of β-MnO_2 in a condition of a precursor to oxidizer ratio of 1:2 is validated by the following reaction:

$$MnSO_4 + 2KMnO_4 \rightarrow 3MnO_2 + K_2SO_4 + O_2 \quad (3)$$

In this reaction, the K^+ required for the stability of the 2 × 2 structure is consumed completely.

Therefore, MnO_6 is rearranged to the stable 1 × 1 structure of β-MnO_2. However, the size of the α-MnO_2 nanorods and β-MnO_2 needles obtained in this work were much smaller than earlier reports using a permanganate oxidizer [6, 8].

The TEM images shown in Figure 2(B) reveal that the microstructure of the hydrothermally prepared samples exhibit nanorod like structure for the entire precursor to oxidizer composition ratios.

The mean diameter of the nanorods for B_1, B_2 and B_3 powders are 95, 80, and 60 nm respectively. The length of the nanorods varies between 1μm to 10 μm.

The experimental results obtained under hydrothermal conditions revealed that all the MnO_2 one-dimensional nanostructures have a common evolving process, which is characteristic of a developing mechanism and phase transformation.

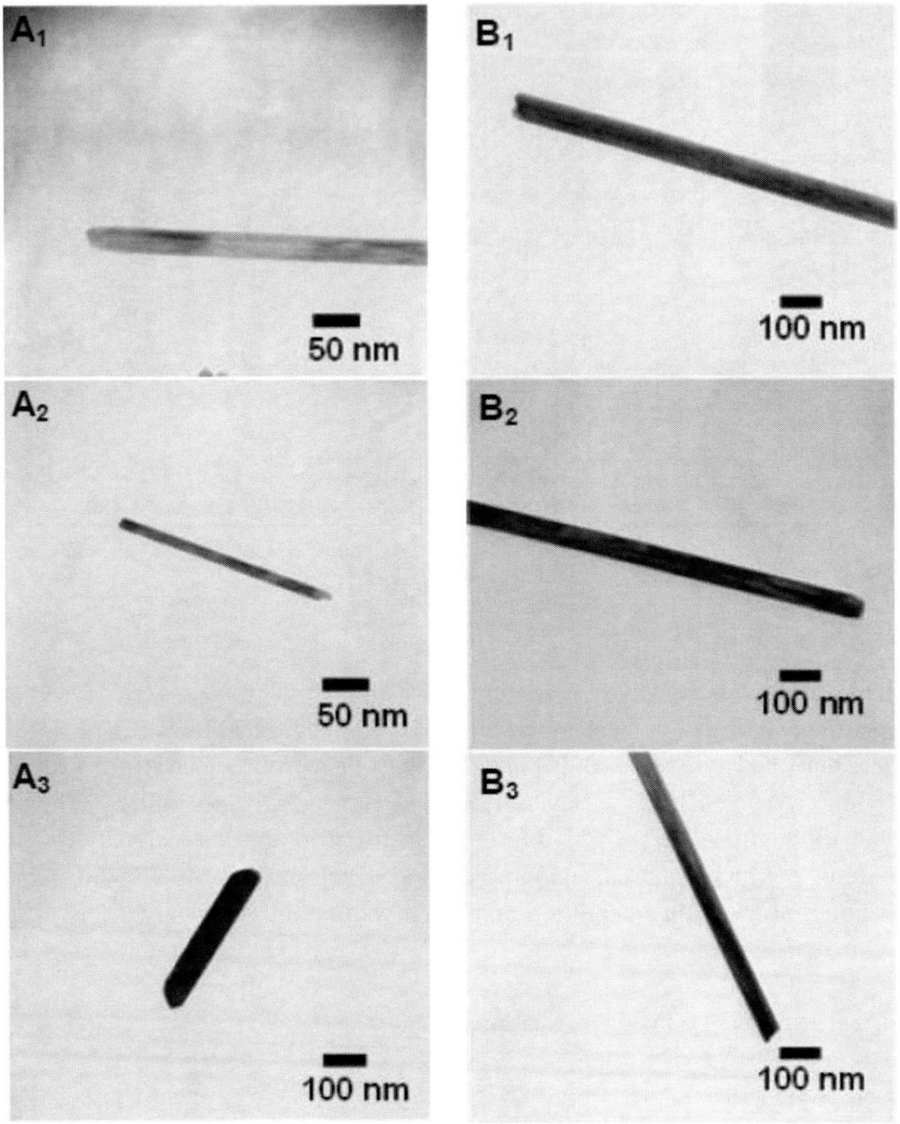

Figure 2. TEM images of the product formed by hydrothermal reaction of $MnSO_4$ and $KMNO_4$ with ratio of (A_1) 1:0.5, (A_2) 1:1, and (A_3) 1:2. TEM images of the product formed by hydrothermal reaction of $MnSO_4$ and NH_4F with ratio of (B_1) 1:0.5, (B_2) 1:1, and (B_3) 1:2.

A plausible mechanism as shown in Figure 3 may be defined as follows:

1 In the initial stage of hydrothermal reaction, the MnO_x units may form in the solution, and followed by condensation reaction to form sheets of layered δ-MnO_2.
2 The layer structure of δ-MnO_2 tends to lock at an elevated temperature and pressure; this might be the decisive step in the formation of one-dimensional MnO_2 nanostructures. Some tubular structure might be formed in the system.
3 The lamellar structure of MnO_2 is in a metastable state. It requires large concentration of cations to make the structure stable. If the cations present in the

structure are not sufficient the formed layer structure may collapse. Depending upon the concentration of cations present in the reaction solution, the transformation of the layered structure into the one dimensional structure may take place in several ways. If the amount of K^+ or NH_4^+ present in the solution is moderate then the layered structure will directly collapse into 2×2 tunnel structure of α-MnO_2; The 1×1 tunnel structure of β-MnO_2 is resulted due to the consumption of the cations in the side reactions occurring in the reaction solution at elevated temperature and pressure or the number of K^+ or NH_4^+ is not sufficient to stabilize the 2×2 tunnel structure.

4 The tubular and curling lamellar structures will get transformed into the nanorod like structure during the phase transformation from layered to one dimensional tunnel structure.

5 In the hydrothermal process, the shorter nanowires may redissolve into the solution phase, and the longer ones grow into much longer ones.

Although other unknown factors may exist that influence the growth of MnO_2 one-dimensional nanostructures, such as the influence of anions, and so forth, the above mechanism is in good agreement with our experiment results.

Figure 4 presents the XPS data for samples A_1, A_2, A_3,B_1, B_2 and B_3. An examination of the Mn 2p spectra of MnO_2 allowed the determination of the oxidation state of Mn at the surface. The peaks observed at binding energies of 642 and 654 eV were attributed to the Mn 2p XPS lines (figure 4(a)) with spin orbit splitting of 12 eV between Mn $2p_{3/2}$ and Mn $2p_{1/2}$ coincident with the previously reported values [9]. The inset shows the deconvoluted XPS spectra of Mn 2p with Gauss-Lorentz distribution function after Shirlay integrated background subtraction, the observed doublets of Mn $2p_{1/2}$ and Mn $2P_{3/2}$ can be assigned to a mixed valent manganese system, most likely Mn^{4+} and Mn^{3+} since the average oxidation state of Mn in the nanostructure form is in between 3.5-4 [10].

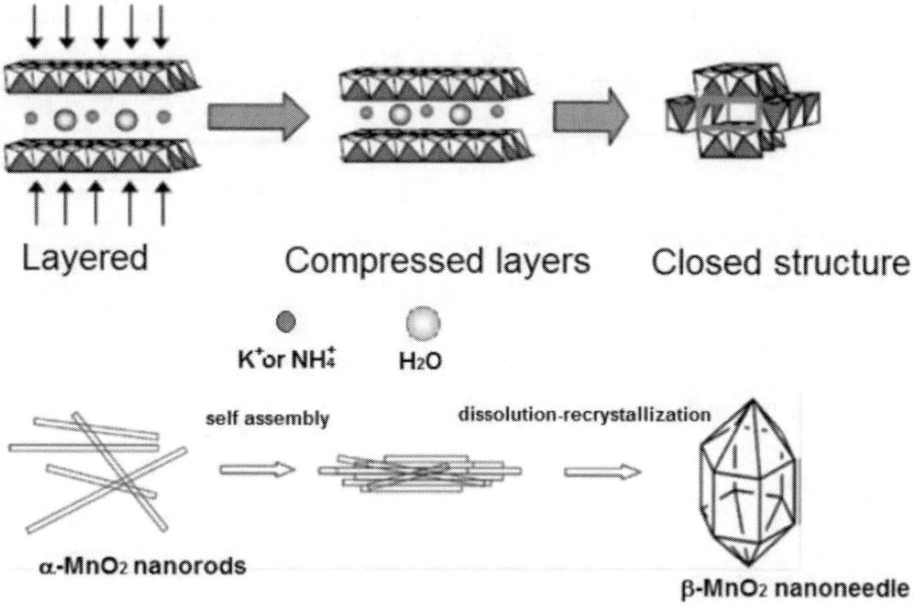

Figure 3. Schematic for structural evolution of MnO_2 polymorphs.

The peak at 641.7 eV suggests that Mn is present in its +4 state. Figure 4(b) shows an interesting evolution of the O 1s peak shape. A peak at 530 eV was related to characteristic metal oxide oxygen, whereas the peak at 532 eV corresponds to an oxygen bond with Mn^{4+} [11].

In addition, the peaks at lower binding energy were due to the oxygen present in the lattice and the peak corresponding to the higher binding energy was for the chemisorbed oxygen [12]. That means the sample A_2, B_1, B_2 and B_3 have the higher concentration of chemisorbed oxygen than that of lattice oxygen while A_1 and A_3 having smaller amount of chemisorbed oxygen than it present in the lattice. Moreover, XPS spectra of Mn 2p and O 1s of A_3 has been shifted to the lower binding energy side compared to that of A_1, this might be the charge induced peak shift due to rearrangement of MnO_6 octahedra to form 1 × 1 tunnel structure of β-MnO_2. The XPS lines for K $2p_{1/2}$ and K $2p_{3/2}$ were observed for samples A_1 and A_2, as shown in figure 4(c).

The much lower intensity of K 2p XPS lines for sample A_3 indicates the absence of K in β-MnO_2. The composition of the nanorods and needles was computed from the XPS data. Oxygen deficient compositions of $MnO_{1.74}$, $MnO_{1.47}$, and $MnO_{1.68}$ were observed for samples A_1, A_2, and A_3 respectively. Further the compositions calculated for the nanostructures in B_1, B_2 and B_3 on the basis of XPS data found to be $MnO_{2.13}$, $MnO_{1.91}$ and $MnO_{2.09}$ having near stoichiometry. This variation of the oxygen content in the structure will affect the properties of these materials.

Figure 5(a) shows a cyclic voltammogram of MnO_2 electrodes measured in a 1 M KOH solution at 25 °C with a scan rate of 1 $mV \cdot s^{-1}$ over the potential range of 100 to -600 mV vs. Hg/HgO. In the absence of O_2 in the KOH solution, the cyclic voltammetry response to the MnO_2 electrodes did not yield any peak, indicating that all the MnO_2 used in this study are electro-inactive in the potential window used. In the presence of O_2, reproducible and well defined cyclic voltammograms were obtained for the A_1 electrode. Two cathodic peaks were observed near -0.072 and -0.46 V vs. Hg/HgO, which were assigned to the two electron reduction process of O_2. O_2 is reduced to HO_2^-, which is further reduced to OH^-. The indirect four electron reduction process of oxygen is given below.

$$O_2 + H_2O + 2e^- \rightarrow HO_2^- + OH^- \quad (4)$$

$$HO_2^- + H_2O + 2e^- \rightarrow 3OH^- \quad (5)$$

Only one peak at -0.083 V vs. Hg/HgO was observed in the cyclic voltammogram for the A_2 electrode. This peak was attributed to the two electron reduction process of oxygen to super peroxide, as given in reaction 4. The disappearance of the second reaction peak for the A_2 electrode can also be explained by the MnO_2 catalyzed chemical conversion reaction of HO_2^-. This suggests that the reduction of O_2 on the MnO_2 electrode occurs through a different pathway from those generally demonstrated. The present case (electrode A_2) follows a catalytic mechanism with an initial two-electron reduction of O_2 to HO_2^- coupled with a chemical reaction of the HO_2^- produced under the heterogeneous catalysis of MnO_2, as shown in reaction 6.

$$2HO_2^- \rightarrow O_2 + 2OH^- \quad (6)$$

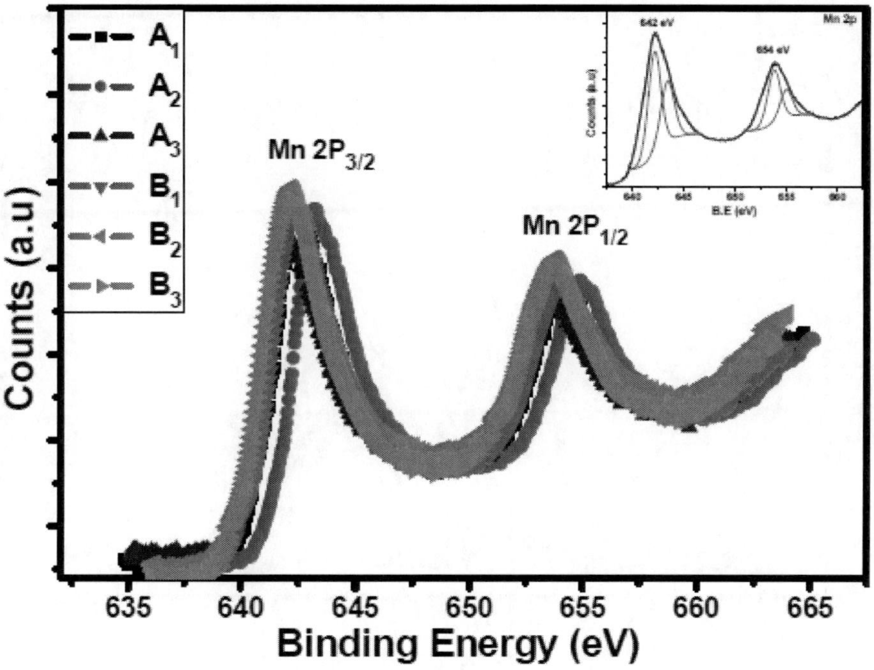

(a)

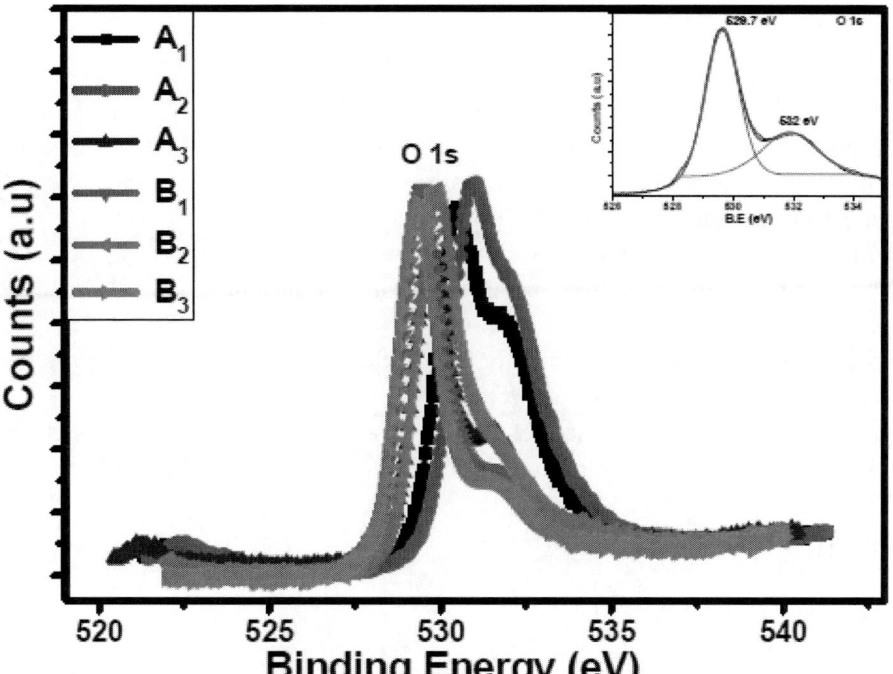

(b)

Figure 4. (Continued).

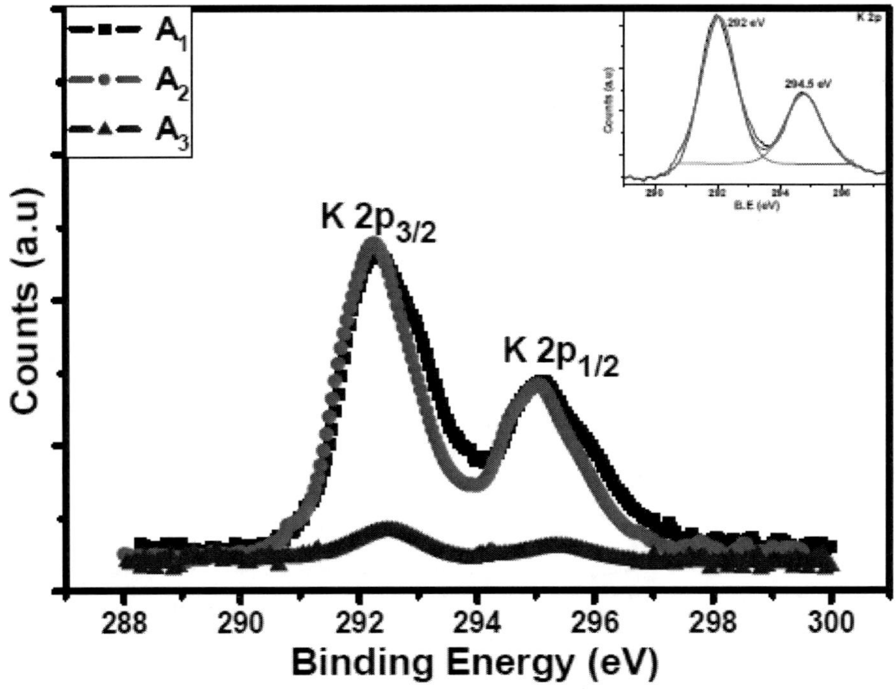

(c)

Figure 4. XP spectra of (a) Mn 2p, (b) O 1s, and (c) K 2p for MnO_2.

This reaction occurs provided the catalyst can dissociate the intermediate of the super peroxide reaction i.e. HO_2^-. The improved catalytic process of the A_2 electrode compared with the A_1 electrode is attributed to the reduced size and enhanced surface area of the catalysts for the oxygen reduction reaction (ORR). As size decreases, the reactivity is enhanced because of large surface-to-bulk ratio and numerous surface defects.

Higher surface areas permit more active sites for the contact between catalyst and electrolyte. In addition, the increased oxygen deficiency in the catalyst may be responsible for the dissociation reaction of peroxide ions on the A_2 catalyzed electrode. An irreversible cathodic peak at approximately -0.245 V for the A_3 electrode is a result of the proton insertion reaction occurring on the electrode.

Owing to the greater size and smaller surface area of the β-MnO_2, protons can be inserted easily into the MnO_2 tunnel structure to form MnOOH, as shown in reaction 7.

$$MnO_2 + H_2O + e^- \rightarrow MnOOH + OH^- \quad (7)$$

$$2MnOOH + O_2 \leftrightarrow 2(MnOOH\cdots O) \quad (8)$$

$$(MnOOH\cdots O) + e^- \leftrightarrow MnO_2 + OH^- \quad (9)$$

Further, oxygen reduction reaction may precede through reaction 8 and 9 in which dissolved oxygen in the electrolyte gets adsorbed on the adjacent MnOOH sites and then reduced by accepting the electron from the electrode.

The absence of an anodic peak corresponding to the reverse reaction in the anodic direction for the de-insertion of protons from MnOOH suggests that the proton insertion reaction has poor reversibility.

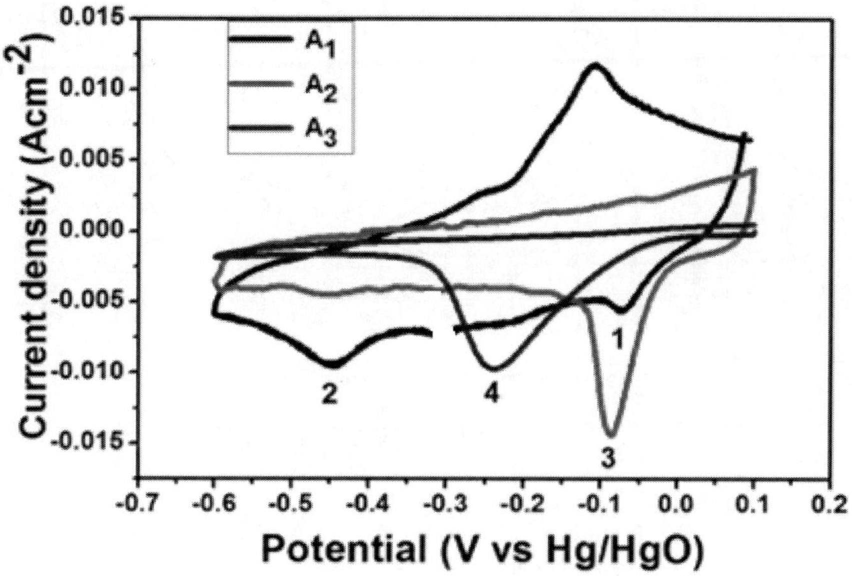

(a)

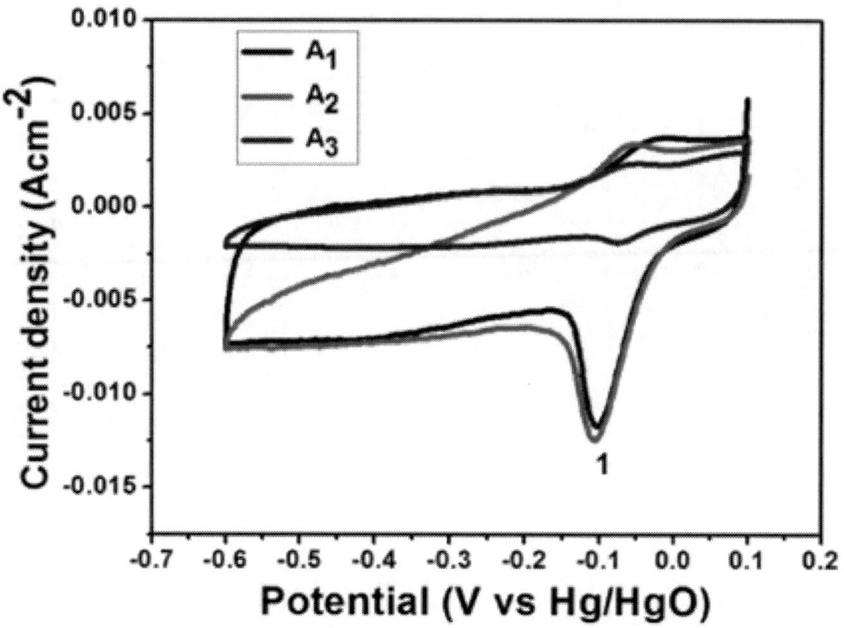

(b)

Figure 5. (Continued).

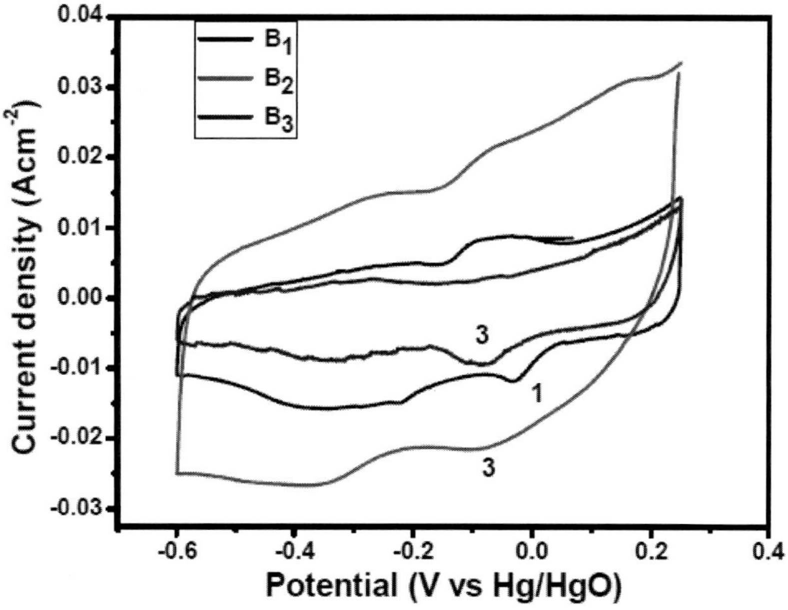

(c)

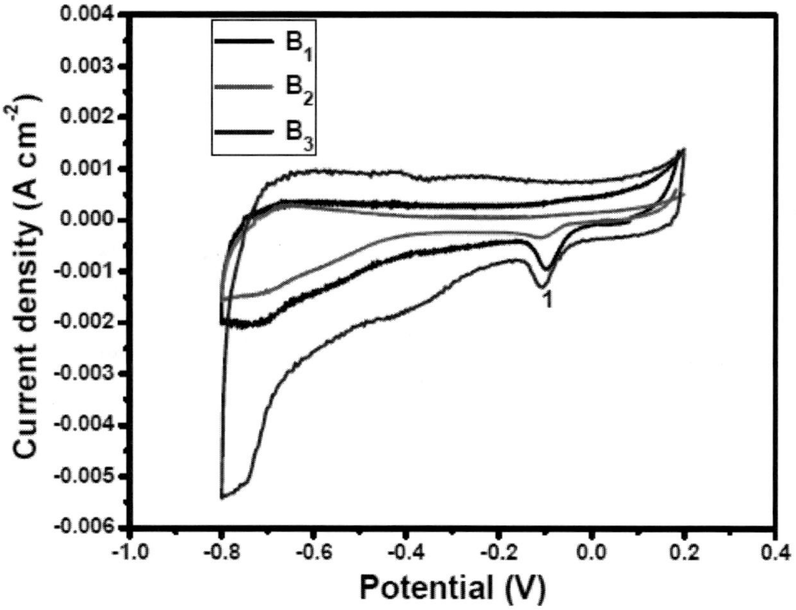

(d)

Figure 5. Cyclic voltammograms recorded for the A_1, A_2, A_3 electrocatalysts in (a) KOH and (b) LiOH solutions. CV for B_1, B_2, B_3 electrocatalysts in (c) KOH and (d) LiOH solutions.

This reaction may induce the continuous reduction of MnO_2 resulting in the electrode becoming inactive for oxygen reduction. This also suggests that the β-MnO_2 exhibits relatively lower electrochemical activity than α-MnO_2.

Figure 5(b) shows the cyclic voltammograms of the MnO_2 electrodes measured in a 1 M LiOH solution at 25 °C. Only one peak was observed for all three electrodes. The peak observed for β-MnO_2 has a lower peak current density than that obtained for α-MnO_2.

The insertion of Li ions in the MnO_2 tunnel structure during the electrochemical reaction might be responsible for the improved catalytic activity of the electrode to give only a single peak for the 2 electron ORR mechanism governed by reaction 4. Li ions can dissolve easily and become inserted into the MnO_2 structure compared to Na^+ and K^+ ions [13]; even the tunnel size of β-MnO_2 is smaller than that of α-MnO_2.

The voltammogram for electrode containing catalyst $B_1$and B_3 as shown in Figure 5 (c) contains only one peak which is due to proton insertion reaction (7) resembling the activity of the catalyst materials A_3.

However the intensity of the reduction peak of B_1 is less than that of A_3. CV plot recorded for electrode B_2 shows single peak arise due to the two electron oxygen reduction reaction (4).

The oxygen reduction occurred on the electrodes B_1, B_2 and B_3 in the LiOH solution is similar to the two electron reduction reaction followed by the disproportionate reaction of the peroxide ions observed for the earlier catalyst materials {Figure 5(d)}.

The discharge characteristics recorded for Li/O_2 cell at discharge current density of 0.2 mA/cm^2 was shown in figure 6. The well- defined discharge curves were observed for the cells having α-MnO_2 (figure 6(a)-(b)) and β-MnO_2 (figure 6(c)) as catalyst in air electrode. Nevertheless, the specific capacity obtained for each cell is different.

The discharge capacities were 1766, 1917 and 1430 $mAhg^{-1}$ for A_1, A_2 and A_3, respectively. This is attributed to the small size of A_2 causing the formation of porous air electrode, which might facilitate the diffusion, adsorption and transport of oxygen gas molecule to the catalytic site. This also suggests that A_2 has the better catalytic activity than other two catalysts in non-aqueous electrolyte. Thus, it is revealed that the catalytic performance of MnO_2 depends strongly on both the crystallographic structure and the morphology. In addition, the catalytic activity of α-MnO_2 is higher than that of β-MnO_2 without taking into account the morphological influence.

Single step discharge curves as shown in Figure 7, were observed for the Li/O_2 cells having nanostructured α-MnO_2 catalyst in the air electrode. The flat discharge voltages of 2.7, 2.9 and 2.75V for B_1, B_2 and B_3 catalyzed cells.

However, the specific capacity obtained for the cell is function of the activity of these catalyst materials. The discharge capacities were 902, 1079 and 1202 $mAhg^{-1}$ for B_1, B_2 and B_3, respectively. The increased capacity is attributed to the decreased size of the MnO_2 nanostructures cause to form the porous air electrode, which might facilitate the diffusion, adsorption and transport of oxygen gas molecule to the catalytic site. It is suggested that the B_3 has the better catalytic activity than other two catalysts in non-aqueous electrolyte.

The supercapacitive behavior of the nanostructured manganese oxide material is investigated using the charge-discharge behavior of the electrode in 1M NaOH solution in the voltage range 0.2 to -0.8 V vs Hg/HgO. The specific capacity of the material is defined as.

$$C = \frac{dQ}{dV} = \frac{I}{\frac{dV}{dt}} \qquad (10)$$

where Q is the charge on the electrode, V is its potential, and I is instantaneous current.

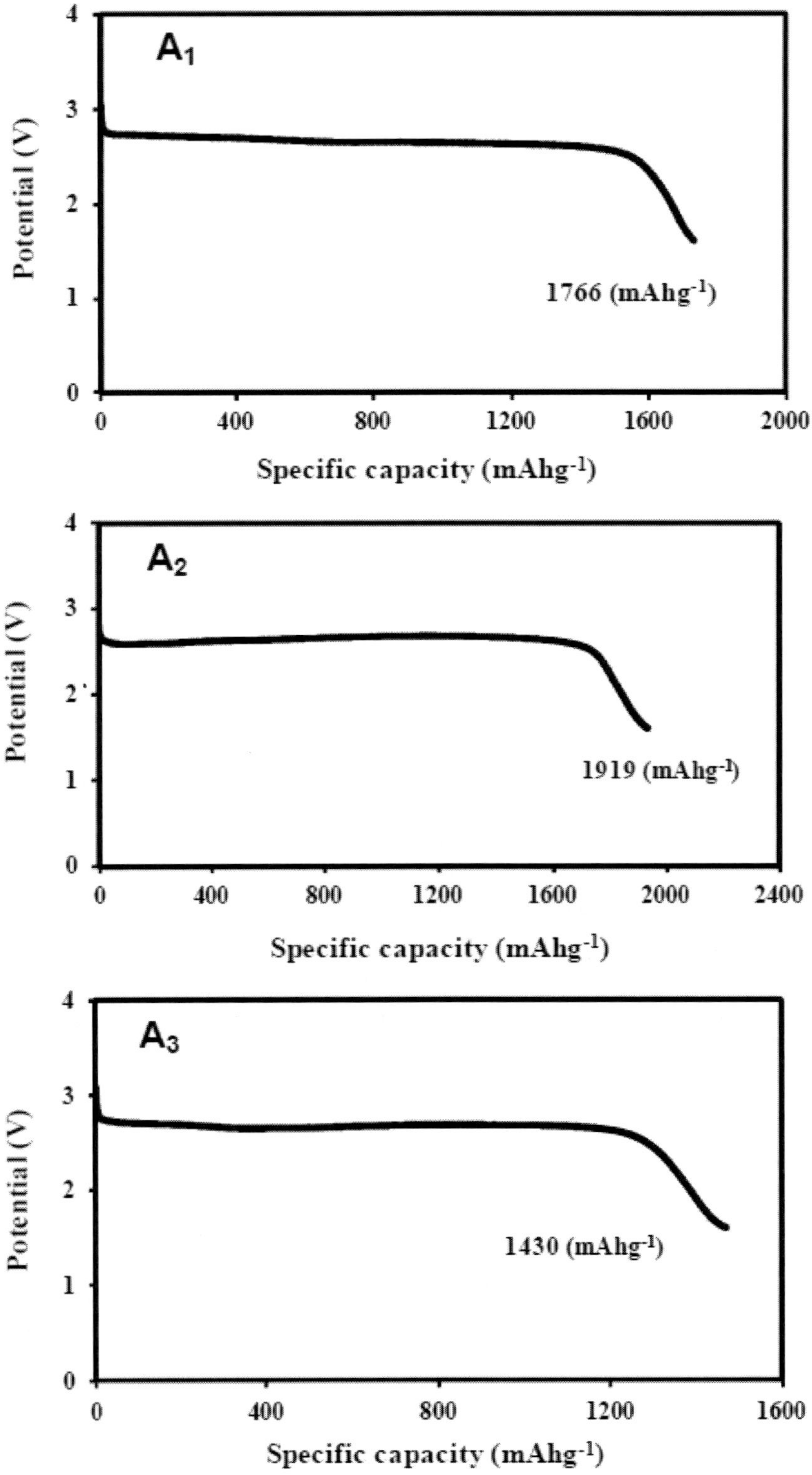

Figure 6. Discharge curve recorded for the Li/O_2 cell using (a) A_1-α-MnO_2 (40nm) (b) A_2-α -MnO_2 (25nm) and (c) A_3- β-MnO_2 electrocatalysts in air electrodes.

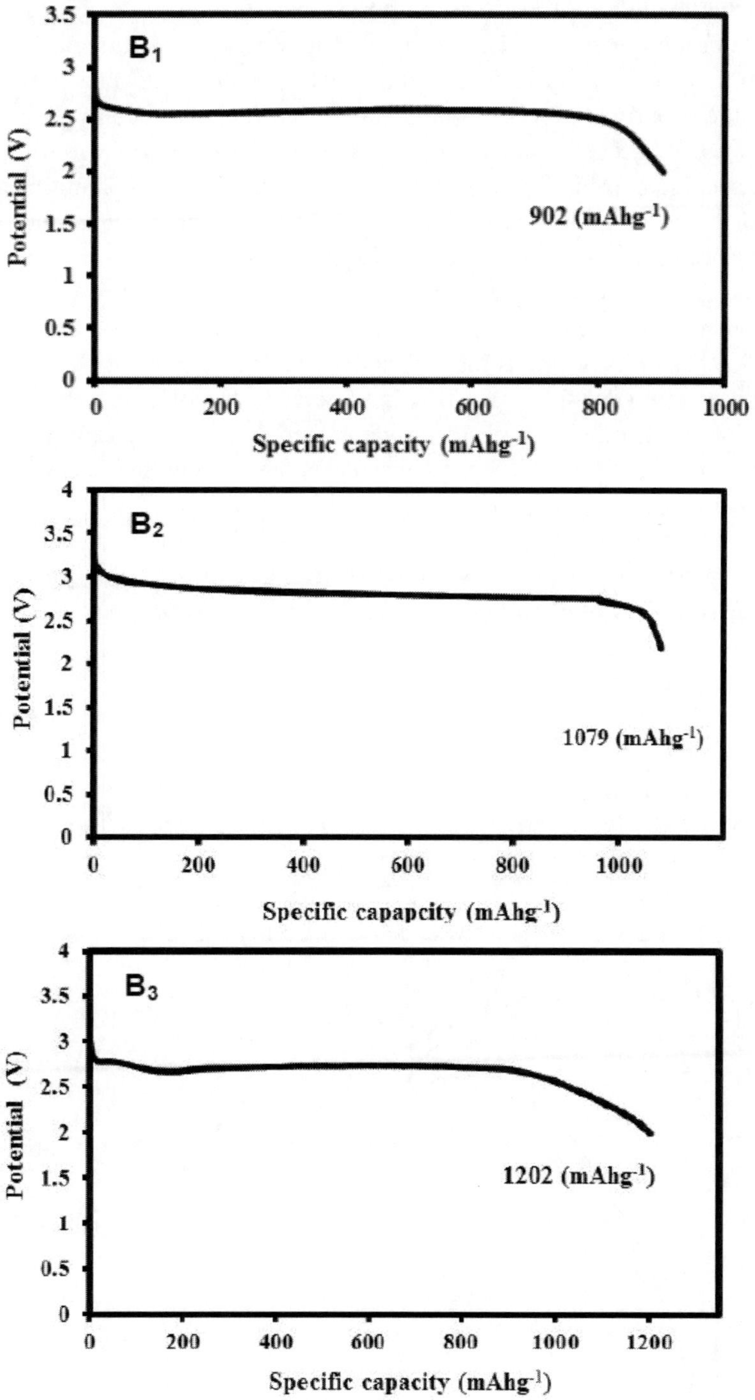

Figure 7. Discharge curve recorded for the Li/O_2 cell using (a) B_1-α-MnO_2 (95nm) (b) B_2-α -MnO_2 (80nm) and (c) B_3- α -MnO_2 (60nm) electrocatalysts in air electrodes.

For systems involving faradaic reactions as well as double-layer effects, an operational definition of an effective capacitance requires the specification of a practical range of potential in which charge-transfer reactions are reversible and do not cause deterioration of the electrode during cycling. For practical purposes, then, the capacitance *C* may be reported as an average value obtained from Eq. 10 applied to various experimental tests within the specified potential range. This definition requires that the value of *C* does not depend greatly on the nature of the experiment and does not change significantly upon cycling. For example, if an electrode is tested by cyclic voltammetry, where the scan rate *dV/dt* is constant, *C* would be given by the average current divided by the scan rate. In a chronopotentiometric discharge test, the value of *C* would be the current divided by the average slope of the potential-time plot over the potential range of interest. If one expresses current as current density based on superficial or geometric electrode area, then *C* would be expressed in units of farads/area. With dispersed or nano particulate electrode materials, it is more common to report specific capacitance, which has units of farads/mass of active material.

The charge storage mechanism occurring in these nanostructures is based on the insertion and deinsertion of the protons in the electrode material during charging and discharging process respectively. The specific capacitance calculated using the charging discharging behavior for one-dimensional MnO_2 nanostructured is shown in Figure 8 (a-b). The specific capacitance values for the NH_4^+ stabilized nanostructures is found to be more compared to the K^+ stabilized one dimensional MnO_2 structures. The highest value of the specific capacitance is obtained for phase pure MnO_2 structure i.e B_3 to be 312 F/g. Specific capacitance for B_1 and B_2 nanostrucres are 281, 265 F/g respectively.

However the capacitance of 125, 164 and 195 F/g are obtained for A_1, A_2 and A_3 MnO_2 structures. The specific capacitance of the MnO_2 electrodes as a function of cycle number is given in Figure 8(c). From the plot, it is revealed that the amount of capacity stored in the MnO_2 electrode over 2000 cycle is slightly decreased than the capacity stored during 1st cycle. All the electrode material shows good storage capability over the thousands of charging discharging cycle in the defined potential window.

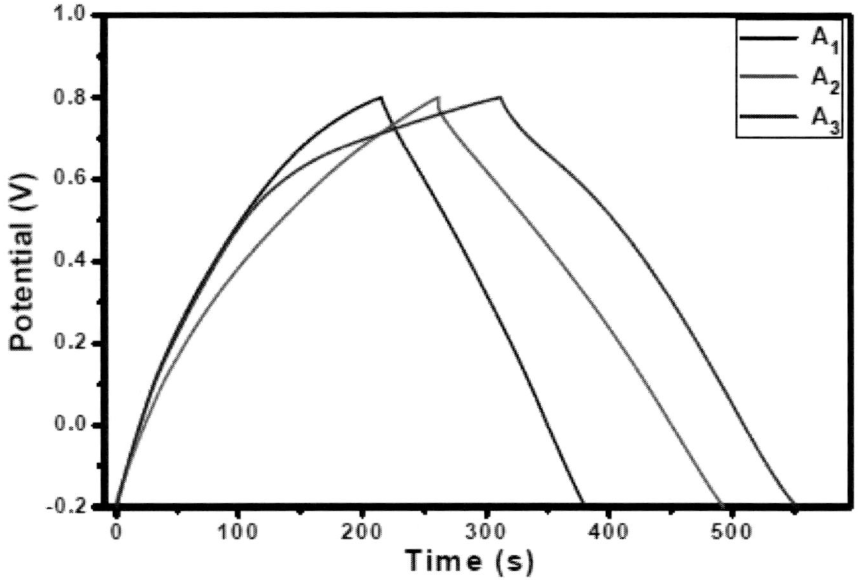

(a)

Figure 8. (Continued).

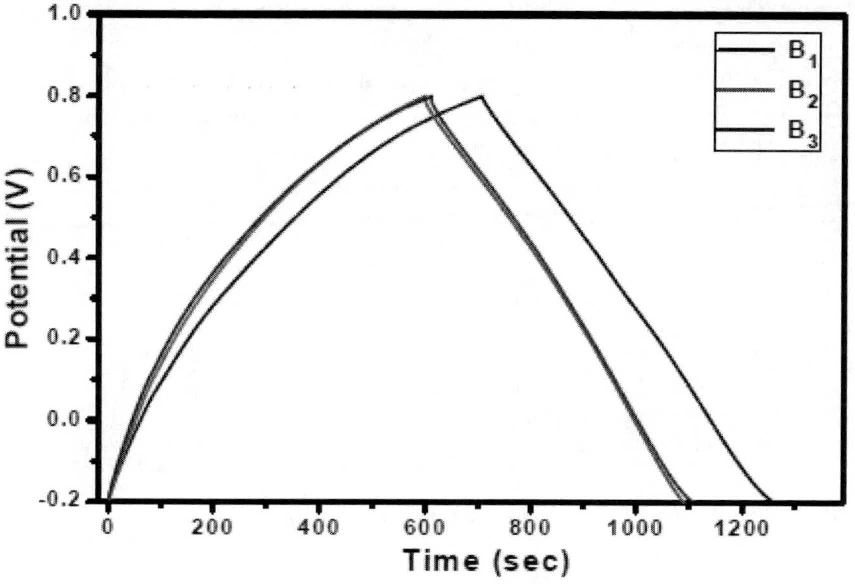

(b)

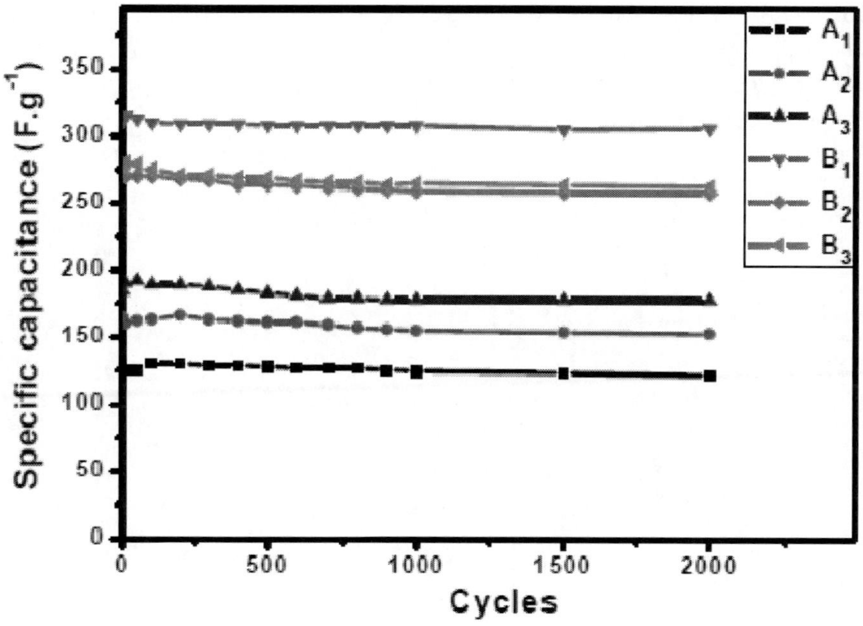

(c)

Figure 8. (a) & (b) Charge discharge curve for MnO_2 nanostructures and (c) Specific capacity plot of the MnO_2 nanostructures as a function of cycle numbers.

The observed capacitance of MnO_2 electrode is believed to be predominantly due to pseudocapacitance as a result of redox processes associated with the reduction of MnO_2 to MnOOH. The initial reduction of MnO_2 involves incorporation of protons and electrons into the MnO_2 lattice. The reduction of MnO_2 proceeds with the dilation of the lattice which leads to the eventual formation of an amorphous MnOOH phase.

This process occurs homogenously and reversibly without any phase change [14-19]. MnO_2 exhibits high charge-storage capacity due to enhanced proton diffusion in and out of MnO_2 lattice in unbuffered solutions [15, 16].

The principal reaction involved in the charging and discharging process of manganese dioxide in neutral electrolytes [15] can be represented as

$$MnOOH_n + \delta H^+ + \delta e^- \leftrightarrow MnOOH_{n+\delta} \tag{11}$$

i.e., when $n = 0$, $\delta = 1$

$$MnO_2 + H^+ + e^- \leftrightarrow MnOOH \tag{12}$$

In alkaline medium, Reaction (12) can be represented as

$$MnO_2 + H_2O + e^- \leftrightarrow MnOOH + OH^- \tag{7}$$

Unlike battery systems, for instance MnO_2/Zn batteries which utilize MnO_2 as the cathode and metallic Zn as the anode, ultracapacitor devices normally use two identical or similar electrodes. We postulate that redox reactions occurring in MnO_2 electrode ultracapacitor devices involve proton "intercalation" or insertion between the positive and negative electrodes which can be represented as follows

Positive electrode

$$MnOOH \underset{\text{discharge}}{\overset{\text{charge}}{\leftrightarrow}} MnO_2H_{1-\delta} + \delta H^+ + \delta e \tag{13}$$

Negative electrode

$$MnO_2 + \delta H^+ + \delta e \underset{\text{discharge}}{\overset{\text{charge}}{\leftrightarrow}} MnO_2H_\delta \tag{14}$$

Overall

$$MnOOH + MnO_2 \underset{\text{discharge}}{\overset{\text{charge}}{\leftrightarrow}} MnO_2H_{1-\delta} + MnO_2H_\delta \; (0 < \delta < 0.5) \tag{15}$$

It is noteworthy that as long as we can maintain the cycling potential and conditions such that only these redox processes can occur on the electrodes, 1 mole of protons out of MnO_2 lattice during each charge and discharge cycle, respectively. Since these processes occur homogeneously without any phase change, high reversibility and long cycle life are assured.

This could be achieved by limiting the charge and discharge cycle to a potential window which corresponds to the highest stability and reversibility of the redox processes. Degradation of films is anticipated from cycling under overcharging or overdischarging conditions. Excessive reaction leads to the formation of complex oxides such as $Mn(OH)_2$, Mn_2O_3, Mn_3O_4, Mn_5O_8, etc., that are known to be electrochemically irreversible.

The higher value of the specific capacitance obtained for the NH_4^+ stabilized one dimentional nanostructure is justified on the base of the charging-discharging behavior of the supercapacitor. The larger size of the 2×2 tunnel structure stabilized by NH_4^+ allows more number of protons to be intercalated and deintercalated during charging and discharing process of supercapacitor than the 2×2 and 1×1 one dimensional structures stabilized by K^+ ion. Hence the capacitance exhibited by the nanostructures A_1, A_2 and A_3 is less compared to B_1, B_2 and B_3.

Conclusion

A hydrothermal reaction is successfully used to prepare tunnel type MnO_2 nanostructures merely by controlling the precursor to oxidizer ratio. α-MnO_2 shows better activity for oxygen reduction than β-MnO_2 in alkaline and non aqueous electrolyte. Potassium used as stabilizing agent enhanced the electrocatalytic properties of the manganese oxide for the oxygen reduction reaction in aqueous as well as in non-aqueous medium. The supercapacitive behavior shown by the NH_4^+ added MnO_2 is higher. For charge storage application NH_4^+ added MnO_2 has shown greater efficiency while K^+ added MnO_2 had possessed higher efficiency for catalytic application. The difference in the electrocatalytic activity of the different structures of MnO_2 highlights the importance of engineering of the tunnel structure, crystal phase and morphology.

Acknowledgments

This work was supported by the Korea Ministry of Education, Science and Technology (MEST) and the National Foundation of Korea (NRF), under the Human Resource Training Project for Regional Innovation (2010-04-Dae-02-013) and by the Korea Ministry of Knowledge Economy (MKE) and the National IT Industry Promotion Agency (NIPA) under the Information Technology Research Center (ITRC) support program (NIPA-2012-H0301-12-1009).

References

[1] Xiao, W., Xia, H., Fuh, J., and Lu, L. 2009 *J. Power Sources* 193 935.
[2] Wang, D., Wang, Q., Wang, T. 2011 *Nanotechnology* 22 135604.
[3] Jiao, F. and Bruce, P. G. 2007 *Adv. Mater.* 19 657.
[4] Sui, N., Duan, Y., Jiao, X., and Chen, D. 2009 *J. Phys. Chem. C* 113 8560.
[5] Wang, X. and Li, Y. 2003 *Chem. Eur. J.* 9 300.

[6] Shen, X. F., Ding, Y. S., Hanson, J. C., Aindow, M., and Suib, S. .L 2006 *J. Am. Chem. Soc.* 128 4570.
[7] Zhang, X., Yang, W., Yang, J., and Evans, D. G. *2008 J. Cryst. Growth* 310 716.
[8] Roche, I. and Scott, K. 2010 *J. Electroanal. Chem.* 638 280.
[9] Foord, J. S., Jackman, R. B. and Allen, G. C. 1984 *Philos. Mag. A* 49 657.
[10] Brock, S. L., Sanabria, M., Suib, S. L., Urban, V., Thiyagarajan, P., Potter, D. I. 1999 *J. Phys. Chem. B* 103 7416
[11] Oku, M., Hirokawa, K. and Ikeda, S. 1975 *J. Electron Spectrosc.* 7 465.
[12] Barr, T. L., Fries, C. G., Cariati, F., Bart, J. C., Giordano, N. 1983 *J. Chem. Soc. Dalton Trans.* 1825.
[13] Li, W., McKinnon, W. R. and Dahn, J. R. *1994 J. Electrochem. Soc.* 141 2310.
[14] Desai, B. D., Fernandes, J. B., Kamat Dalal, V. N. 1985 *J. Power Sources.* 16 1
[15] Kelliher, E. M., Rose, T. 1986, *Mater. Res. Soc. Symp. Proc.* 55 287
[16] Era, A., Takehara, Z., Yoshizawa, S. 1967 *Electrochim. Acta*, 12 1199
[17] McBreen, J. 1975 *Electrochim. Acta* 20 221
[18] Schumm, J.r. B., Middaugh, R. L., Grotheer, M. P., Hunter, J. C. 1985 "Manganese Dioxide Electrode: Theory and Practice for Electrochemical Application", PV 85-4, The Electrochemical Society Proceedings Series, Pennington, NJ
[19] Ruetschi, P., Giovanoli, R. 1988 *J. Electrochem. Soc*, 135 2663

In: Doping
Editor: Lixin Yu
ISBN: 978-1-62618-097-0

Chapter 3

GROWTH, CHARACTERIZATION AND MAGNETIC PROPERTIES OF POLYCRYSTALLINE FIBERS OF ELECTRON DOPED MANGANITES

C. A. Cardoso,[1] S. Sergeenkov,[1] M. R. B. Andreeta,[2] A. C. Hernandes,[2] E. R. Leite,[3] O. F. de Lima[4] and F. M. Araujo-Moreira[1]

[1]Departamento de Física, Grupo de Materiais e Dispositivos/CMDMC, UFSCar, São Carlos, SP, Brazil
[2]Instituto de Física de São Carlos, Grupo de Crescimento de Cristais e Materiais Cerâmicos/CMDMC, USP, São Carlos, SP, Brazil
[3]Departamento de Química, LIEC/CMDMC, UFSCar, São Carlos, SP, Brazil
[4]Instituto de Fisica "Gleb Wataghin", UNICAMP, Campinas, SP, Brazil

ABSTRACT

In this Chapter we review some of our latest results regarding the growth, characterization and magnetic properties of bulk samples (composite fibers) of the optimally doped $La_{1-x}Ce_xMnO_3$ manganites by using the Laser Heated Pedestal Growth technique. It has been widely recognized that, in comparison with the hole doped manganite (like, e.g., $La_{1-x}Ca_xMnO_3$), its electron doped counterpart $La_{1-x}Ce_xMnO_3$ is a metastable phase. That is why, before optimizing the growth conditions, we were able to grow polycrystalline multiphase fibers. Their composition and granularity were verified through Scanning Electron Microcopy with Energy Dispersive X-ray analysis. The magnetic behavior of the fibers was probed by AC susceptibility and magnetization measurements as a function of temperature. Several ferromagnetic transitions were identified. In addition to a weak transition at T_{C0}=45K (due to MnO_2 impurities), three more transitions with Curie temperatures T_{C1}=126K, T_{C2}=180K, and $T_{C,max}$=300K were observed and linked to fiber regions with cerium concentration of x=0.063, x=0.13, and x=0.3,respectively.

Keywords: Electron-doped manganites, cerium compounds, laser heated pedestal growth technique, magnetization, ferromagnetic transitions, Curie temperature

PACS: 75.47.Gk, 75.47.Lx, 81.10.Jt.

INTRODUCTION

Magnetic manganites of the type $R_{1-x}A_xMnO_3$ [with R={La, Pr, Nd} and A={Ca, Sr, Ba}] constitute one of the most fascinating classes of materials, exhibiting a wide variety of structures and properties [1, 2], including colossal magnetoresistence (CMR), due to the presence of the mixed valence of manganese as Mn^{3+} and Mn^{4+}. Although a complete understanding of the basic mechanisms involved is still lacking [1], a massive amount of information has been gathering in the last years. For instance, it is well known that the Curie temperature (T_C) in those materials is determined by Mn–O–Mn bond angle [2]. In fact, the Curie temperature decreases with the reduction in the ionic radius of the divalent metal A^{2+} or of the trivalent rare earth R^{3+} since in both cases there is an increasing distortion in the MnO_6 octahedra [3-7]. Recently, some effort has been dedicated to dope manganite systems with tetravalent ions, for instance with cerium [8-15]. Such materials would be electron doped and their magnetic properties should arise from Mn^{2+} and Mn^{3+} interactions. Considering that Mn^{2+} and Mn^{4+} are both non Jahn-Teller ions whereas Mn^{3+} is an Jahn-Teller ion, one can argue that electron doped manganites should present magnetic properties different from their hole doped relatives. The possibility to synthesize electron and hole doped manganites would open the possibility of building dipolar devices where both spin and charge could be used, which is of great interest in the field of spintronics. The first reports on cerium doped manganites [8, 9] claim their polycrystalline samples are single phase $La_{0.7}Ce_{0.3}MnO_3$ and that this phase presents magnetic properties very similar to the hole doped system $La_{0.7}Ca_{0.3}MnO_3$. Such coincidence was interpreted as an evidence that electron and hole doping are equivalent in terms of inducing the CMR effect on manganites. However, these results were strongly criticized by several authors [12, 16, 17]. First of all, a closer inspection of the X ray diffraction (XRD) patterns revealed the presence of the most intense peaks from CeO_2, indicating that the samples were not single phase $La_{0.7}Ce_{0.3}MnO_3$. More than that, it was demonstrated [16, 17] that the magnetic response measured in those samples may be explained by considering an inhomogeneous material formed by CeO_2 and lanthanum deficient $La_{0.7}MnO_3$. In fact, some authors question whether it is even possible to synthesize Ce doped manganites because of the small size of Ce^{4+} ions which would lead to a very large distortion of the MnO_6 octahedra, let alone if cerium would have enough reduction power to induce the 2+ valence in the manganese ions [17, 18]. Although such arguments were widely accepted for polycrystalline bulk samples, researchers keep pursuing the synthesis of $La_{0.7}Ce_{0.3}MnO_3$ but now in the form of thin films [12, 14, 15, 19-21]. In fact, thin films of $La_{0.7}Ce_{0.3}MnO_3$ grown by the very energetic pulsed laser deposition (PLD) technique do not present the CeO_2 peaks in XRD measurements [15, 19].

In this Chapter we report our recent results on the growth and characterization of polycrystalline fibers of nominal composition $La_{0.7}Ce_{0.3}MnO_3$ (LCMO) which clearly demonstrate that it is indeed possible to grow cerium doped manganite even in bulk form, although this chemical phase seems to be metastable.

SAMPLES GROWTH AND CHARACTERIZATION

The fibers employed in this work were grown by using the *Laser Heated Pedestal Growth* (LHPG) technique. The same technique was previously used [22] by us to grow high-quality single crystals of $La_{0.7}Ca_{0.3}MnO_3$. The LHPG is a variation of the Floating Melt Zone technique, and its setup is presented in Figure 1. A powerful CO_2 laser has been used to melt a polycrystalline pedestal. Namely, the highly focused laser beam heats the pedestal to very high temperatures (up to 2600°C).

These high temperatures may be useful in breaking apart even very stable oxides as well as helping the formation of less stable structures. In addition, the small size of the melt zone reduces the influence of convection and the evaporation of elements.

Therefore LHPG may be a useful tool in the preparation of material that cannot be obtained by other techniques.

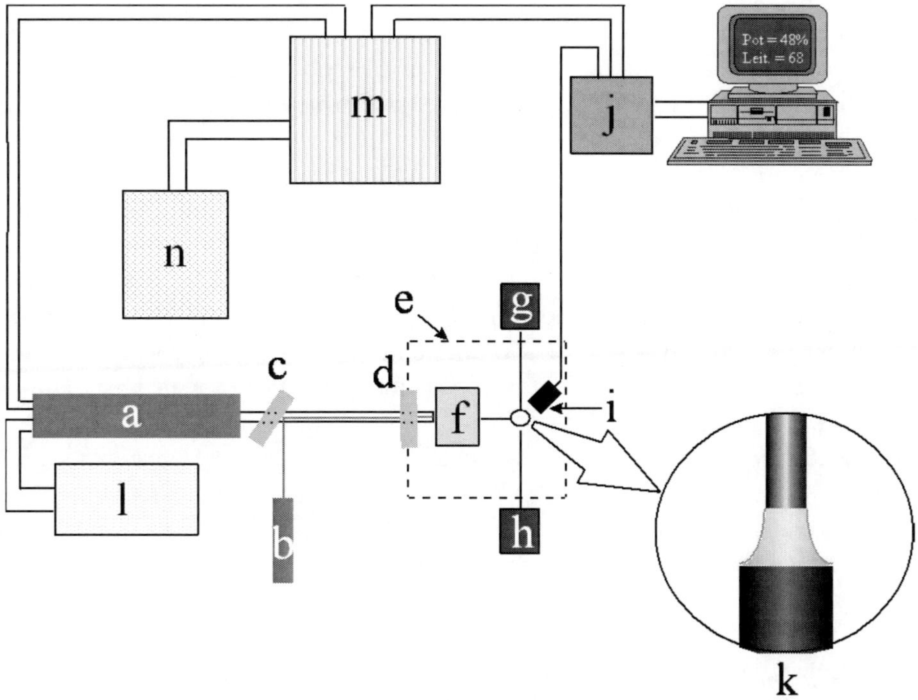

Figure 1. Schematic representation of a Laser Heated Pedestal Growth setup. (a) CO_2 laser, (b) He-Ne laser, (c) ZnSe Beam splitter, (d) ZnSe window, (e) Growth chamber, (f) Focusing system, (g) Fiber pulling system, (h) pedestal moving system, (i) phototransistor, (j) laser control system, (k) representation of the melting zone, (l) Cooling system, (m) and (n) RF source, (o) CCD video camera.

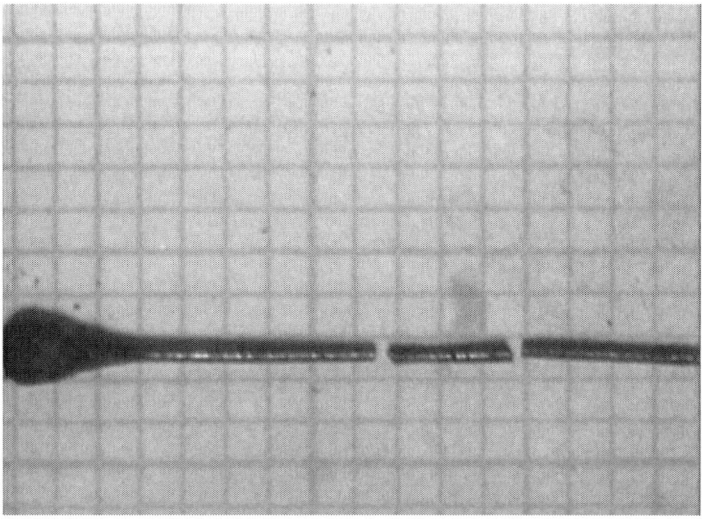

Figure 2. Photography of manganite fiber, showing part of the pedestal (on the left) and pieces of the crystalline fiber.

On the other hand, such high thermal gradient induces a large stress in the crystal thus limiting the fiber diameter. In our case the diameter of the obtained crystalline fiber is around 0.7 mm (see Figure 2).

In order to prepare the $La_{0.7}Ce_{0.3}MnO_3$ fibers, source pedestals of the same composition were initially prepared by mixing in the stoichiometric proportion of high purity La_2O_3, CeO_2 and MnO powders. No pre-reaction was done since previous tests on $La_{0.7}Sr_{0.3}MnO_3$ (LSrMO) indicated that crystals with similar quality were obtained no matter if the pedestal was pre-reacted or not.

Attempts to prepare single phase LCMO powder by both solid state reaction and polymeric precursor method failed, and clear CeO_2 peaks were detected in the XRD patterns of all powders. The powder used for pedestal preparation was cold extruded. The growth process was performed in the conventional LHPG system presented in Figure 1, with a CO_2 laser (Synrad, 60-1 -125 W, CW) [23]. The fiber and pedestal pulling speeds, as well as the laser power, were modulated by an automatic diameter control system [23]. The average pulling speed was 18 mm/h.

The obtained fibers are polycrystals with black opaque surfaces, as shown in Figure 2. All fibers have a diameter below 1 mm and are up to 30 mm long. A longitudinally polished fiber was examined in a scanning electron microscope (SEM) and its composition was verified by EDX (*Energy Dispersive X-ray Analysis*). Images obtained using backscattered electrons revealed the fiber was formed by 3 main compositions. From the seed end of the fiber, we observe a grey matrix with a white precipitate first appearing periodically and later in a continuous way, but always in the form of a small reticulate, as shown in Figure 3.

As we move away from this end of the fiber, we start to observe the appearance of a third phase which is black in the SEM images. Curiously this black phase occurs basically in the central core of the fiber, with some elongated regions growing towards the fiber surface.

In this region we also identify the white phase also in the form of elongated regions pointing in the diagonal from the center to the surface of the fiber, getting closer to the surface as the growth of the fiber continued.

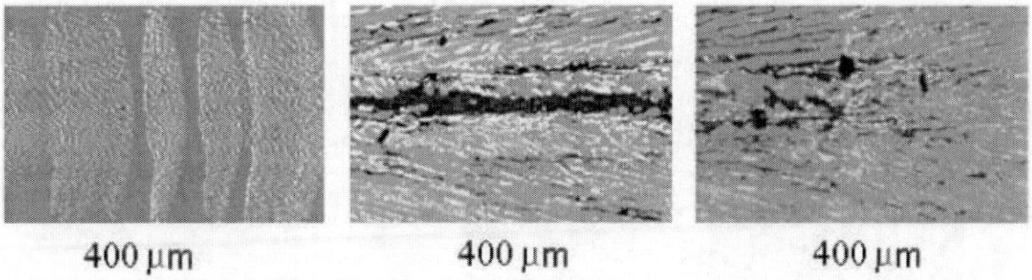

Figure 3. SEM images (200x magnification) of three different regions of a LCMO fiber, showing the evolution of the growth process (from the seed end at the left to the pedestal end at the right).

Finally, in the pedestal end of the fiber, the three phases seems to be mixed in a more homogeneous way (see Figure 3). It is important to notice that the gray phase is the dominant one along the entire fiber. In order to identify the composition of these three phases, we performed EDX measurements in different regions of the fiber.

In Figure 4 we present a SEM image of a 30×40 μm region in the center of the fiber along with the compositional maps for the different elements present in the fiber. It can be seen that the white region is rich in cerium, while the black one is rich in manganese.

A quantitative analysis indicate that the white region is indeed cerium oxide, while the black one should be some manganese oxide, with only traces of lanthanum and no cerium at all. More interesting is the result for the dominant grey region. The quantitative analysis indicates a proportion of 49:39:12 for Mn, La and Ce respectively, pointing to a stoichiometry of $La_{0.78}Ce_{0.24}Mn_{0.98}O_x$ in the pedestal extremity of the fiber and 49:42:09 (or $La_{0.84}Ce_{0.18}Mn_{0.98}O_x$) in the seed end of the fiber. The quantification measurements do not present reliable information about the oxygen content of the fiber, so we cannot point out in an affirmative way the exact content of oxygen in each of these phases.

The obtained stoichiometries for the grey regions indicate that it is indeed possible to grow the cerium doped LCMO phase even in a bulk configuration, although to obtain single phase LCMO it is necessary to optimize the growth conditions.

Our results agree with previous works [12] about the necessity of using a highly energetic technique to synthesize LCMO, like PLD and LHPG.

The appearance of CeO_2 clusters when LCMO film is submitted to thermal treatments [21] indicates that this material is a metastable phase and by adjusting the growth condition it may be possible to obtain pure LCMO in bulk form, just like it was achieved for thin films grown by PLD [12].

Our results indicate the occurrence of some substitution of Mn by Ce besides the desirable occupation of Ce in the A site (substituting the La ions), a possibility already indicated as most probable in a previous work [17]. Therefore, the general stoichiometric formula for this compound should be written as $La_{1-x}Ce_{x+y}Mn_{1-y}O_3$.

Interestingly, the amount of Ce substitution in the B site remains constant during all the fiber pulling process, while the substitution of La by Ce increases along the fiber (from the seed end to the pedestal end). Another possible interpretation of the fact that Mn/(La+Ca) ratio is less than unity is to assume the presence of vacancies in the Mn sub-lattice. In fact, the structure of LCMO allows for the existence of cation vacancies [24] in both La and Mn sub-lattices up to 2%at. The presence of Mn vacancies in fibers grown by LHPG was previously reported [22, 25].

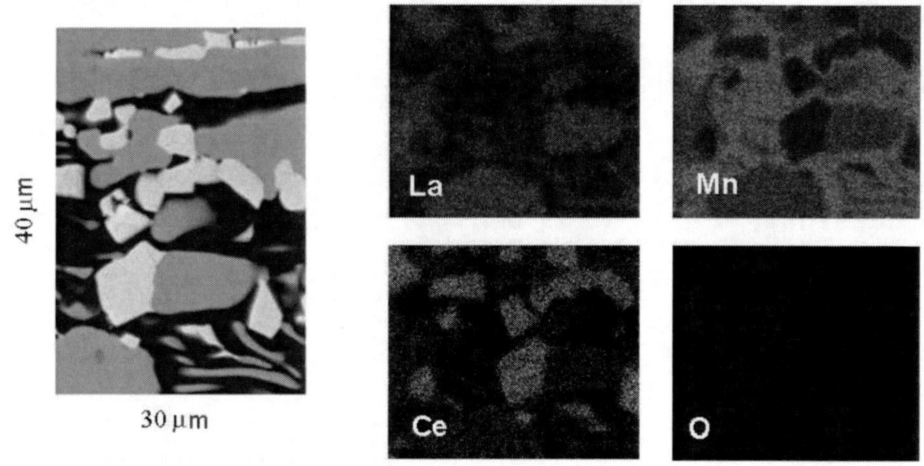

Figure 4. Left: SEM image (1000x magnification) of a 30×40 μm region in the center of the fiber. Right: Compositional maps for the different elements present in the fiber. Quantitative measurements identified the black regions as MnO_2, white ones as CeO_2 and the grey matrix are the desired cerium doped manganite.

The amount of manganese and cerium oxides segregated during the pulling process seems to happen to accommodate the excess of those elements which could not be incorporated in the grey LCMO phase. It is important to stress that we do not detect by EDX the self doped composition, $La_{0.7}MnO_3$, which was pointed out to be responsible for the ferromagnetic transition around T_C~ 250 K in previous works.

MAGNETIC PROPERTIES

Now let us turn our attention to the magnetic properties of our fiber. A small piece (3mm long) was cut from the center of the fiber and its magnetic response was measured as a function of temperature.

According to Figure 5, it is possible to identify three major magnetic transitions around 45K, 126K, and 180K indicating the presence of different phases within the measured part of the fiber. Most probably, the transition at 45K corresponds to the low-temperature FM phase (created by Ce-free regions) affected by the presence of a weak antiferromagnetic transition [11] due to MnO_2 impurities (revealed by EDX measurements and corresponding to the black regions in our SEM images).

Next, we find a large broad transition at 180 K which presents an initial step from this temperature up to 126K. It is worth to mention that the transitions observed in the temperature range 126–181K cannot be assigned to either CeO_2 or MnO_2 impurities because they have never been observed before in other samples where these impurities were detected. According to the proposed [19] phase diagram for LCMO thin films, we identify the transitions at 126K and 180K with LCMO phases corresponding to x=0.063 and x=0.13, respectively. Both transitions may be seen in the AC susceptibility measurements, as shown in the inset of Figure 5. And finally, we identify the broadest transition at 300K as a magnetic response from the (highly inhomogeneous) nominally doped phases with x=0.3.

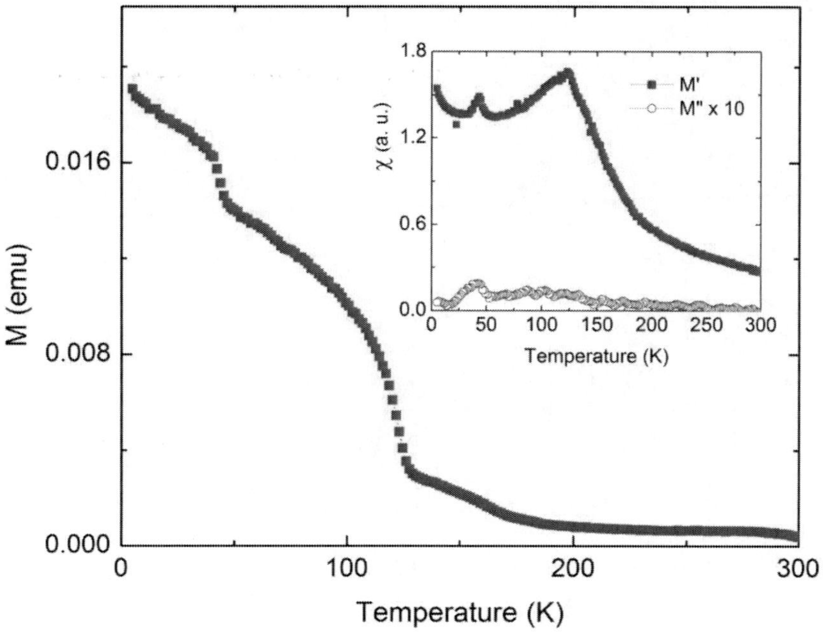

Figure 5. Magnetization versus temperature for the central region of one LCMO fiber. Inset shows the real and imaginary parts of AC susceptibility as a function of temperature for the same sample.

To better understand the origin of the above-mentioned magnetic transitions in the observed behavior of the spontaneous magnetization in our sample (see Figure 5), we adopt the spin polaron scenario for doped manganites and assume that the temperature and x dependence of the magnetization is governed by the classical Curie-Weiss contribution [26, 27]:

$$M(T,x) = M_0(x)\tanh\left\{\sqrt{\left[\frac{T_C(x)}{T}\right]^2 - 1}\right\} + M_R(x) \qquad (1)$$

here $T_C(x)$ is the doping dependent Curie temperature, $M_0(x)$ accounts for the deviation of the saturation magnetization of the undoped material in the presence of Ce atoms, and $M_R(x)=M(T_C(x),x)$ is Ce induced residual (paramagnetic) contribution. In fact, save for the $M_R(x)$ term, Equation (1) is an analytical (approximate) solution of the well-known Curie-Weiss mean-field equation on spontaneous magnetization. Figure 6 presents the best fits according to Equation (1) for experimental data on magnetization (shown in Figure 5) using the following set of model parameters: $T_{C1}=T_C(0.063)=126K$, $T_{C2}=T_C(0.13)=180K$, $T_{C,max}=T_C(0.3)=300K$, $M_0(0)=0.99M(0,0)$, $M_0(0.063)=0.76M(0,0)$, $M_0(0.13)=0.32M(0,0)$, $M_0(0.3)=0.05M(0,0)$, $M_R(0)=0.3M(0,0)$, $M_R(0.063)=0.15M(0,0)$, $M_R(0.13)=0.05M(0,0)$, and $M_R(0.3)=0.001M(0,0)$ with $M(0,0)=0.02emu$. Figure 7a shows the deduced doping dependence of the Curie temperature $T_C(x)$ which follows a linear law $T_C(x)=T_C(0)(1-x)$ with $T_C(0)=90K$.

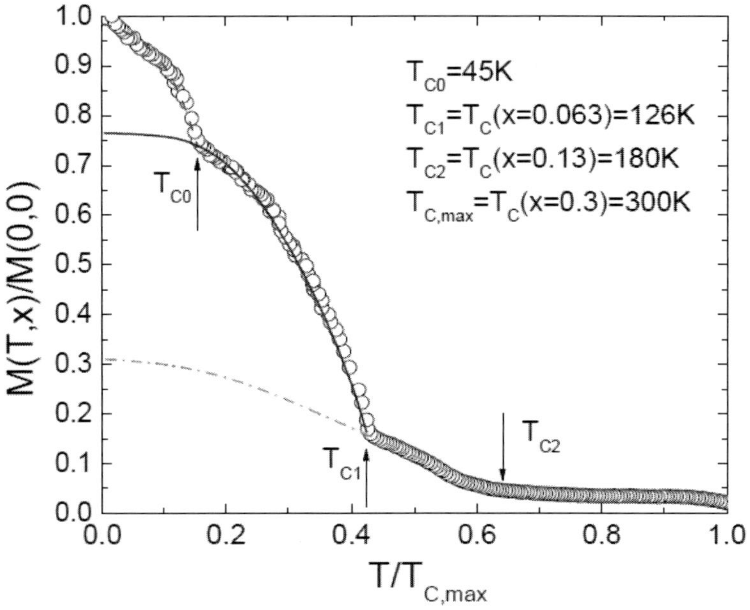

Figure 6. The best fits of the normalized magnetization M(T,x)/M(0,x) versus reduced temperature $T/T_{C,max}$ according to Equation(1) for the data from the central region of LCMO fiber shown in Figure 5. Four magnetic transitions can be identified with the Curie temperatures T_{C0}=45K, T_{C1}=126K, T_{C2}=180K, and $T_{C,max}$=300K corresponding to the concentration of Ce atoms equal to x=0, 0.063, 0.13, and 0.3, respectively.

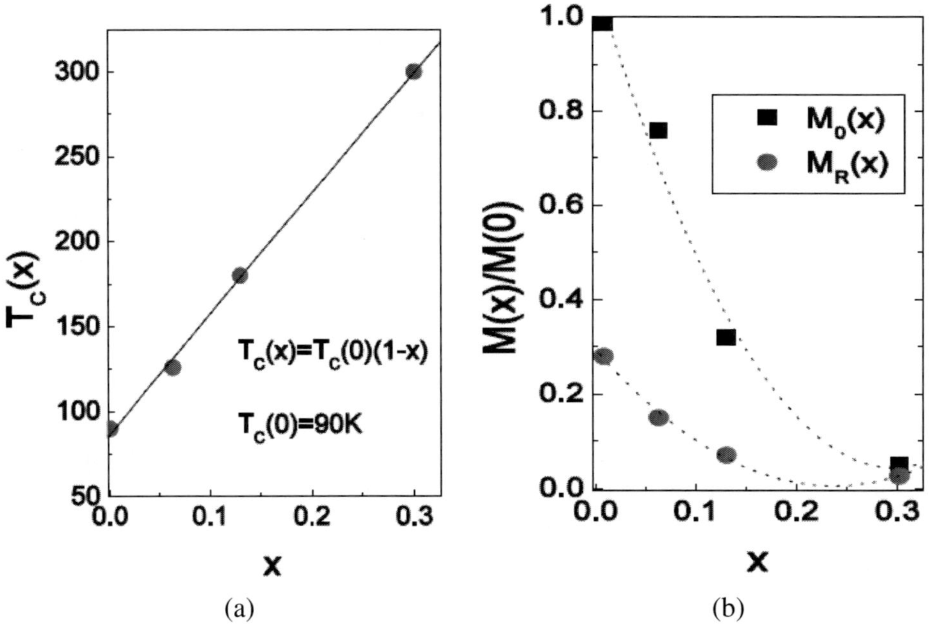

Figure 7. (a) The doping dependence of the Curie temperature $T_C(x)$; (b) Ce induced modification of the saturation magnetization $M_0(x)$ and residual (paramagnetic) contribution $M_R(x)$ deduced from the fits shown in Figure 6. In turn, Figure 7b demonstrates Ce induced modification of the saturation magnetization $M_0(x)$ and residual (paramagnetic) contribution $M_R(x)$.

It is interesting to point out that within this scenario we can treat a weak transition at T_{C0}=45K on the same footing as Ce mediated phase transitions by noting that with good accuracy [19] T_{C0}=0.5 T_C(x=0).

CONCLUSION

We reported our latest results on the growth of LCMO manganite fibers by the LHPG technique. Although the fibers present multiple phases, we could clearly identify by EDX measurements the presence of a cerium doped manganite phase. The LCMO phase is metastable and can only be achieved by using highly energetic techniques, like LHPG or PLD. We also identify the presence of CeO2 and MnO2 impurities in our samples.

The optimally cerium doped manganite phase exhibits a ferromagnetic transition at the inflection point 126K with a tail going up to 180K. The proportion of La, Ce and Mn observed in our samples also indicates the possible substitution of Ce by Mn or, alternatively, the presence of Mn vacancies. Based on the presented here results, we may conclude that it is quite possible to obtain single phase LCMO by proper optimization of the growth conditions.

ACKNOWLEDGMENTS

The authors gratefully acknowledge the financial support from Brazilian Agencies FAPESP, CAPES and CNPq.

REFERENCES

[1] *Colossal Magnetoresistive Oxides;* Tokura, Y., Ed.; Gordon and Breach, NY, 2000.

[2] *Colossal Magnetoresistance, Charge Ordering and Related Properties of Manganese Oxides;* Rao, C. N. R., Raveau, B., Eds.; World Scientific, Singapore, 1998.

[3] Zhou, J. P., McDevitt, J. T., Zhou, J. S., Yin, H. Q., Goodenough, J. B., Gim, Y., Jia, Q. X. *Appl. Phys. Lett.* 1999, 75, 1146-1149.

[4] Moutis, N., Panagiotopoulos, I., Pissas, M., Niarchos, D. *Phys. Rev. B* 1999, 59, 1129-1136.

[5] Rubinstein, M., Gillespie, D. J., Snyder, J. E., Tritt, T.M. *Phys. Rev. B* 1997, 56, 5412-5419.

[6] Hwang, H. Y., Cheong, S.-W., Radaelli, P. G., Marezio, M., Batlogg, B. *Phys. Rev. Lett.* 1995, 75, 914-917.

[7] Nie, J. C., Wang, J. H., Zhao, B. R. *J. Magn. Magn. Mater.* 1999, 192, L379-L384.

[8] Mandal, P. Das, S. *Phys. Rev. B* 1997, 56, 15073-15080.

[9] Philip, J., Kutty, T. R. N. *J. Phys. Condens. Matter* 1999, 11, 8537-8542.

[10] Raychaudhuri, P., Mukherjee, S., Nigam, A. K., John, J., Vaisnav, U. D., Pinto, R., Mandal, P. *J. Appl. Phys.* 1999, 86, 5718-5724.

[11] Gebhardt, J. R., Roy, S., Ali, N. *J. Appl. Phys.* 1999, 85, 5390-5398.

[12] Mitra, C., Raychaudhuri, P., John, J., Dhar, S. K., Nigam, A. K., Pinto, R. *J. Appl. Phys.* 2001, 89, 524-530.
[13] Mitra, C., Raychaudhuri, P., Kobernik, G., Dorr, K., Muller, K.-H., Schultz, L., Pinto, R. *Appl. Phys. Lett.* 2001, 79, 2408-2411.
[14] Mitra, C., Raychaudhuri, P., Dorr, K., Muller, K.-H., Schultz, L., Oppener, P. M., Wirth, S. *Phys. Rev. Lett.* 2003, 90, 017202.
[15] Mitra, C., Hu, Z., Raychaudhuri, P., Wirth, S., Csiszar, S. I., Hsieh, H. H., Lin, H.-J., Chen, C. T., Tjeng, L. H. *Phys. Rev. B* 2003, 67, 092404.
[16] Ganguly, R., Gopalakrishnan, I. K., Yakhmi, J. V. *J. Phys. Condens. Matter* 2000, 12, L719.
[17] Joly, V. L. J., Joy, P. A., Date, S. K. *J. Magn. Magn. Mater.* 2002, 247, 316-319.
[18] Hu, Z., Meier, R., Schüβler-Langeheine, C., Weschke, E., Kaindl, G., Felner, I., Merz, M., Nücker, N., Schuppler, S., Erb, A. *Phys. Rev. B* 1999, 60, 1460-1470.
[19] Raychaudhuri, P., Mitra, C., Mann, P. D. A., Wirth, S. *J. Appl. Phys.* 2003, 93, 8328-8334.
[20] Han, S. W., Kang, J.-S., Kim, K. H., Lee, J. D., Kim, J. H., Wi, S. C., Mitra, C., Raychaudhuri, P., Wirth, S., Kim, K. J., Kim, B. S., Jeong, J. I., Kwon, S. K., Min, B. I. *Phys. Rev. B* 2004, 69, 104406.
[21] Yanagida, T., Kanki, T., Vilquin, B., Tanaka, H., Kawai, T. *Phys. Rev. B* 2004, 70, 184437.
[22] Cardoso, C. A., Araujo-Moreira, F. M., Andreeta, M. R. B., Hernandes, A. C., Leite, E. R., de Lima, O. F., Mombrú, A. W., Faccio, R. *Appl. Phys. Lett.* 2003, 83, 3135-3137.
[23] Andreeta, M. R. B., Caraschi, L. C., Hernandes, A. C. *Materials Research* 2003, 6, 107-112.
[24] Tofield, B. C., Scott, W. R. *Solid State Chem.* 1974, 10, 183-196.
[25] Sergeenkov, S., Cardoso, C. A., Andreeta, M. R. B., Hernandes, A. C., Leite, E. R., Araujo-Moreira, F. M. *Phys. Status Solidi A* 2011, 208, 1704-1707.
[26] Sergeenkov, S., Bougrine, H., Ausloos, M., Cloots, R. *JETP Lett.* 1999, 69, 858-863.
[27] Sergeenkov, S., Mucha, J., Pekala, M., Drozd, V., Ausloos, M. *J. Appl. Phys.* 2007, 102, 083916.

In: Doping
Editor: Lixin Yu

ISBN: 978-1-62618-097-0

Chapter 4

DOPING: ION IMPLANTATION

Seongjae Cho[1] and Byung-Gook Park[2]

[1]Department of Electrical Engineering, Stanford University, Stanford, CA, US

[2]Department of Electrical Engineering and Computer Science, Gwanak-gu, Seoul, Republic of Korea

ABSTRACT

In this section on ion implantation, one of the most famous doping techniques, information associated with both theoretical and experimental aspects will be provided. For the former aspects, we address the basic equipment anatomy associated with the major parts of the ion implanter, the tools related with ion implantation other than the implanter, and the applications of ion implantation. For the latter, experimental sources are provided and we take a closer look into the physics of ion implantation and criteria for process control.

The experimental resources have been obtained from the periodic semiconductor process studies held for graduate students and field engineers at the Inter-university Semiconductor Research Center (ISRC) of Seoul National University, Seoul, Korea. These have been conducted for more than 10 years and are still being conducted today. The ion implanter and the other process tools installed in the ISRC have been used throughout the chapter.

INTRODUCTION

One of the most attractive properties of semiconductor materials is that their conductivity can be controlled as needed. Semiconductor materials frequently requested for device fabrication, such as silicon, germanium, and even compound semiconductor materials with two or more atomic species, are doped with n-type or p-type dopants in the integrated circuit (IC) fabrication. There are two classical methods to dope semiconductors, diffusion and ion implantation, and other techniques have been incessantly developed for nanoscale device fabrication.

Ion implantation is still the dominant technique in the IC fabrication, through which it can be distinguished as one of the core process in modern very large scale integration (VLSI) technologies.

Ion implantation is a process by which dopant atoms, in the form of an energetic ion beam injection, are added in the semiconductor. The dopants are accelerated by an electric field and thrown into the target in the form of ionized atoms or molecules.

Adding ions changes the electrical properties in the local area on a semiconductor wafer or piece. The properties are closely related with the conductivity (equivalently resistivity) of the material in that region or threshold voltage (V_{th}) when it comes to complementary metal-oxide-semiconductor (CMOS) devices. Therefore, it can be said that ion implantation is the key to control the electrical characteristics in ICs as the engineers design.

1. Brief History

Although the concept of doping by ion implantation to change semiconductor conductivity was disclosed in the original 1954 patent by Shockley, doping was mostly conducted by the diffusion process in high-temperature furnaces before the mid-1970s. Before ion implantation became very popular, solid-source diffusion and gas-source doping were the preferred doping methods.

In solid-source diffusion, the silicon (Si) surface is first coated with a thin film (of a SiGe alloy or silicon dioxide, for example) containing dopants either as deposited or due to subsequent implantations of dopants into this film (and leave the damage in it). Dopants are diffused into Si and the doping film may be removed by wet etching.

In practice, gas-source doping is used to dope Si with phosphorus (P) only. There are no convenient gas sources for arsenic (As) or boron (B). It is carried out in a furnace similar to that used for oxidation. The N_2 carrier gas would pass through a bubbler containing phosphorus oxychloride ($POCl_3$, often pronounced "pockle") that is a liquid at room temperature.

The N_2 carries the vapor of the source into the furnace tube. The reaction with Si or other gases liberates phosphorus atoms which diffuse into Si. Following the doping step, a high-temperature drive-in process diffuses the dopant atoms into the substrate to the depth required by the device design, which usually accompanies an additional oxidation process to prevent an out-diffusion of the impurities already doped into the substrate. After all these processes, the passivation oxide layer is stripped by a wet etch process, known as a deglaze. Pockle gas-source doping is still in use for some cases, such as doping the gate of CMOS devices, which needs minimized surface damage to maintain a high quality of gate material for better direct-current (DC) and radio-frequency (RF) device performances.

The most critical issue in the diffusion process is the gate alignment, which posed a great challenge when the feature size began to shrink. The application of ion implantation solved the gate alignment problem by using the so-called self-aligned source/drain (S/D) doping process. In this case, the gate oxide is grown and polycrystalline silicon (shortly polysilicon in usual) is deposited, patterned, and etched.

After photoresist (PR) stripping, a high-current ion implantation is used to form the S/D junctions. Since the polysilicon gate structure and field oxide block the ions, the source and drain are always aligned with the polysilicon gate. Some of the terminologies used here can be unfamiliar to some readers not participating in electrical engineering, but ion implantation is easily explained without them and they will be clearer in the later sections, although detailed explanations on the operation of a metal-oxide-semiconductor field-effect transistor (MOSFET) device and its physics will not be made.

The first true commercial ion implanters were manufactured in the early 1970s. These early implanters were typically capable of producing a few microamperes of beam current with energies of up to 200 keV. In 1978, the first true high-current ion implanter was introduced.

It was capable of producing 10 mA of beam current at energies up to 80 keV. Complex implanted dopant structures are required for modern integrated circuits.

For example, deep buried layers for device isolation can extend to depths of approximately 1 μm, necessitating the use of MeV-implantation, whereas MOSFET S/D extension implants require only sub-keV energies and halo implants require implantation into tilted wafers. This broad diversity of demands has driven the continuous evolution of ion implanters.

2. Ion Implantation Equipment

2.1. Overview

Ion implantation equipment, the ion implanter, is usually a very large piece of equipment, most probably the largest process equipment that can be found in the semiconductor fabrication facility. It has several subsystems, including an ion source, mass analyzer, acceleration tube, scanners, load-lock chamber for the end station, and a couple of vacuum systems. An ion implanter is shown schematically in Figure 1. A gas containing the desired impurity is ionized within the source and is then extracted into a mass analyzer to ensure that only the desired ion species enter the drift tube. After the mass separation, the ions are passed to an acceleration tube where the desired kinetic energy is obtained. In many ion implanters the mass analyzer occurs before the ions are accelerated to high energy as in the case in Figure 1, but the positions of two parts are exchanged in some ion implanters. The ion beam is then focused and scanned electrostatically over the surface of the wafer in the target chamber. Repetitive scanning in a raster pattern provides an exceptionally uniform doping of the wafer surface. The target chamber commonly includes automatic wafer-handling facilities to speed up the process of implanting many wafers per unit of time.

2.2. Ion Source

Dopant ions are generated in an ion source through the ionization discharge of atoms or molecules of a dopant vapor or a gaseous dopant chemical compound. A hot-filament ion source is the most commonly used ion source.

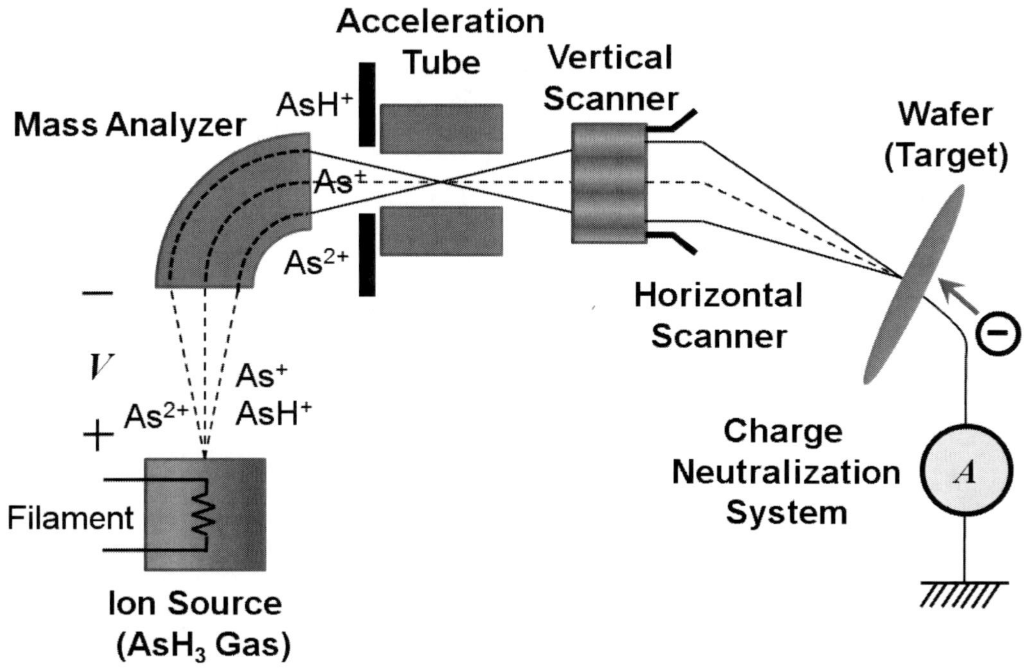

Figure 1. Schematic of an ion implanter (Arsenic gas has been taken as an example).

In this kind of ion source, large electrical currents from the filament power supply heat the tungsten filament and causes thermal electron emission from the red-hot filament surface. The thermal electrons are accelerated by the arc power supply with high enough voltage to dissociate and ionize the dopant gas molecules and dopant atoms.

Great demand is placed on the ion source because, unlike most other accelerators where the ion source is required to produce only a single ion species, the ion implanter requires an ion source capable of the contamination-free, on-demand production of a variety of different species. The difficulty of this task has historically limited the suitable source types to one called Freeman and Bernas, and the latter is the one that is currently prevalent .

2.3. Mass Analyzer

In almost every ion implanter, a mass analyzer is applied to precisely select the desired ion species and filter out unwanted ion species. In a magnetic field, charged particles rotate by the magnetic force, which is always perpendicular to the direction of the charged parties.

For the fixed magnetic field strength and ion energy, the gyroradius is related only to the mass/charge ratio, or simply m/q, of the charged particle. The mass analyzer passes the wanted atoms at high precision based on this principle.

The magnetic force on a beam of moving ions is given by the following vector equation, the Lorentz' equation, when the electric field is ignored:

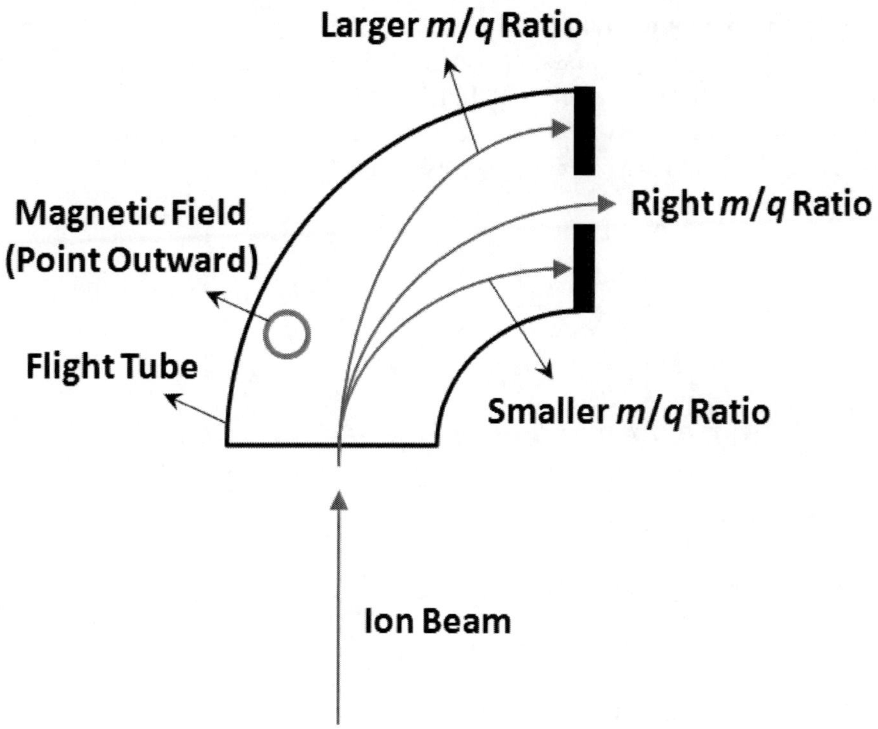

Figure 2. Schematic view of the mass analyzer of an ion implanter.

$$F = q\, v \times B \tag{1}$$

where F is the magnetic force, q the charge on the ion, v is the velocity vector, and B the magnetic field vector. Thus, F is represented by a scalar of the cross product between v and B, by which the direction of the magnetic force is also determined. As shown in an exemplary situation in Figure 2, given that a positive charge is moving upward and the magnetic field points outward from the figure surface, the vector operation directs the resulting magnetic force rightward. The magnetic force is mutually perpendicular to the velocity and magnetic field directions which would tend to force the ion into a circular orbit, the radius of which is determined by equality between the centripetal magnetic force and the inertial centrifugal force (in terms of only the vector magnitudes):

$$\frac{mv^2}{R} = qvB \tag{2}$$

where R is the radius of curvatures of the ion trajectory. In addition, the ion velocity is related to the energy imparted by the extraction potential by the equation below:

$$E = \frac{1}{2} mv^2 \tag{3}$$

Therefore, according to Equations (2) and (3), the radius of the ion trajectory is uniquely determined by the ion charge, mass, energy, and magnetic field intensity. A resolving slit at the exit of the analyzing magnet will block all species except those having the proper trajectory radius *R* for a given magnetic field intensity *B*. The product, *RB,* called the magnetic rigidity, quantitatively defines these trajectory requirements and may be derived from two equations above:

$$\frac{mv^2}{R} = qvB, \; RB = \frac{mv}{q} = \frac{m}{q}\sqrt{\frac{2E}{m}} = \sqrt{\frac{2mE}{q^2}} = \sqrt{\frac{2mV}{q}}$$

$$\text{or equivalently, } R = \sqrt{\frac{2mE}{q^2B^2}} = \sqrt{\frac{2mV}{qB^2}} \qquad (4)$$

where *V* is the potential difference that accelerates the charge, previously shown in Figure 1. Before the ions enter the analyzer, their energy is determined by the potential difference between the ion source and the extraction electrode, which normally is set at about 50 kV. The energy of the single charged ions after the extraction is 50 keV. Knowing the ion *m/q* ratio and ion energy, an on-board computer can calculate the required magnet-field strength for the ion trajectory passing through the narrow slit. Adjusting the current in the magnet coils enables the mass analyzer to precisely select the desired dopant ions. In practice, the magnetic field generated by the mass-resolving magnet is adjusted until the desired ion species travels along that trajectory radius that allows its transport through the mass-resolving slit. A detector placed at the site of the resolving slit would then typically show an ion distribution that has a Gaussian-like profile, having a peak at the radius position *R*, corresponding to an ion mass *M* and some half-width, equivalent to a mass range ΔM. Within the limits of *M* and ΔM, the mass resolving slit serves to block the transport of all ion beams except the beam of the desired specie which is allowed to pass and travel onto the wafer implantation site.

2.4. Acceleration Tube

After the analyzer selects the correct ion specie, the ions go into the acceleration tube, where the beam current and the final ion energy are controlled. In some textbooks, this "real" acceleration process is termed as "postacceleration" to be distinguished from the acceleration that occurred between the ion source and the extraction electrode before the mass analyzer. The ion beam current is controlled by a pair of adjustable vanes and the ion energy by acceleration electrode potential. Ion beam focus and beam shape are also controlled in this part by the defining apertures and electrodes.

The high-energy ion implanters mainly used in the well and buried layer implantation process in CMOS technology requires several high-voltage acceleration electrodes connected in series along the beam line to accelerate ions to several MeV energy levels. So that ion implanters could be used for ultrashallow junction implantation, especially for the p-type with boron implantation, the electrode in acceleration is connected in a reversed way such that the ion beam is decelerated instead of accelerated when passing the electrode. A pure ion beam with energy as low as 0.5 keV can be generated in this manner.

2.5. Beam Scanner and Wafer Handler

The beam scanning system provides the methods for uniformly spreading the desired ion species over the entire surface of the target wafer. Scanning may involve moving the beam relative to the wafer, or moving the wafer relative to the beam, or moving both of them in combination. As may be expected, all these techniques can result in a variation of the incidence angle of the beam on the wafer. For a nanoscale device, even a small angle variation can cause undesirable device variability. Consequently, most modern implanters also provide some ways of beam angle control incorporated into the scanning system.

Low-perveance beams, such as those present in medium-current ion implanters, may be scanned by passing the beam between a pair of electrodes that are electrically activated by a scanning generator. The generator imposes a high frequency alternating potential to the electrodes which causes beam deflections. Some attempts have been made to use alternating magnetic dipole fields to scan high-perveance beams. The large inductive load of such a scanner limits the fast scan frequency to approximately 150-200 Hz, which could lead to implant dose micro-non-uniformities (striping) for low-dose implants where the number of beams passing over the wafer is few.

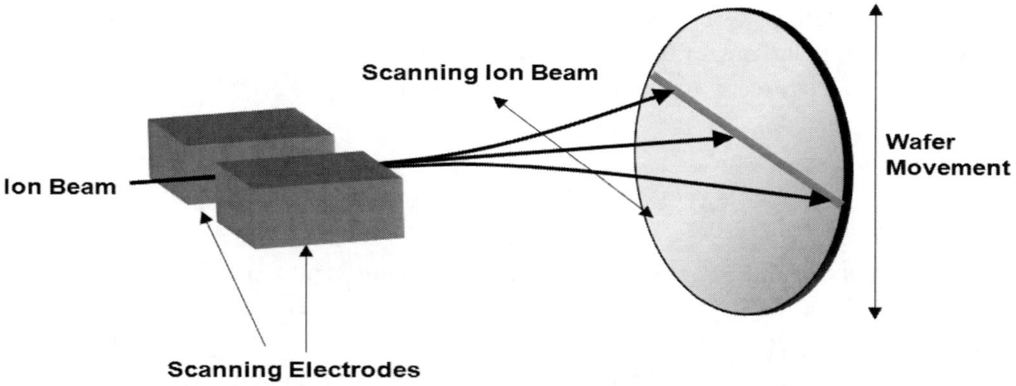

Figure 3. Schematic of the beam scanner.

Combined with the functions of the beam scanner, the most important duty of the wafer handler is to achieve uniform implantation profiles across the wafer. The diameter of the ion beam is about 25 mm (~1”). It needs to move either the ion beam or wafer, or, for some implanters, both, to scan the ion beam uniformly across the wafer, which has a diameter up to 300 mm. The spin-wheel system spins at a high rate. When a wafer passes through the ion beam, ions are implanted into part of the wafer, in the form of an arch band with the width of the ion beam diameter. The center of the wheel swings back and forth, which allows the ion beam to scan uniformly across the whole of each wafer on the spin-wheel. The spin-disk is similar to the spin-wheel, the difference being that instead of swinging the whole disk, it scans the ion beam to achieve uniform ion implantation across the wafer.

Another kind of wafer-handling system of ion implanters is shown in Figure 3. It combines ion beam scanning and wafer movement. Varying the applied bias potential between the scanning electrodes enables the ion beam to scan back and forth on the x-axis, while the wafer can be uniformly implanted in this way. This kind of scanning technique can be used in a single-wafer implantation system.

2.6. Charge Neutralization System

When ions are implanted into the Si substrate, they carry a positive charge to the wafer surface. Making the atoms ionized is essential in ion implantation systems since they can be sorted and controlled in ion states by a mass analyzer and accelerator, respectively. However, if the positive charge accumulates in the substrate, it could cause a wafer-charging effect. The positively charged wafer surface tends to expel positive ions, which could cause a so-called beam blowup and non-uniform ion implantation, resulting in a non-uniform dopant distribution across the wafer. Even worse, when the surface charge concentration is too high, the charge-induced electric field could be high enough to break down the thin-gate oxide layer and significantly affect the IC chip yield. When accumulated positive charges rise to a certain level, they could discharge in the form of arcing, and the arcing spark could cause severe defects on the wafer surface.

To address the wafer-charging problem, large amounts of negatively charged electrons are needed to neutralize positive ions on the wafer surface. There are several approaches to wafer neutralization, and one of the methods was schematically shown in Figure 1, where the electrons are supplied directly to the backside of the wafer.

3. Applications of Ion Implantation

Pure crystalline (single-crystal) Si rarely exists in nature and thus it is made artificially. Man-made crystalline Si has a very high resistivity. Si is in the group IV column on the periodic table, and the conductivity of crystalline Si can be controlled by adding dopants such as boron (B), indium (In), phosphorus (P), arsenic (As), or antimony (Sb). For an n-type (electron donor) doping process with atoms from the group V column, P is a light dopant and As and Sb are heavy dopants. For p-type (electron acceptor) doping with group-III species, B is a light dopant, and a heavier molecule, BF_2^+, is doped for easy ion extraction with higher energy.

3.1. Source/Drain Implantations

Source/Drain (S/D) implants are used to form highly doped (>10^{20} cm^{-3}) source and drain regions adjacent to the relatively lightly doped (~10^{18} cm^{-3}) active channel and well regions of the device. Figure 4 shows the cross-sectional view of the CMOS transistors fabricated on the Si substrate. Most applications for CMOS technology is summarized in this figure. The type of dopant in S/D junctions of an NMOSFET is an n-type so that a dopant can act as a donor of an electron. This is a counter-type of the substrate doping. In other words, an NMOSFET is fabricated on the p-type region and has n-type S/D regions. Likewise, a PMOSFET is fabricated on the n-type region and has p-type S/D junctions. When doping one of these different types of S/D regions, the other site is screened by photoresist (PR) to selectively open the regions to be implanted. S/D structures must be heavily doped to supply enough electrons for higher current drivability, and minimize MOSFET parasitic resistance and contact resistance to the metal contact. These days, to suppress the punch-through and the dependence of the device threshold voltage (V_{th}) on channel length (short-channel effect), the S/D junctions tend to be very shallow. Regarding S/D implantations, S/D extension (SDE) implantation is a frequently used technique. SDE is a highly doped (~10^{20} cm^{-3}) region under the sidewall spacer between the S/D regions and the channel, where lightly doped drain (LDD) structures used to be.

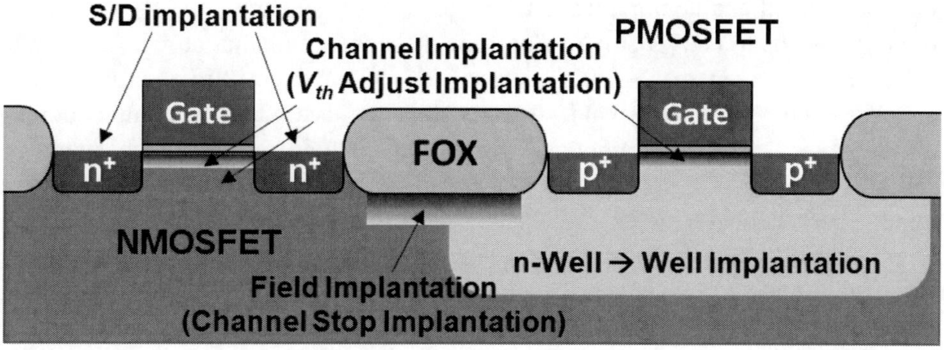

Figure 4. Cross-section of the CMOS on the Si substrate.

The LDD structure forced a trade-off between the hot carrier immunity and transistor current drivability. In spite of this trade-off relation, SDE implants are adopted to suppress short channel effects such as punch-through in the recent nanoscale CMOS devices.

3.2. Well Implantation

An NMOSFET where the carriers are electrons donated by donors is fabricated on a p-type Si substrate. On the other hand, a PMOSFET where the carriers are holes supplied by acceptors in the source junction needs to be implemented on the n-type substrate. This arrangement blocks the leakage current under a forward bias between S/D junctions and the substrate, by which the only current path is formed between the source and the drain junctions.

Grafting two different types of substrates for CMOS operations is not cost and area-effective at all. Thus, for a PMOSFET to be integrated on the p-type substrate with a NMOSFET, an n-type substrate-like atmosphere around the PMOSFET should be prepared. For this, well implantation is performed with high acceleration energy as can be seen in Figure 4. In this n-type doped region, the PMOSFET operates as if it were fabricated on an n-type Si substrate.

3.3. Channel Implantation

Ion implantation is also carried in the channel beneath the gate oxide. Its intent is to control the V_{th} of the device as the designer wants. So, it can also be called as the V_{th} adjusted implantation. In terms of the doping type, the channel is a part of substrate and it prevents current conduction when the inversion layer is not yet formed.

If there is no channel with a dopant type opposite to S/D junctions (i.e, the S/D junctions and channel are in the same dopant type to make a simple resistor with n-type doping), once the drain is applied by a positive bias with the source grounded, the current flows from the source to the drain regardless of gate voltage. The latter means that doping the channel with dopants of this type is the same as doping with S/D dopants, both lowering the device V_{th}.

On the other hand, the doping type is the same with the substrate; the V_{th} is raised. For an NMOSFET, V_{th} needs to be a positive value. Thus, channel implantation with n-type dopants reduces V_{th} and can even make it a negative value. Nonetheless, implanting with p-type dopants makes it a higher positive value. The V_{th} shift in a PMOSFET by channel doping with two different types and their amounts can be understood in the same manner. This implantation is performed to prevent short channel effects in a nanoscale CMOS device with an extremely short channel length.

Higher channel doping (counter-type to the S/D junctions) helps in securing the channel from being fully depleted, by which the punch-through between S/D junctions through the depletion region is prohibited. Since direct punch-through ignoring the gate control makes the device V_{th} undesirably low or even negative, the channel implantation is indispensable.

3.4. Field Implantation

Anb electrical isolation between MOSFET devices should be well established. Before the invention of planar technology, transistors and diodes were usually fabricated as mesa structures. However, surface leakage currents were a big problem in these devices. Local oxidation of silicon (LOCOS) is a famous process to electrically isolate devices.

After growing a thin oxide (10-20 nm), a layer of LPCVD Si_3N_4 (100-200 nm) is deposited. The nitride is then patterned to form the active device regions. Because the Si_3N_4 is resistant to oxidation, oxide will only grow in those regions that have no nitride present. Before growing the field oxide, boron is implanted (1×10^{13} – 4×10^{13} cm^{-2}) into the field regions in order to assure an acceptable field threshold. Subsequently, the field oxide is grown and the isolation structures are completed by stripping the Si_3N_4 film in hot H_3PO_4 and the pad oxide in a buffered HF (BHF) solution. Figure 5 shows the process flow for isolation by LOCOS.

The thick oxide regions in the field remaining after all the processes are called field oxides (FOX). The effective length of a path to possibly connecting devices is made long enough by the FOX, which cuts off the unwanted leakage path among devices so the field implantation can be omitted in the LOCOS process. However, implantation makes leakage harder to occur when combined with a FOX formation during the LOCOS process. Thus, it is also called a channel-stop implantation, as previously shown in Figure 4.

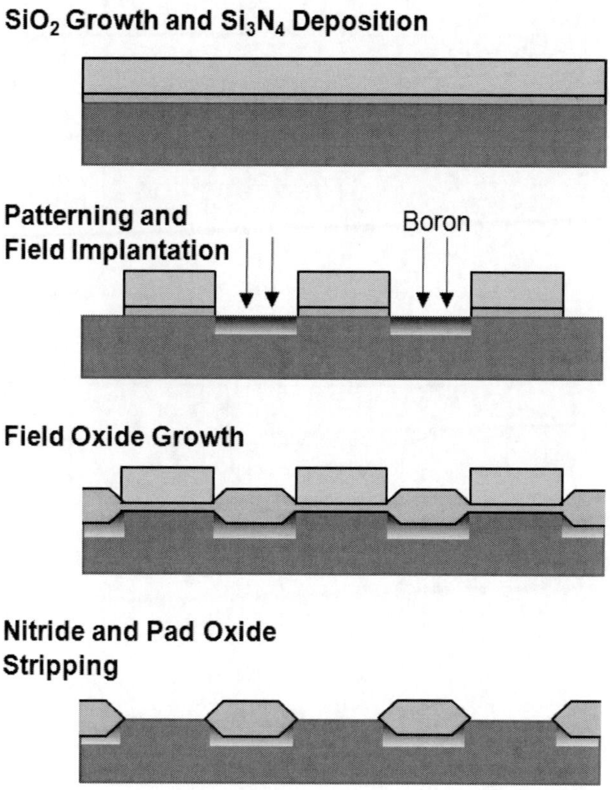

Figure 5. LOCOS process flow.

4. EXPERIMENTS

In this section, experimental results on ion implantation and discussions are given. As^+ ions were implanted into a p-type Si substrate having two process variables: acceleration energy and dose. Then, an ion implantation annealing process was performed to activate the implanted dopants. The tendency in sheet resistance change by process conditions is traced and the underlying causes are analyzed. Some tools, concepts, and terminologies specifically used in relation with ion implantation are also briefly surveyed.

4.1. Advantages of Ion Implantation

It was mentioned that there are solid-source and gas-source diffusion methods, other than ion implantation, to dope some impurities into the Si substrate. Figure 6 demonstrates the processes for $POCl_3$ doping by gas-source diffusion. Drawbacks of a doping process by diffusion are first visited.

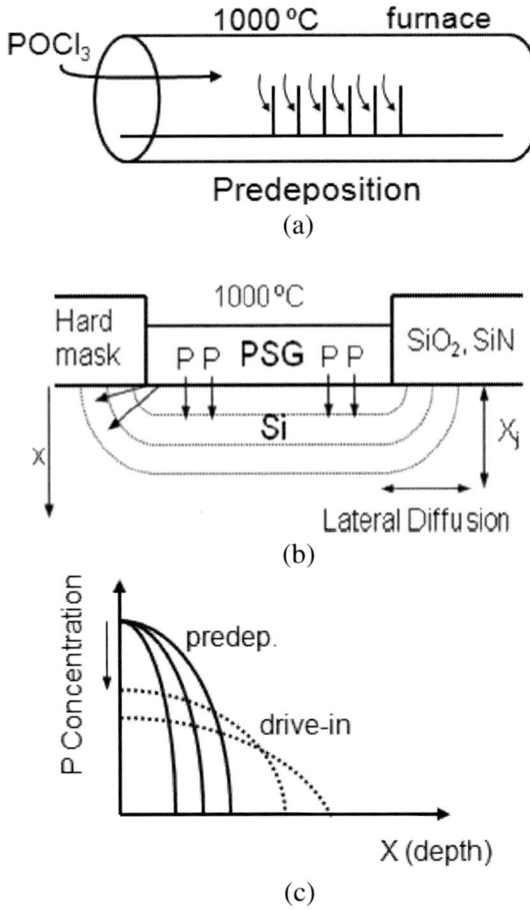

Figure 6. $POCl_3$ doping. (a) Predeposition. (b) Phosphosilicate glass (PSG) passivation and drive-in. (c) Doping profile change during drive-in.

A constant surface concentration is always expected for doping by diffusion. At a given temperature (>900 °C), the solid solubility of species in the Si substrate is determined. In other words, the Si substrate has an upper limit in accepting the dopants at the temperature, which gives a constant surface concentration. Although the doping profile changes as time goes by during the drive-in process, the overall shape cannot be tailored from the complementary error function as shown in Figure 6 (a) through (c). Once the passivation oxide is deposited, there is no way-in or way-out of the dopants in the Si substrate, so the total number of dopants is fixed and unchangeable. Another disadvantage is that there is a lateral diffusion as much as the diffusion in the depth direction, since the thermal diffusion is a process occurring with regard to the dopant concentration. Doping by diffusion usually takes place in a furnace which has a very long process time, making it difficult to form a very shallow junction. Other than the diffusion process itself, it inevitably accompanies other processes to prepare hard masks made from SiO_2 or Si_3N_4.

The advantages of ion implantation are opposite to the above drawbacks of conventional doping processes by diffusion. Ion implantation is a fast, uniform, and repetitive process. The number of dopants implanted into the Si substrate is precisely controlled. The purity of doping is very high, mainly due to the help of a mass analyzer, a part that an ion implantation genuinely has. Depth control is very easy since the acceleration energy is controllable. Non-Gaussian dopant distributions are achievable by multiple ion implants with different doses, energies, and various species with different atomic masses. Also, the lateral diffusion can be minimized. As will be explained in a later part, the ion implantation needs a thermal process known as annealing, but it can be conducted for a very short time compared to the furnace process. Moreover, the initial depth can be determined to be as deep as wished by controlling the acceleration energy so the effect of lateral diffusion can be negligible.

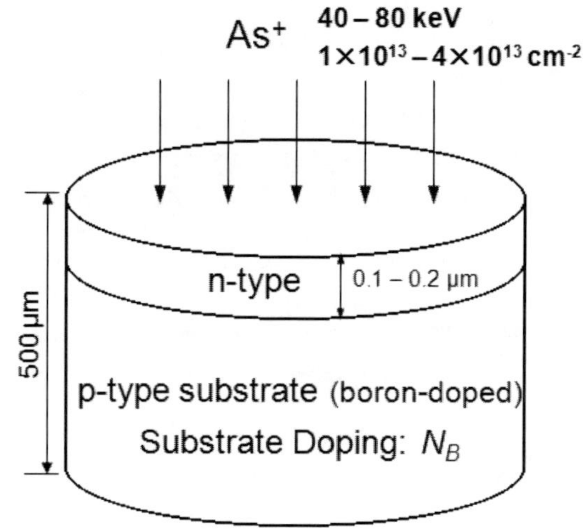

Figure 7. Ion-implantation onto a 4” wafer.

One of the major advantages is that ion implantation is a low-temperature process. It does not need a hard mask grown at a high temperature that endures the high temperature during a diffusion-based doping process. Instead, PR is used to form the mask for ion implantation. Since the ion implantation is carried out at room temperature (although there can be wafer heating during the process, PR and Si can be sustained without deformations), it does not require hard masks grown at high temperatures in general. Although a drawback of ion implantation is that the process results in lattice damages and needs an additional annealing process, it is usually resolved by either a subsequent process known as rapid thermal annealing (RTA) or other advanced annealing techniques. More about the annealing process will be explained in a later part.

4.2. Experiment Design

The experiments were performed with 5 slices of 4” p-type (100) Si wafers. The initial resistivity was checked by the sonogauge to confirm that the substrate doping concentration was the same as given in the wafer specification. As^+ ions were implanted, and the process variables were dose and acceleration energy. For the annealing process, the doped wafers were put into the furnace, and then final sheet resistances were measured by a four-point probe. The energy splits were 40, 60, and 80 keV, and the implanted doses were 1×10^{13}, 2×10^{13}, and 4×10^{13} cm^{-2}. The annealing process was performed at 900 ℃ for 4 minutes. This amount of thermal budgeting, designed for efficient use of time, is not sufficient for full dopant activation, but it is effective in making significant activation results and analyzing the results in a coherent tendency.

Making a brief investigation on the ion-implanted 4" wafer, as shown in Figure 7, the thickness of 4" wafer is usually 500 – 550 μm and the depth of implanted ions reaches 0.1 – 0.2 μm. Thus, the ion implantation is completed in a very shallow region compared to the whole wafer thickness. The substrate doping concentration can be checked by a sonogauge, from a simple relation of:

$$\rho = \frac{1}{q\mu_p N_B} \tag{5}$$

where ρ is the resistivity of material, μ_p is the hole mobility at a given doping concentration, and N_B is the doping concentration which is the substrate doping, in this case.

4.3. Tools for Experiments

As the first step in semiconductor processing, the wafers or pieces need to be identified with serial numbers when the different process splits are made on each wafer/piece. Figure 8(a) shows the diamond pencil by which the serial numbers can be easily carved on the wafer/piece with the diamond of the strongest hardness among most solids. Figure 8(b) shows the laser maker operated automatically by the serial numbers entered at the computer terminal and robot arms, which is better for handling wafers, especially when they are large in number. After the identification, the wafers are put into an initial cleaning state. The sonogauge pachymeter, or simply sonogauge, is used to measure thickness, sheet resistivity, and/or bulk resistivity. This system is only intended for use on wafers without junctions. The non-contact sensing head contains concentric elements which simultaneously measure thickness and sheet resistance at the same point. Differential thickness is measure by sonic (related with sound waves, where "sono" in the name comes from) elements in the upper and lower heads.

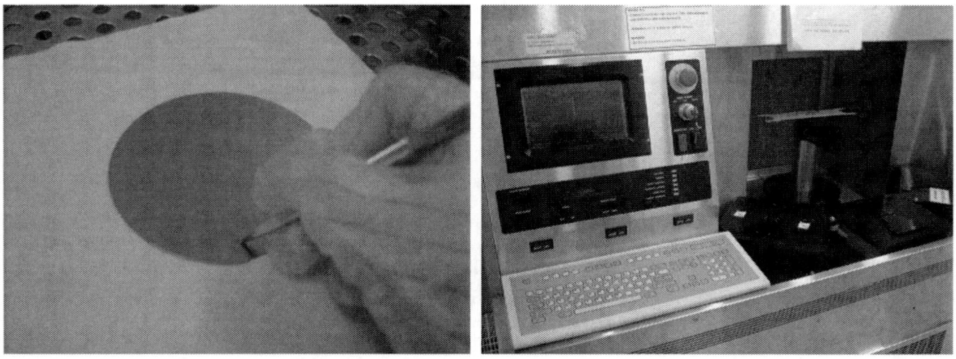

(a) (b)

Figure 8. Wafer identification tools. (a) Diamond pencil. (b) Laser marker.

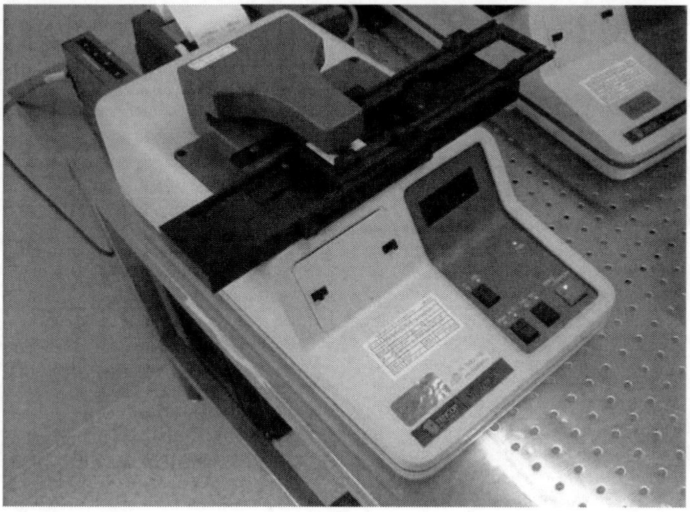

Figure 9. Sonogauge.

A high-quality RF tank circuit measures sheet resistance. Resistivity is computed automatically from the thickness and sheet resistance measurements taken at the same point. The relation among resistivity, thickness, and sheet resistivity will be made known later on. To improve accuracy, the RF tank circuit automatically re-zeroes each time the wafer carriage starts from the load position on a new measurement sequence. At the same time, the sonic elements automatically measure a reference wafer to determine probe-to-probe spacing, which is used in wafer thickness determinations. This tool is used before the ion implantation to verify initial conditions of the bulk wafer.

Figure 10. Medium ion implanter.

(a) (b)

Figure 11. Annealing tools. (a) Conventional furnace (n^+ annealing tube). (b) RTA.

Figure 10 shows the front panel of the medium ion implanter (Varian E220) used for the experiments. The process pressure is maintained at a high vacuum of 10^{-6} Torr to eliminate particles, reducing the mean free path of accelerated ions. B^+, BF_2^+, As^+, P^+, and Ar are the available species and the uniformity of doping concentrations along the laterally formed beam (Figure 3) should be within 3% to commence implants. The energy range is 5 – 200 keV and tilt and twist angles are 0° – 90°±1° and 0° – 360°±1.8°, respectively. To secure the required vacuum conditions during the process, cry pumps, turbo pumps, and dry pumps are operated. The implant dose range is 1×10^{11} – 1×10^{16} cm^{-2} (the maximum value for B^+ is 5×10^{15} cm^{-2}, exceptionally).

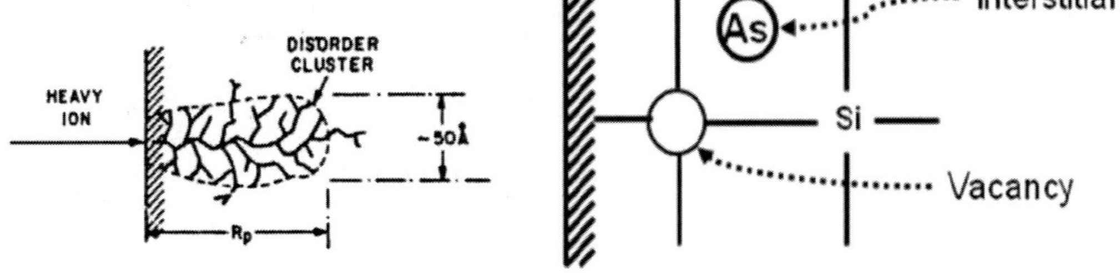

(a) (b)

Figure 12. Purpose of annealing process. (a) Surface destruction after ion implantation. (b) Occupation process of a vacancy by an interstitial ion.

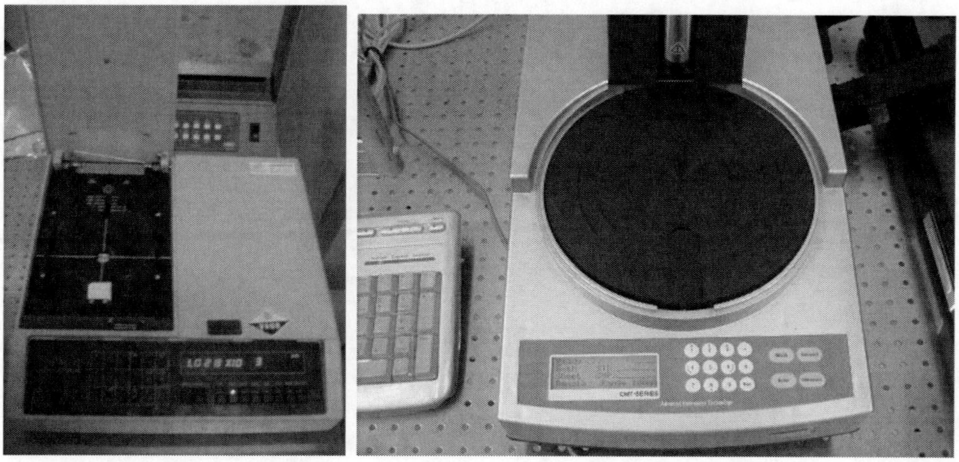

(a) (b)

Figure 13. Four-point probe. (a) Manual-type and (b) automatic four-point probes.

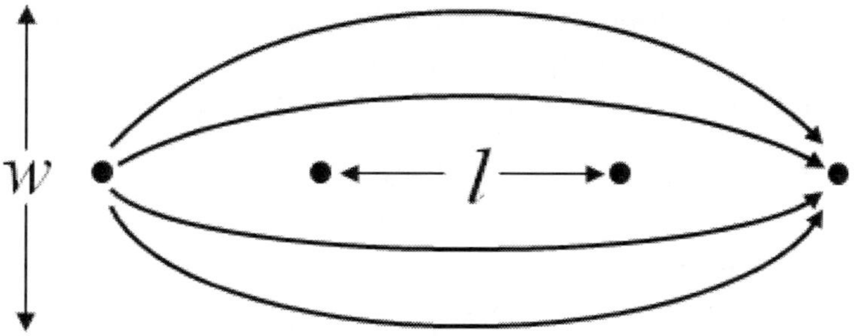

(a)

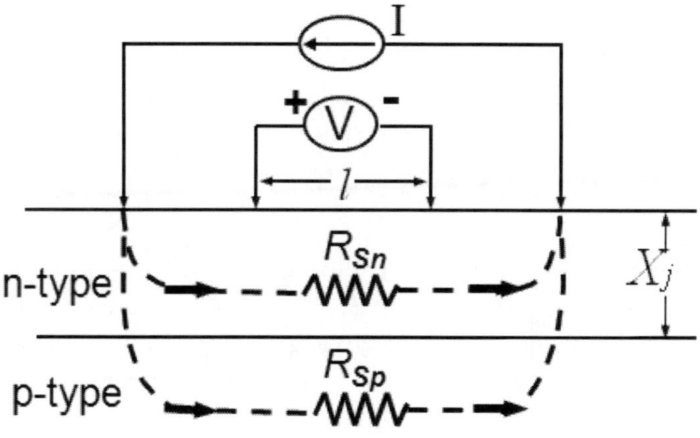

(b)

Figure 14. Four-point configuration. (a) Collinear arrangements of the probes. (b) Role of each probe (X_j: junction depth of the n-type region on the p-type Si substrate).

After ion implantation, the wafers are annealed in either the furnace or RTA tool, shown in Figure 11(a) and (b). The annealing process is performed in an inert ambient location so that there is no surface reaction with gas, by which any loss of implanted dopants is expected.

The lattice structure of the Si wafer surface is totally broken so that the crystalline structure is turned into an amorphous state. The sheet resistivity is immeasurably high at this point.

Two main purposes of the annealing process can be summarized as damage cure and dopant activation. The amorphized wafer surface recovers the original crystalline structure through the annealing process. The Si atoms are aligned in positions with a minimum bonding energy that eventually restores the crystalline Si.

However, this is not enough for obtaining the resistivity lowered by doping impurities into the semiconductor. The intersitial atoms, the implanted dopants, also need thermal energy to find the energy-stable positions where they participate in the covalent bonding with Si atoms so that implanted impurities behave as donors and acceptors, a process known as dopant activation. This process is accompanied by the damage cure during the annealing process.

After the annealing process, the sheet resistance of the ion-implanted wafer is measured. Usually, a four-point probe is used for this purpose. Figures 13(a) and (b) show the images of manual-type and automatic four-point probes, respectively. The latter is more advanced and provides us with more kinds of information from the measurement, but the fundamental principle is the same with the former one.

The four-point probe is commonly used to measure the semiconductor resistivity. It is an absolute measurement without recourse to calibrated standards and is sometimes used to provide standards for other resistivity measurements. Two-point probe methods would appear to be easier to implement because only two probes need to be manipulated. However, by using four rather than two probes, the parasitic voltage drops can be eliminated, even though the voltage probes (inner two probes) contact the device on the same contact pads as the current probes (outer two probes). Such four contact measurements are frequently referred as Kelvin measurements, after Lord Kelvin.

By Figures 14(a) and (b), the basic principle of a four-point probe can be explained more explicitly. The mathematical expression for the four-point probe resistivity is derived as follows:

$$\rho = 2\pi l F \frac{V}{I} \tag{6}$$

where l is the distance between probes (all the distances are assumed to be same as l) and F is the factor that corrects the resistivity without F that is derived by assuming a wafer with an infinite radius to a realistic value. If the effective spreading width (w) is much smaller than the distance between the outer two probes (for current measurement) and the thickness (t) of conductive region (ion-implanted region) is much smaller than l (this is true in most cases since t is much thinner than 1μm and l is in order of millimeter), F is reduced to $F = (t/l)/2\ln(2)$ and Eq. (6) is also simplified as:

$$\rho = \frac{\pi}{\ln(2)} t \frac{V}{I} = 4.532 t \frac{V}{I} \tag{7}$$

where V and I are voltage and current measured by two inner probes and two outer probes, respectively. Eventually, the four-point probe displays the sheet resistivity (R_{sh}):

$$R_{sh} = \frac{\rho}{t} = 4.532\frac{V}{I} \tag{8}$$

Now, let us find out the relation between resistivity, conductor thickness, and sheet resistivity. When the thickness is too thin to probe the two sides at the end of a conductor, sheet resistivity (sometimes simply called sheet resistance) provides a very useful physical quantity. Along with Figure 15(a) where the current flows in the longitudinal direction, the resistance of a conductor (the ion-implanted region in the Si substrate is also categorized as a conductor) is defined as:

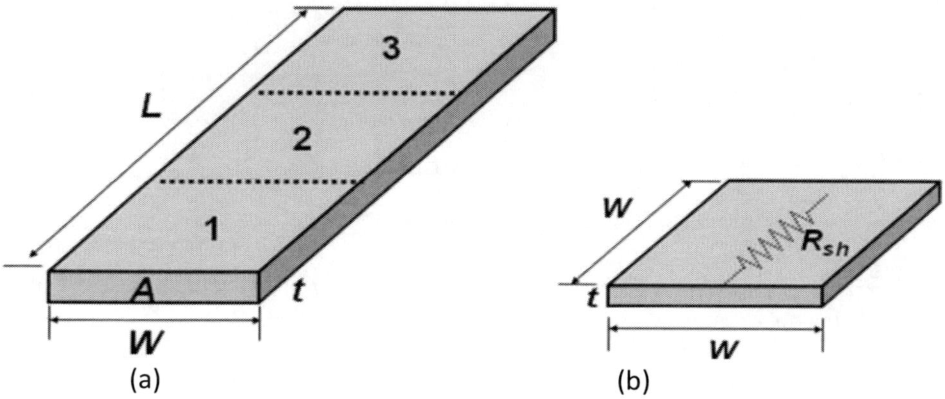

Figure 15. Definition of sheet resistivity. (a) Conductor strip. (b) Unit square for sheet resistivity.

$$R = \rho\frac{L}{A} = \rho\frac{L}{Wt} \tag{9}$$

where ρ is the resistivity, L is the length of conductor, A is the cross-sectional area the current passes through, expressed by Wt. The ratio of L/W indicates how many squares with a side of length L are included in the conductor in the length direction. By dividing both sides by L/W, sheet resistivity is obtained as follows:

$$R_{sh} = \frac{R}{L/W} = R/\square = \frac{\rho}{t} \tag{10}$$

where □ is the number of squares, L/W, counted in the length direction. So, a new physical quantity not related with conductor width, W, has been obtained. Strictly speaking, the unit of sheet resistivity is ohm (Ω) since the length units for L and W are cancelled out, which is the reason that it is simply referred to as sheet resistance.

To revive the meaning of sheet resistivity, resistivity per unit thickness as can be implied by Eq. (10) to distinguish it from resistance, it is common to say ohm-per-square (Ω/Υ) rather than ohm.

It cannot be overemphasized that R_{sh} does not depend on W. No matter how long either L or W is, R_{sh} is invariable with those values as long as the ratio of L/W is maintained to be constant in the same material system.

The circuit designers find the concept of sheet resistivity to be very handy in drawing the circuit layouts. The only thing that they need to do is connect the square tiles in a series to construct metal lines once they know the width that a single square mandatorily has by the design rule, the resolution of the photolithography tool that the fabrication facility is equipped with.

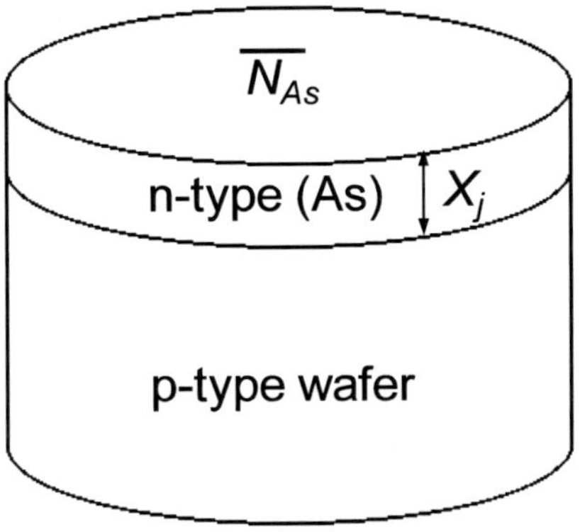

Figure 16. Schematic of a Si wafer implanted with As^+ dopants (X_j: junction depth, $\overline{N_{As}}$: average doping concentration of As atoms).

4.4. Experiment Results

Before investigating the experimental results and discussions, one needs to be familiar with the concept of "dose" which is frequently used in ion implantation. Dose is also a concept associated with the number of particles, which indicates the number per unit area. This is differentiated from "concentration" which is number per unit volume.

The electrical characteristics of a fabricated device are determined by concentration rather than dose. However, the ion implanter is only concerned with the number of ions it shoots onto the substrate/piece it is facing in the nearly normal direction without knowing the depth of the junction being formed. This is a physical quantity based on a two-dimensional manipulation just as sheet resistivity is, whereas concentration and resistance are viewed in a three-dimensional manner. One can guess there can be a relation between sheet resistivity and dose from this analogy.

$$\rho = \frac{1}{q(\mu_n n + \mu_p p)} \approx \frac{1}{q\mu_n n} \quad (\because n \gg p \text{ in the } n^+ \text{ junction})$$

$$R_{sh} = \frac{\overline{\rho}}{t} = \frac{\overline{\rho}}{X_j} = \frac{1}{X_j} \frac{1}{q\,\mu_n \overline{n}} = \frac{1}{q\mu_n \overline{N_{As}} X_j}$$

$$R_{sh} = \frac{1}{q\mu_n D} \quad (\because \overline{N_{As}} = D/X_j) \tag{11}$$

where q=1.6×10^{-19} C, $\mu_{n/p}$: electron/hole mobility, n/p: mobile electron/hole concentration, X_j: junction depth, upper bar: average value, and D: dose. Therefore, one can conclude that sheet resistivity (R_{sh}) is proportional to the inverse of D. This is not strictly right due to the mobility dependence on doping concentration, but keeping this relation in mind would be helpful in interpreting the experimental results.

4.4.1. Dose vs. Sheet Resistivity

In the first experiment, the acceleration energy was fixed to be 60 keV and dose was controlled to be 1×10^{13}, 2×10^{13}, and 4×10^{13} cm^{-2}. Figure 17(a) shows the process simulation results under these conditions. It is confirmed that the location of junction depth (X_j) is almost the same at the constant acceleration energy, but the area under curves expands as the dose increases. A higher dose results in a higher R_{sh} as shown in Figure 17(b) and as can be confirmed by Eq. (11) as well. The red dotted line indicates points assuming a constant μ_n. However, a line from the experimental result (black line with error bar) deviates from the hypothetical line. The factor making the difference is electron mobility, which monotonically decreases as the number of dopants increases, as can be confirmed in Figure 18(a). Therefore, the actual curve does not pace up with the curve, presumably ignoring the effect of electron mobility. At a same X_j, a higher D means a higher concentration (larger number of mobile electrons), and consequently, resistivity (ρ) shows a monotonic decrease, of which results are coherent with the curve in Figure 18(b). In most cases, the electrical characteristics are determined by the net sum of donors and acceptors ($|N_A$-$N_D|$), but it is noticeable that mobility is affected by the total number of impurities (N_A+N_D) since an electron regards both donor and acceptor atoms as particles distracting its traveling.

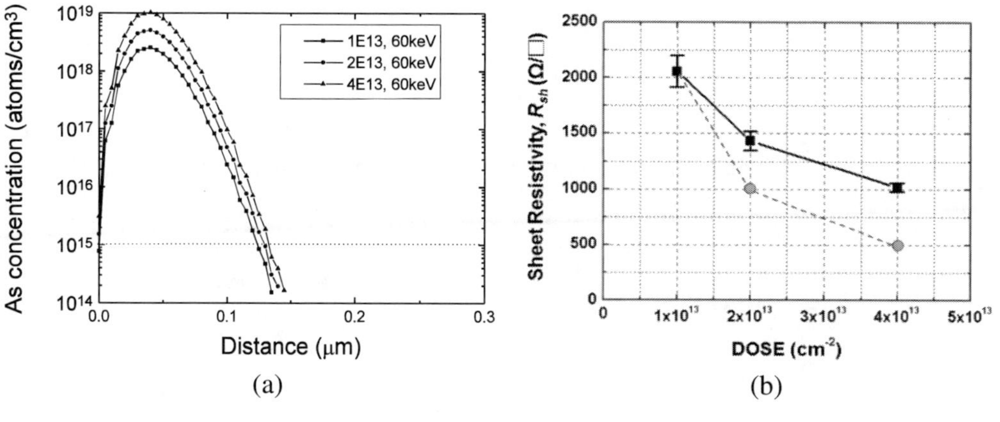

Figure 17. Ion implantation with dose variation. (a) Doping profile. (b) Dose vs. sheet resistivity.

4.4.2. Acceleration Energy vs. Sheet Resistivity

In the second experiment, the acceleration energy was controlled to be 40, 60, and 80 keV, while the dose was fixed at 2×10^{13} cm^{-2}. By a fixed dose, it is meant that the total number of impurity atoms is constant in all cases. As the acceleration energy increases, the junction is formed at a deeper location as shown in Figure 19(a). Consequently, electron mobility is enhanced since its mean free path between collisions with dopants is increased due to less impurity scattering as the result of a reduced number of implanted atoms in a unit volume. The sheet resistivity is reduced as mobility increases as shown in Figure 19(b), which is confirmed by referring to Eq. (11) where R_{sh} and μ_n are inversely proportional.

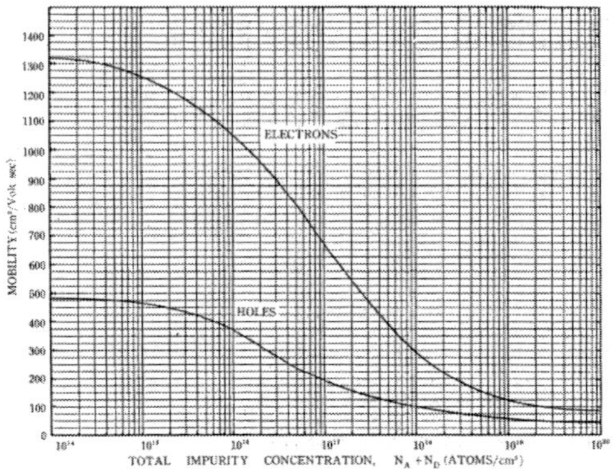

(a)

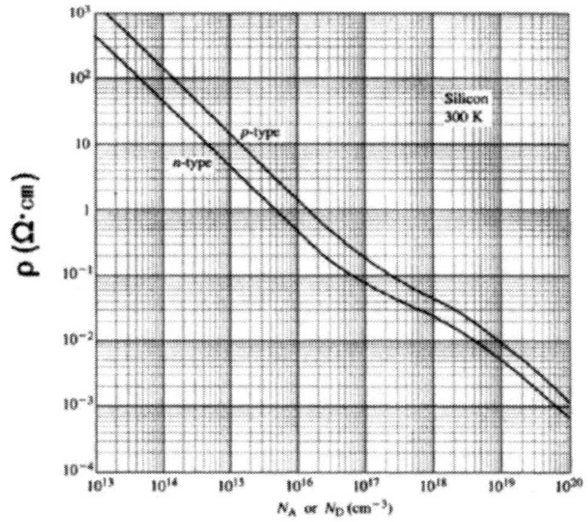

(b)

Figure 18. Effect of impurity number on (a) electron and hole mobilities and (b) resistivity.

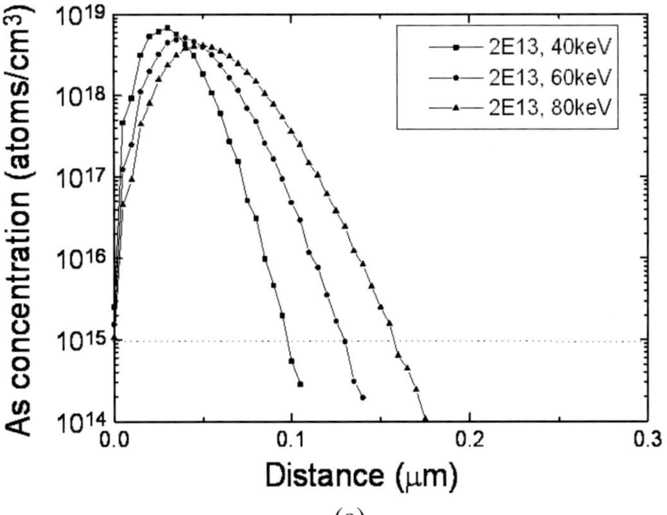

(a)

Figure 19. (Continued).

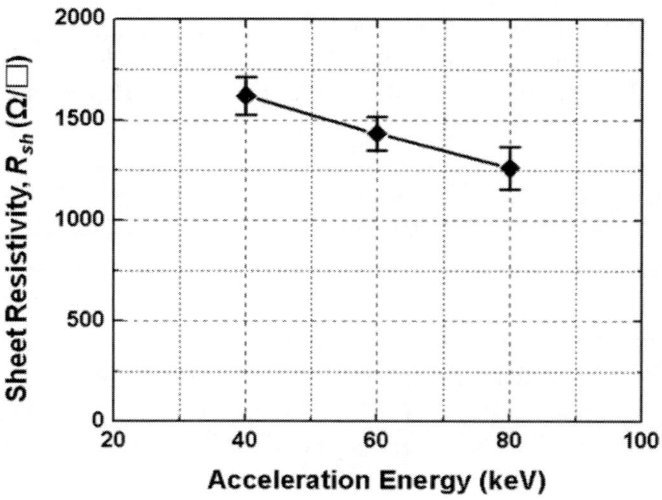

(b)

Figure 19. Ion implantation with acceleration energy variation. (a) Doping profile. (b) Acceleration energy vs. sheet resistivity.

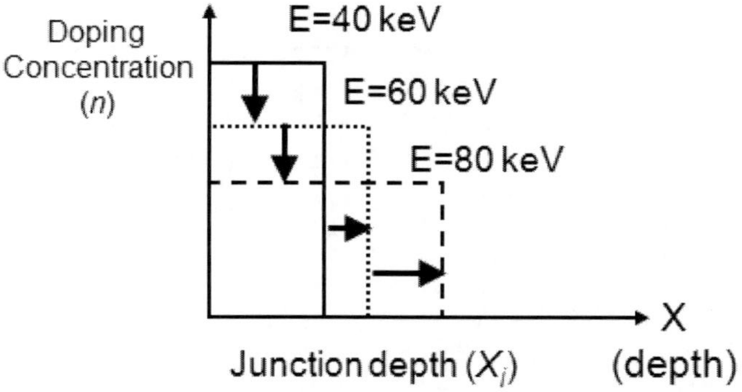

Figure 20. Ion implantation with variable acceleration energy and constant dose.

The relation among X_j, n, and D are clarified by the schematic in Figure 20. These parameters are mathematically formulated as follows:

$$D = \int_0^{X_j} n(x)dx \tag{12}$$

Dose can be understood as the area under an arbitrarily shaped curve in the depth-concentration plane, and it is simply the area of a rectangle, which is concentration multiplied by junction depth in Figure 20.

4.4.3. Annealing vs. Sheet Resistivity

In the third experiment, dose and acceleration energy were fixed to be 4×10^{13} cm^{-2} and 60 keV, respectively. Taking a look at a single curve with an annealing time in Figure 21, as the annealing temperature is increased, sheet resistivity monotonically decreases.

Also, making comparisons among R_{sh} values at a fixed temperature, R_{sh} shows a monotonic decrease as annealing time increases. It is noticeable that once the annealing temperature gets high enough, like at 1,000°C, the R_{sh} will not vary with annealing time any longer. This supports that the annealing temperature is more critical than annealing time in dopant activation. In Eq. (11), there is no significant change in parameters affected by annealing conditions other than D. In the previous experiments, D had been assumed to reflect the whole number of implanted dopants.

However, in reality, the actual D value or activated D, D_{ACT}, can be less than the designed D depending on the annealing conditions. Thus, creating higher annealing temperatures and longer annealing times are the efforts taken to make D_{ACT} approach D as long as the thermal budget is in a permissible range for device performance and reliability.

The annealing process was performed by RTA to create precise control over annealing time. In order to rapidly heat the Si wafer up (and to cool it off) for short-duration annealing, a special heating technique other than the furnace process is required.

In RTA, the Si wafer is heated to a high temperature in seconds by a bank of maps. Cooling off is also fast because the thermal mass of the entire system is small. Similar systems can be used for rapid thermal oxidation and rapid thermal chemical vapor deposition (CVD). Together, they are called rapid thermal processing (RTP).

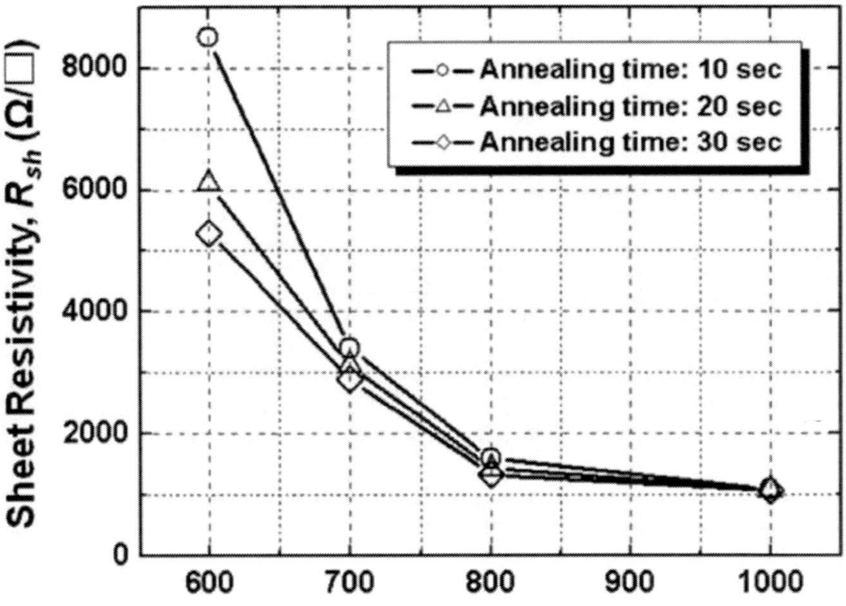

Figure 21. Sheet resistivity as a function of annealing conditions.

Conclusion

In this chapter, ion implantation as the conventional and still prevalent doping technique in modern CMOS technology has been reviewed. Novel technologies for nanofabrication will be researched and developed for the unceasing demands as the semiconductor device shrinkage is driven.

Through surveying the brief history, mechanisms by parts, applications in CMOS technology, and hands-on experiments, it is hoped to harden the understanding that ion implantation will serve as one of the unchangeably essential semiconductor process tools in the technology evolution.

ACKNOWLEDGMENTS

The authors would like to thank Prof. Jong Duk Lee for his dedication and endeavors for the ISRC, and Drs. Cheon An Lee, Won Bo Shim, and Jung Han Lee for their constructive discussions, effort, and time spent in establishing the research associated with the semiconductor process at ISRC. Also, the authors thank all the support unsparingly given by the ISRC staff and engineers.

REFERENCES

B. G. Streetman and S. Banerjee, *Solid State Electronic Devices, 5/e*, Prentice Hall, New Jersey, 2000.

C. C. Hu, *Modern Semiconductor Devices for Integrated Circuit, Prentice Hall*, New Jersey, 2010.

D. K. Schroder, *Semiconductor Material and Device Characterization, 3/e*, Wiley Inter-Science, New Jersey, 2006.

H. Ryssel and H. Glawischnig, *Ion Implantation Techniques*, Springer-Verlag, Berlin, 1982.

H. Xiao, *Introduction to Semiconductor Manufacturing Technology*, Prentice Hall, New Jersey, 2001.

R. Doering and Y. Nishi, *Handbook of Semiconductor Manufacturing Technology, 2/e*, CRC Press, Florida, 2008.

R. J. Baker, *CMOS: Circuit Design, Layout, and Simulation*, Wiley Inter-Science, New Jersey, 2008.

Y. Taur and T. H. Ning, *Fundamentals of Modern VLSI Devices*, Cambridge University Press, Cambridge, 1998.

In: Doping
Editor: Lixin Yu

ISBN: 978-1-62618-097-0

Chapter 5

PHOTOEMISSION STUDIES OF THE ELECTRONIC STRUCTURE IN ALKALI METAL DOPED MOLECULAR ORGANIC SEMICONDUCTORS

Huanjun Ding and Yongli Gao[*]
Department of Physics and Astronomy,
University of Rochester, Rochester, NY, US

ABSTRACT

The development of electronic devices based on molecular organic semiconductors, such as organic light emitting diodes (OLEDs), organic thin film transistors (OTFTs), and organic photovoltaic cells (OPVs), has been intensively explored in the last two decades. To employ successfully organic molecules as the semiconducting materials in the devices, it is essential to be able to modify the electronic structures of these molecules in a controlled way. As for inorganic semiconductors, doping has proven to be one of the most efficient techniques to adjust the electronic properties of the organic semiconductors. Doping of organic materials often causes significant changes in these materials and it is of both fundamental and practical interest. However, due to the localization of the electronic wavefunctions and the resulting strong correlation effects between the electrons and the nuclei, doping theories developed in inorganic semiconductors may not be applied without careful justification. It therefore remains a fundamental and important issue as how the energy level shift depends on the doping density in organic semiconductors. In particular, it is important to investigate a guest-host system whose properties are otherwise well studied, in order to reveal fundamental patterns and rules governing the electronic structure of doping in organic semiconductors. In this context, alkali metals are good candidates for prototypical studies due to their simple electronic structures and low electron affinity. In this chapter, we will first review the works about alkali metal doping-induced electronic structure modification in several molecular organic semiconductors. Based on experimental studies, we proposed a two-stage model to describe the evolution of the energy levels for the doped-organic

[*] Corresponding author: Yongli Gao. E-mail: ygao@pas.rochester.edu.

molecules. Density functional theory (DFT) calculation will be employed to explain the experimental results and the proposed model. Further, the reversal of the energy level shift induced by metal-deposition or oxygen exposure for the doped-organic molecules will be discussed. Finally, as a particularly interesting case, we will discuss the role of thin LiF insertion layer in metal/ tris-(8-hydroxyquinoline) aluminum (Alq) interface, which has been suggested to be a critical factor that determines the device performance.

INTRODUCTION

The development of electronic devices based on organic semiconductors has been intensively explored in the last two decades. To employ successfully organic molecules as the semiconducting materials in the devices, such as organic light emitting diodes (OLEDs), organic thin film transistors (OTFOs), and organic photovoltaic cells (OPVs), it is essential to be able to modify the electronic structures of these molecules in a controlled way. As for inorganic semiconductors, doping has proven to be one of the most efficient techniques to adjust the electronic properties of the organic semiconductors. Doping of organic materials often causes significant changes in these materials and it is of both fundamental and practical interest. Theoretically, it is representative of low-dimensional guest-host systems with strong correlation and anisotropic interactions. Practically, doping is widely used to improve organic semiconductor devices analogous to that in the inorganic semiconductor industry, and proven to be one of the most effective methods to improve the performance of the devices. One of the examples is the emission wavelength shift and the emission efficiency improvement caused by doping appropriate molecules into classical OLED materials such as Alq. [1, 2] Besides the use for energy transfer, doping can also improve the lifetime of OLED by enhancing the stability of the organic materials. [3] In addition, doping is expected to modify the properties of the metal-organic interface by adjusting the energy level alignment. Kido et al. report that doping of Alq with Li near the cathode interface significantly improves the device performance, and the highest luminescence is observed with the ratio Li:Alq=1 (atom: molecule). [4] Lous et al. and Heller et al. observe that doping of C_{60} can result in the Schottky barrier formation. [5, 6] Yang et al. and Antoniadis et al. successfully used doped polyaniline as a buffer layer between the anode and a hole transport layer, which results in efficient suppression of leakage currents. [7, 8] Despite these exciting observations of the doping effects in organic semiconductors, the basic relationship between guest and host energy levels is not yet clearly understood. It is difficult to provide a general theory for doping in organic materials, partly because of the localization of the electronic wavefunctions and the resulting strong correlation effects between the electrons and the nuclei. The energy level position is generally speculated to depend on the doping density the same way as that well established in inorganic semiconductors. However, such a speculation has not been verified experimentally. It therefore remains a fundamental and important issue as how the energy level shift depends on the doping density in organic semiconductors. In particular, it is important to investigate organic semiconductors whose properties are otherwise well studied, in order to reveal fundamental patterns and rules governing the electronic structure of doping in organic semiconductors. Several possible dopants, including organic molecules, [9] oxygen, [10] and alkali metals, [11-16] have been discussed in the literature. Among them, alkali metals are widely used in device fabrication as electron donors and are good candidates for prototypical studies due to their simple electronic structures and low electron affinity. Alkali metal doping in commonly used organic semiconductors, such as copper phthalocyanine (CuPc) [12, 17], tris(8-hydroxyquinoline) aluminum (Alq) [18, 19], perylene-tetracarboxylicacid-dianhydride (PTCDA) [20, 21] and oligophenylene [22, 23] has been investigated by several groups. [10-12, 17-19, 24-99]

This chapter will start with a brief introduction on the electronic structure of molecular organic semiconductors and the surface analytical tools involved in the studies discussed in this chapter. Then, we will review the works about alkali metal doping-induced electronic structure modification in several molecular organic semiconductors, including Alq, CuPc, PTCDA and oligophenylene. Based on experimental studies, a two-stage model is proposed to describe the evolution of the energy levels for the doped-organic molecules. Density functional theory (DFT) calculation is employed to explain the experimental results and the proposed model. Further, the reversal of the energy level shift induced by metal-deposition or oxygen exposure for the doped-organic molecules will be discussed. Finally, as a particularly interesting case, we will discuss the role of thin LiF insertion layer in metal/ tris-(8-hydroxyquinoline) aluminum (Alq) interface, which has been suggested to be a critical factor that determines the device performance.

1. Electronic Property of Π–Conjugated Molecular Organic Semiconductors

Molecular organic semiconductor refers to the kind of molecules with a well-defined molecular weight, including the oligomers. [100, 101] The studies discussed in this chapter are limited to small molecule organic systems.

In order to understand the electronic structure of organic semiconductors, let start with the basic knowledge about the molecular bonding between carbon atoms, which are the most essential building block in organic semiconductors. The element carbon is one of the most versatile elements on the periodic table in terms of the number of compounds it can form. This is largely due to the types of bonds it can form and the number of different elements it can join in bonding. Bonding in any element will take place with only the valence shell electrons, which are found in the incomplete, outermost shell. A carbon atom has four valence electrons. In the ground state of an isolated carbon atom, two of the valence electrons occupy the 2s sub-shell and the other two are in the 2p sub-shell. The 1s electrons are considered to be core electrons and are not available for bonding. Since energy is released when bonds form, as shown in Figure 2-2, it would be to carbon's benefit to try to maximize the number of bonds it can form. In other words, an excited state is favorable for a carbon atom in a multi-atom molecule by promoting one of its 2s electrons into its empty 2p orbital. Furthermore, three types of hybridizations may be introduced by mixing the 2s orbital with all or part of the three 2p atomic orbitals. In organic semiconductors, the most important type of hybridization is sp^2-hybridization, where the 2s orbital is hybridized with two 2p orbitals to form three equivalent sp^2 hybrid orbitals in a symmetric triangular plane geometry.

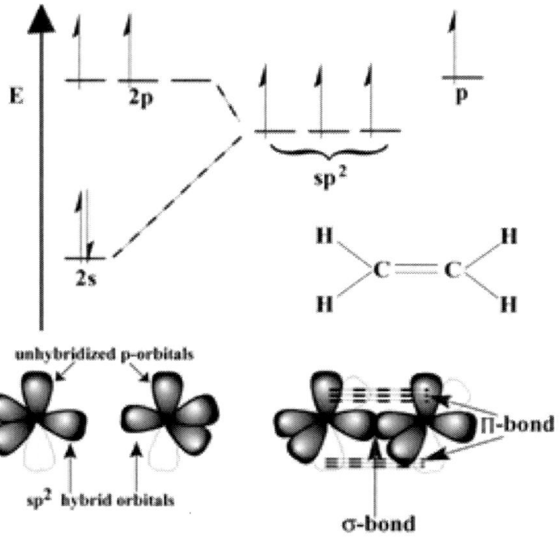

Figure 1. Illustration of sp^2-hybridization in conjugated organic semiconductors and formation of and bonds.

The remaining p_z orbital is perpendicular to this plane and free to overlap with adjacent carbon atoms, as shown in Figure 1. When the adjacent carbon atoms form molecular bonds in an organic semiconductor, sp^2 hybrid orbitals overlap and couple, they produce covalent σ and σ* energy levels for bonding and antibonding molecular orbitals. At the same time, the overlap of the p_z orbitals, which are all standing out-of-plane, yields π and π* bonds. As a result, the bonding in the molecule or the polymer backbone always alternates between single and double carbon-carbon bonds, which is referred to as conjugation. The σ-bonds are strong and stable, which keep the chemical structure of the molecules and the polymers, avoiding destructive processes during charge injection or exciton formation. The π-bonds give rise to states that are delocalized over the whole molecule, or a large segment of the polymer chain. The filled π-orbitals are usually called highest occupied molecular orbital (HOMO), while the unoccupied π*-orbitals are termed as lowest unoccupied molecular orbital (LUMO). A simplified picture of the electronic structure of a polyatomic molecule or a single polymer chain is schemed in Figure 2 (a). The essential difference between the organic semiconductors and their inorganic counterparts is introduced in the next step, when molecules are condensed into solid phase. The wavefunction of electrons in the molecular lattice is described by Bloch function, in the form of [102]

$$\Psi_k(r) = u_k(r)\exp(ikr) \tag{1}$$

where *exp*(*ikr*) is a typical plane wave envelope function, and $u_k(r)$ is a periodic function that has the same periodicity as the potential in a form

$$u_k(r) = u_k(r+R) \tag{2}$$

for all Bravais lattice vector *R*.

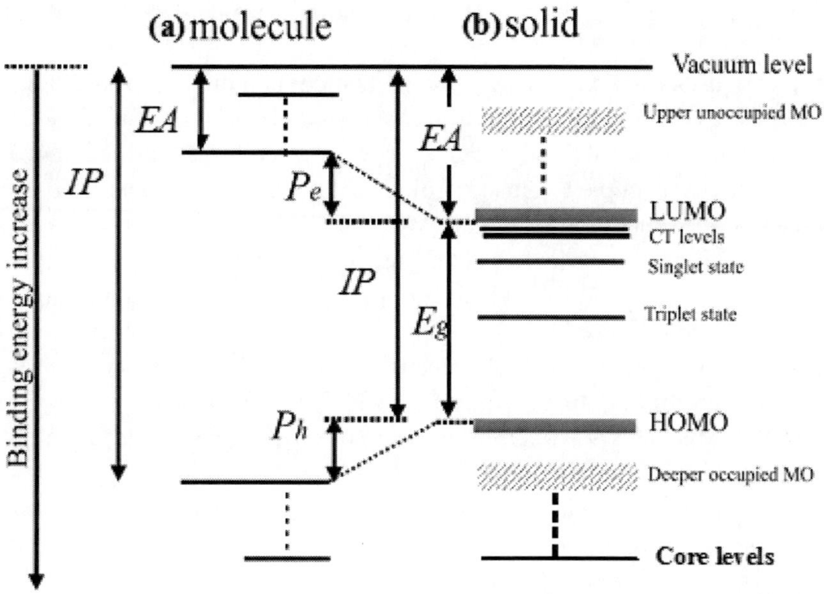

Figure 2. Energy level of (a) an isolated molecule and of (b) a solid. The vacuum level is used as a reference. The parent molecular states at left are lined by dashed lines to the corresponding solid states.

Bloch's theorem implies that for an electron in a periodic electronic potential lattice, the electron wavefunction is also in a periodic form. For inorganic semiconductors, the interaction between molecules is dominated by covalent bonding, leading to a strong overlap of the electron wavefunction. Therefore, the occupied and unoccupied molecular orbitals form valence and conduction bands, respectively. The charge transport can be approximately described by delocalized Bloch wave, and the electron-lattice interactions are generally weak. However, organic solids differ from inorganic ones essentially in the aspect that the electronic interaction between the molecules is predominately Van der Waals force, which is much weaker than the covalent bonds in inorganic solids. The weak interaction between the molecules constituting the lattice suggests that the wavefunction overlap is small and the transport bands are narrow. The typical intermolecular bandwidths for organic semiconductors are usually less than 0.1 eV. [103] With such a narrow bandwidth, the mean free paths of the charge carriers and excitons are of the order of the lattice spacing, which renders the localized nature of those carriers on an individual molecule. [104] In other words, the electronic structure of an organic solid preserves largely that of a molecule, as shown in Figure 2 (b), and the application of band theory developed for classical inorganic systems is usually limited. Here the highest occupied level in the valence region and the lowest unoccupied level in the conduction region are still called HOMO and LUMO, respectively, reflecting the correspondence with the molecular state. In this aspect, several most widely used parameters related to the molecular orbital should be introduced, in order to better describe the electronic structure of organic materials.

First of all, consider an isolated electron sitting at rest in a vacuum, the energy of that electron is defined as the vacuum energy level at infinite distance from the system. However, the real measurement of a solid sample involves electrons that are at rest just outside the sample surface, and are still affected by the potential of the solid. [105] It is thus a different value from the vacuum level of infinite distance, and is variable depending on the measured system. [103]

The next two important parameters are ionization potential (IP) and electron affinity (EA). IP is defined as the energy needed to remove an electron from the highest filled energy level in a neutral molecule; it is also the energy released when an electron at infinity is brought into the singly occupied valence orbital of a positive ion. Hence, IP can be simply deduced as the energy separation of the HOMO and the vacuum level. On the contrary, the energy released when an electron is placed into the lowest unoccupied state in a neutral molecule from vacuum is called EA, which is defined as the energy separation of LUMO and vacuum level. One should note here that the values of IP and EA are different for isolated molecules and for solids due to the multi-electronic effect. During the condensation process, the electronic structure of an ionized molecule will be significantly affected by the electronic polarization in the molecules surrounding it. The polarization energy leads to an increase of the LUMO binding energy and a decrease of the HOMO binding energy. It is important to note that in solid phase, HOMO is closer to the vacuum level, as it is easier to ionize a neutral molecule when the remaining positive charge is stabilized by its polarization energy. For the same reason, it is more difficult to remove an excess electron to the vacuum level from an ionized molecule due to the stabilization induced by the surrounding molecules. While the polarization energy of a positive charge carrier (P^+) can be deduced from the difference between the IP in gas phase and in solid, respectively, the polarization energy of a negative charge carrier (P^-)' cannot be measured directly, but has to be calculated. To the zero-order approximation, P^+ equals P^-, being typically around 1.3 eV. But the inclusion of the charge-quadrupole interactions introduces energy asymmetry such that P^+ is larger than P^-. The energy separation between the HOMO and LUMO is defined as the band gap energy (E_g), which is the energy released when an electron and hole recombine from their lowest energy states. It is thus also the least energy needed to produce an uncorrelated pair of carriers inside the crystal. The polarization effect in the condensed phase causes Eg to decrease. [106] In solid phase, this energy separation is different depending on whether it is measured for a neutral molecule or ionized one. The difference is the correlation energy between an electron and a hole, which is required to break an exciton into free charges.

Another important parameter relevant to the work discussed in this thesis is the work function (Φ), which is introduced as the difference in energy between an electron at rest in vacuum at infinite distance and an electron at the Fermi level inside the solid. For metals, the work function depends critically on the surface atomic geometry, and is the most important parameters to describe the ability of charge injection across the interface. For organic semiconductor, due to the presence of the band gap, work function is used as a concept during interface analysis, and can be very sensitive to charge doping.

Finally, the other important aspect of the electron-lattice interaction is the polaron states. The localized nature of the molecular orbital leads to a considerably smaller number of free (delocalized) electrons in organic solid in comparison to the case of metal or even inorganic semiconductors. As a result, the screening of an excess charge introduced into the solid is incomplete, resulting in the polarization of the medium. Considering the case when an electron is introduced into a molecule, the added electron has to be placed into the unoccupied states under the rigid-band assumption. The localization of the electron wavefunction and the polarization effect imply that the correlations between the excess charge and the crystal lattice are much stronger than the case of inorganic semiconductors. Therefore, the added charge is no longer a free electron, but an electron that interacts with the surrounding molecules. This ionized species is then named electronic polaron, first introduced by Toyozawa. [107] If two electrons are bound together in the same molecule, leading to a doubly charged polaron, it is known as a bipolaron. The charge carriers in the organic molecular solid are strongly localized and move in the lattice in the form of polarons. As a result, the corresponding conduction levels of charge carriers are determined by the actual conduction levels of the electronic or molecular polarons. The energy separation of the positive and negative non-relaxed electronic polaron states is named as optical energy gap, whereas, such energy separation for relaxed states of molecular polarons is called adiabatic energy gap. Polarons of opposite signs may interact with each other through Coulomb attractions, forming bonded (non-conducting) polar states. For non-relaxed electronic polaron pair, it is customary to call it charge-transfer state. The pair of relaxed molecular polarons is usually named as charge pair state, which is a characteristic feature of organic molecular solid and behaves as an intermediate state in many electronic processes.

2. Photoemission and Inverse Photoemission Spectroscopy

Electrons and Photons are the most easily available particles with which to probe matter. The electrons that are ejected from a solid subsequent to the absorption of high energy photons contain considerable information about the static energy level structure of the solid and the density of states at these energies. The principle of photoemission spectroscopy (PES) is illustrated in Figure 3, where a photon impinges on a sample and an electron is excited, which then escapes into vacuum and finally is detected by an energy analyzer. The basic principle of the technique is based on the photoelectric effect explained by Einstein in 1905, [108] which relies on the creation of photoelectrons via interaction between the irradiating photons and the sample. An important point in this process is that the obtained PES spectra contain information only from those energy states whose final energy lies above vacuum level of the system.

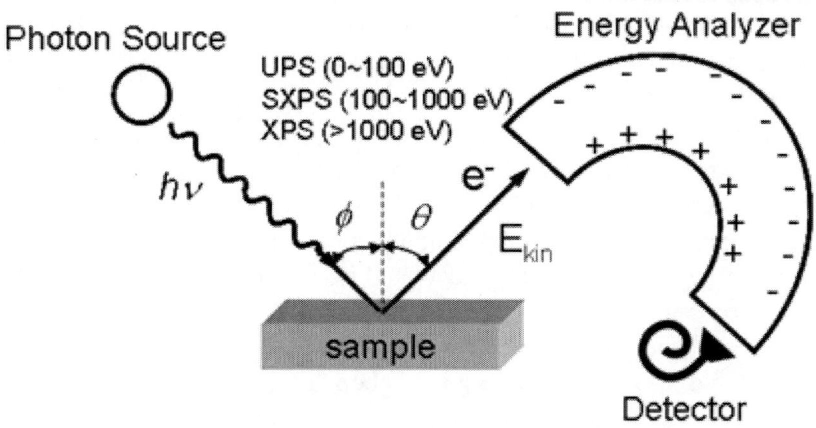

Figure 3. Sketch of modern PES experiment.

Due to possible scattering during the travel inside the solid, the total current of the photoemitted electrons will be composed of primary electrons that have suffered no energy loss and secondary electrons that have experienced inelastic collisions. [109] Generally, in the measured PES spectra, the primary electrons contribute to the main features, while the secondary electrons form the parabolic background. The detailed discussions about the secondary electrons will not be included in this chapter. They can be found in the referred books. [109-111] The following discussion will thus focus on the energetic of the primary electrons, as they carry the most direct information about the position and the density of the energy levels in a solid. In a PES experiment, a hole is left on the molecule after an electron is emitted. In a rigorous view, the photoemitted electron and the sample should be considered as a joint excited state. According to the energy conservation rules, the energy difference between such excited state and the ground state, at which sample has not been hit by the photon yet, must equal the energy of the annihilated photon. However, this rigorous view is practically very hard to apply. Very often, researchers consider the photoemission process in a one-electron picture. In this simplified model, the emitted electron comes from a one-electron orbital within the sample and escapes into vacuum without suffering energy losses. In this case, the energy of the emitted electron equals the photon energy minus the initial binding energy of the corresponding bound electronic states. Therefore, the analysis of the energy distribution of the photoemitted electrons provides information about the energies of the occupied one-electron states. This simple relation can be expressed by following equation, [112]

$$E_k = h\nu - E_B - \Phi \tag{3}$$

where E_k is the kinetic energy of photoemitted electron, $h\nu$ the incident photon energy, E_B the binding energy of the initial state with respective to the Fermi level, and Φ the work function of the sample. In a PES experiment, photons with known energy are used to excite electrons out of the sample surface, whose kinetic energy and momentum can be measured in an electrostatic analyzer placed nearby the sample surface, with a cylindrical, spherical, or ellipsoidal field pattern. [113, 114] The schematic of this relation is presented in Figure 4, where photoelectrons originate from the occupied energy levels, including the valence band in inorganic semiconductors or HOMO for organic semiconductors as well as the core-levels, which correspond to closed atomic shells.

In an ideal case, there is a one to one correspondence between the peaks in the photoemission spectra and the atomic or molecular energy levels in the neutral ground state, as suggested in Koopmans' theorem, [115] which proposes that the ionization energy of an orbital is equal in magnitude to the orbital binding energy. In this theory, Koopmans assumed that the binding energy difference can be calculated from Hartree-Fock wave functions for the state with *N* electrons and the state with *(N-1)* electrons. However, in practice, many factors including both inter- and intra-molecular relaxation effects contribute to the difference between the observed spectra and Koopmans' theorem. One of the major effects is the relaxation process during the photoemission event. In a simple classical picture, this effect can be understood as follows: an electron in photoemission process typically leaves a molecule within 10^{-15} s. The nuclear geometric relaxation time is about 10^{-13} s, which corresponds to an optical phonon frequency. At the same time, the corresponding electronic

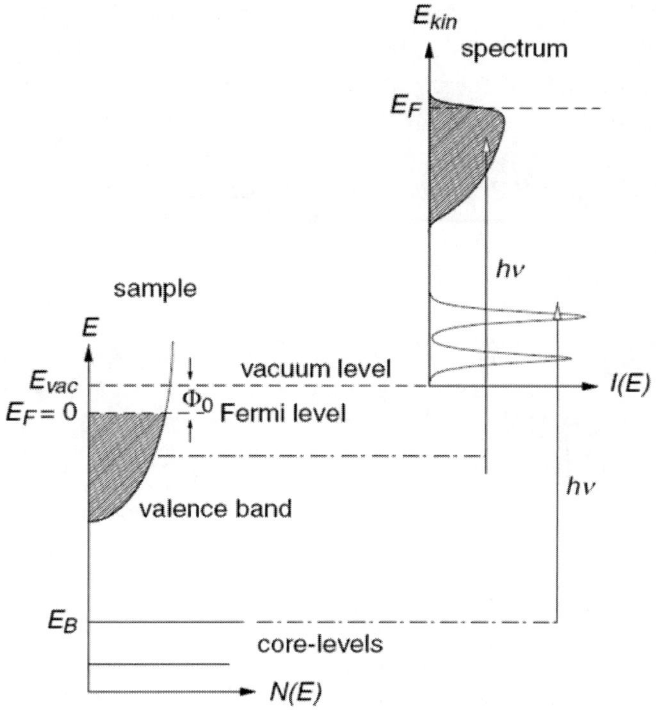

Figure 4. A schematic representation of photoemission spectroscopy.

relaxation time is around 10^{-16} s. As a result, the electrons have time to relax and the hole left by the photoemitted electron is fully screened. However, the nuclei are frozen during the process. Therefore, in the system where the ionized final-state energy can be significantly affected by geometric relaxation, which is most likely the case for π-conjugated organic molecules, PES measurements will give an overestimation of the binding energy. Besides relaxation effects, several other processes may also occur during the photoemission event, which are summarized in Figure 5. [116] Photoionization is the main process in PES as we have learned from previous discussion. The shake-up and shake-off processes are similar, which happen when the escaping electron perturbs the outer electrons of the atoms. If the secondary electron gains enough energy to escape the solid, it is a shake-off process. Otherwise it is a shake-up process when the perturbed electron is excited to a higher unoccupied orbital. In these two processes the photoelectron loses energy, resulting in a shallow peak at the higher binding energy side of the main peak. Finally, the Auger process occurs when the de-excitation energy obtained from the relaxation of an ionized atom is transferred to another electron (Auger electron), which is then emitted by the atom. These processes are normally easy to be observed in XPS where the photon energy is large.

Finally, the other important aspect about PES measurement is its surface sensitivity. In any systematic surface studies, one has to recognize the fact that the number of surface atoms scales as $N_A^{2/3}$, where N_A is the number of atoms in bulk. As such, the signal from the bulk atoms will quickly overwhelm that originating from the surface, if the probing depth of the experimental technique exceeds several atomic layers. In this context, PES turns out to be one of the best surface analytical tools, as it utilizes photoemitted electrons as the probing particles, which have a strong interaction with the rest of the solid, leading to a relatively short escape depth, or more generally named mean free path (MFP). The MFP λ is defined simply as the distance an electron can travel before suffering an inelastic scattering, which depends strongly on the kinetic energy of the photoelectron and is relatively independent of the material. The universal curve for MFP in Å as a function of the electron kinetic energy is shown in Figure 6, [117] where one notices that within the energy range of interest, typically between 10 to 2000 eV, the MFP is only on the order of a few monolayers.

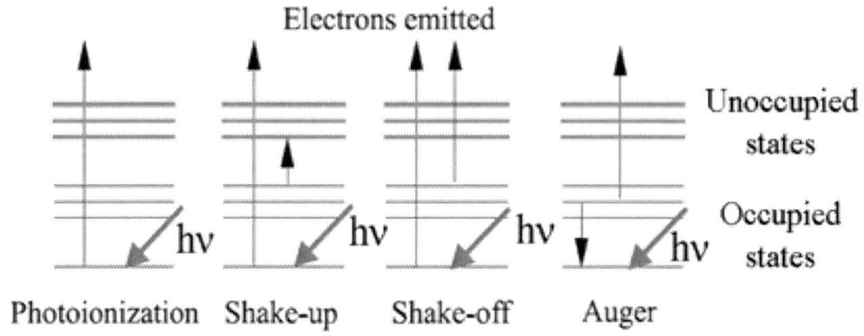

Figure 5. Different processes occur during the photoemission event.

The energy loss caused by inelastic collisions induces an attenuation of the photoelectron intensity, which depends on the depth of the source. When photoelectrons are ejected normal to the surface, the relation between signal intensity and the source depth d is given by

$$I = I_0 \exp(-d / \lambda) \tag{4}$$

where I0 and I are the original and attenuated photoelectron intensity, respectively. For samples containing two perfect layers, the PES signal intensity $I_o\,(\theta)$ along the surface normal of a uniform overlayer film is given by

$$I_o(\theta) = I_o(\infty)[1 - \exp(-\theta / \lambda)] \tag{5}$$

where $Io(\infty)$ is the signal intensity when the film is infinitely thick and θ is the film thickness. At the same time, the signal intensity of the substrate $Is\ (\theta)$ attenuated by a laminar growth of the overlayer is given by

$$I_s(\theta) = I_s(0)\exp(-\theta / \lambda) \tag{6}$$

where $I_s\,(0)$ is the intensity of substrate signal without any overlayer. From these equations, an empirical value of λ can be obtained. Consequently, the intensity variation of the photoemission signal as a function of overlayer thickness can be obtained, which provides information about the growth morphology of the overlayer.

In PES experiments, photons with various energies are used as excitation source. Thus, PES can be categorized into several classes, according to the light source used to generate photoelectrons. The common light sources include x-ray, ultraviolet, synchrotron radiation, and most recently developed pulsed laser beam. The wide energy range provided by different kinds of light source enables the investigation of both core-level and frontier orbital in solids.

2.1. X-Ray Photoemission Spectroscopy

The most commonly used x-ray sources for X-ray photoemission spectroscopy (XPS) are the unmonochromatized *Kα* radiation resulting from collision of high energy electrons onto Mg or Al anode. In order to improve the energy and spatial resolution, the x-ray source can be further monochromatized through specially-designed quartz crystal. The typical photon energies for Al *Kα* and Mg *Kα* lines are 1486.6 eV and 1254.6 eV, respectively. With such high energy photons, XPS is capable of probing the core-level electrons, whose binding energies depend on the elements from which the photoelectrons originate. Thus, XPS is also widely known as electron spectroscopy for chemical analysis (ESCA), due to its wide application in elemental identification in a given sample.

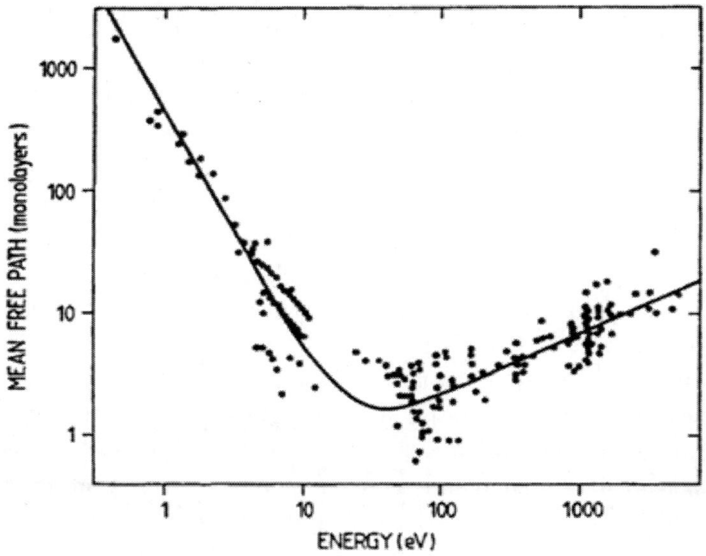

Figure 6. Universal curve of electron mean free path in solids.

Further, the binding energies of the core-level electrons are very sensitive to any changes in the chemical environment. Such changes can be revealed by small chemical shift of the core-level peaks for a given element. Through the analysis of the chemical shift of the peaks, the chemical state of the atom may be determined. Understandings about the chemical shift in XPS and the corresponding change in the chemical environment provide fundamental knowledge about the chemical structure of the sample.

Finally, another important application of XPS involves quantitative chemical analysis from the core-level peak intensities. The photoemission intensity of core levels can be employed as a means to determine the concentration of an element in a sample. This is, however, not straightforward, as the observed intensity of a core-level peak is the sum of the primary and the secondary spectrum. The separation of the two is, in most cases, very difficult. In an ideal case, the primary spectrum of a given core level can be considered as a Lorentzian centered at energy *E (i, x)*, where *i* is the index of the core level and *x* that of the element. If the total integrated experimental intensity of the core level is *I (i, x)*, one has:

$$I(i,x) = B\sigma(\hbar\omega,\alpha,E(i,x))\lambda_{tot}(E(i,x))T(i,x)n(x) \tag{7}$$

where *B* is the instrumental factor which contains properties such as the total x-ray flux, the angular acceptance of the electron energy analyzer, and the total transmission factor of the analyzer. The second quantity, $\sigma(\hbar\omega,\alpha, E(i, x))$, is defined as:

$$\sigma(\hbar\omega,\alpha,E(i,x)) = \sigma(\hbar\omega,E(i,x))f(i,x,\alpha);$$

$$f(i,x,\alpha) = 1 + \frac{\beta(i,x)}{4}(1-3\cos^2\alpha) \tag{8}$$

where $\beta(i, x)$ is the asymmetry parameter as calculated by Reilman at al., [118] $\sigma(\hbar\omega, E(i, x))$ is the total photoabsorption cross section of the core shell of the element, calculated by Scofield for free atoms at $\hbar\omega$ = 1487 eV (Al K_α) and 1254 eV (Mg K_α), [119] and α is the angle between the incident photon and the emitted photoelectron. The third quantity in Equation 7, $\lambda_{tot}(E(i, x))$ is the total escape depth of the photoexcited electrons, which can be approximated by $CE(i, x)^{0.71}$, where *C* is a constant weakly dependent on the electronic properties of the sample. *T (i, x)* is the transmission coefficient of the electrons through the surface, and finally, *n (x)* is the atomic density of element x in the sample. It is necessary to point out that the application of core-level intensity analysis is possible with good accuracy only if one has a symmetric peak shape, which more or less implies a single final state without the presence of other more complicated situations, such as satellites, plasmons, and electron-hole pair excitations. In most real applications, one is interested mostly in the relative density ratio between two elements, *I (i, x)/I (j, y)*. In this case, the instrumental factor *B* can be eliminated. If the kinetic energies of the photoelectrons from the two elements do not differ too much, $\lambda_{tot}(E(i, x))$ and *T (i, x)* usually show a small variation, and thus can be neglected as well. Thus in the simplest approximation, the relative concentrations of the constituents in a sample are only associated with the photoabsorption cross sections and the measured intensities, and can be expressed by:

$$\frac{n_x}{n_y} = \frac{I_x/\sigma_x}{I_y/\sigma_y} \tag{9}$$

This equation will be frequently used in the discussions of alkali metal doping density in organic semiconductors. Generally, the elemental sensitivity and accuracy thus obtained are about 1%.

2.2. Ultraviolet Photoemission Spectroscopy

Although in principle, XPS can also be used to investigate the electronic structure of the valence region of a solid, it faces several obstacles. The most severe problem is that the photoionization cross section for the valence electrons is much lower than that of the core electrons, when using high energy photons. Thus it is common to study the electronic structure of the valence region using ultraviolet light generated by a He discharge lamp as the photon source, which has a high cross section for the electrons in the valence region, allowing improved statistics for these measurements and thereby shorter measurement time. This technique is known as UPS. Another advantage is that UPS provides a better spectrum resolution due to the smaller linewidth of the UV light source, which is only about 20 meVs for the most used He I line at 21.22 eV and He II line at 40.6 eV. Finally, UPS is typically less damaging to the sample than XPS, a feature especially valuable to organic semiconductors that are usually vulnerable to photo induced damages and chemistry. The probing depth of UPS is typically 15 Å.

UPS has been extensively used to study the energy level alignment at the interfaces. The energies of several important molecular levels can be directly obtained from the UPS spectrum by simple analysis, including the vacuum level, the Fermi level (for metallic surfaces) and the HOMO or valence band maximum (for semiconducting surfaces). Figure 7 shows a typical UPS spectrum of a clean Au surface. The Fermi edge, derived from emission from occupied states just below the Fermi level, can be clearly observed, and is usually used as a reference point to calibrate the obtained binding energies of other energy levels. The Fermi level of the whole system should be aligned under the thermal equilibrium conditions when all parts are in electrical contact. It is important to note here that the Fermi level position obtained from the metallic surface remains fixed and can be used for semiconducting samples, although the Fermi edge cannot be directly observed for such surfaces because the density of states at the Fermi level is essentially zero. The strong features close to the Fermi edge are the signals from the occupied 5d states in Au. At the left side of the spectrum, a sharp cutoff can be identified, which is called the low energy secondary electron cutoff.This abrupt cutting edge stems from the fact that only those electrons, whose binding energy is smaller than the photon energy (21.22 eV), can be ejected out of the sample surface. Therefore by simply adding 21.22 eV to the cutoff energy, the binding energy of the vacuum level relative to the Fermi level can be obtained, as shown in the figure. The direct measurement of above mentioned major energy levels is critical to sketch the electronic structure at the surface and interface, because this allows calculation of several important parameters of the sample, such as the WF and IP. The equations for WF and IP calculations are frequently used in the following chapters and are given as follows:

$$WF = h\nu - (E_{Fermi} - E_{cutoff}) \tag{10}$$

$$IP = h\nu - (E_{HOMO} - E_{cutoff}) \tag{11}$$

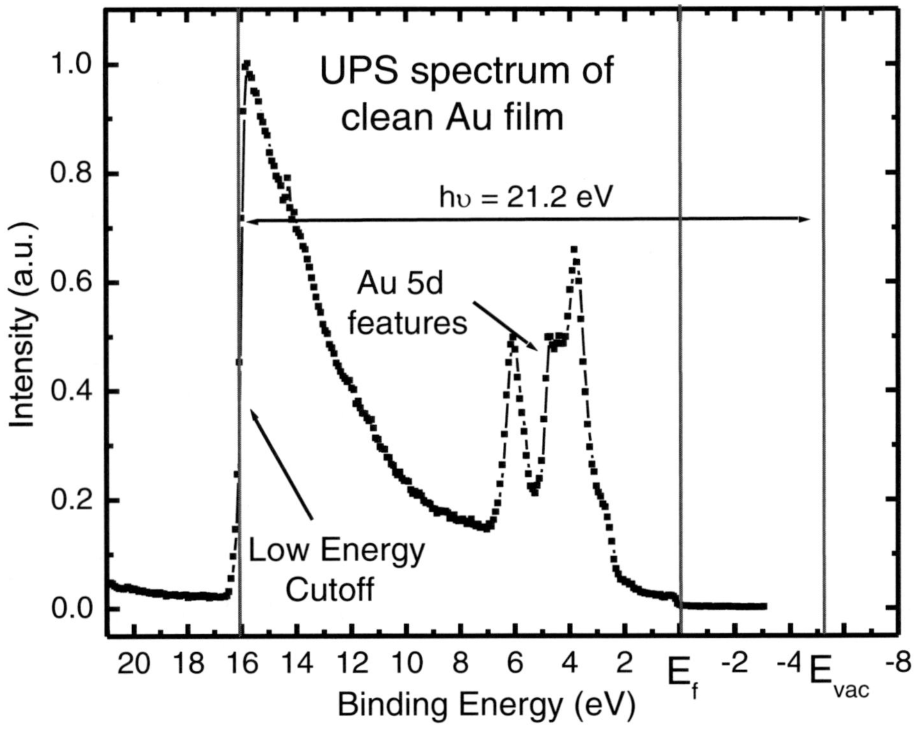

Figure 7. UPS spectrum of a clean Au surface, showing the major features of UPS measurement.

where *hυ* is the photon energy, in most cases, 21.22 eV. E_{Fermi}, E_{cutoff} and E_{HOMO} refer to the kinetic energy obtained directly from the measurement.

2.3. Inverse Photoemission Spectroscopy

Inverse Photoemission Spectroscopy (IPES) is the time reversed photoemission spectroscopy. [120-124] The surface under investigation is bombarded with electrons of a known kinetic energy, usually less than 20 eV. The relaxation of the incident electrons into the unoccupied energy levels generates photons through radiative decay. The ejected photons are then collected, thus providing a direct measure of the unoccupied energy levels of the surface. IPES is usually applied in two different methods. The first one fixes the kinetic energy of the incident electrons and directly measures the spectrum of emitted photons using a spectrometer. Whereas the second approach measures at single photon energy, and obtains the information of the unoccupied states by varying the kinetic energy of the electrons arriving at the sample surface.

Given the energies of the incident electrons this technique has approximately the same surface sensitivity as UPS. Typical energy resolution ranges from 0.3-0.8 eV depending on the instrumentation. [120-124] In the investigation of organic semiconductors, the IPES measurement is frequently combined with UPS measurement to describe the electronic structure of both the unoccupied and occupied energy levels, as shown in Figure 8. In the measured spectra, the width of the peaks is presumably due to inhomogenuity of the amorphous film. Therefore, the features are usually fitted with Gaussian peaks, whose center energy corresponds to the energy of the most populous molecule. Analogous to the earlier discussion of PES, IPES measures the molecular anions induced by the injected electrons in its relaxed (polarized) state, instead of the neutral state of the molecule. Thus the energy separation of the HOMO-LUMO peaks obtained by UPS and IPES is the energy difference of relaxed positive and negative polarons separated at infinity, sometimes referred as transport gap (E_t) of the organic material. However, the charge injection into solid does not necessarily occurs at the most populous or average species. It is most favorable to inject into the molecules with lowest charge injection barrier, which is defined by the onsets of the HOMO and LUMO peaks, as indicated in Figure 8. In most cases, the injection of hole and electron occur at different molecules. It is thus a general practice to define the injection gap (E_g) as the energy separation of the HOMO and LUMO onsets, which can be obtained by linear extrapolations of the leading edges of the peaks. It should also be noted that in many cases the obtained injection gap is very close to the optical band gap (E_{opt}), obtained from optical absorption measurement.

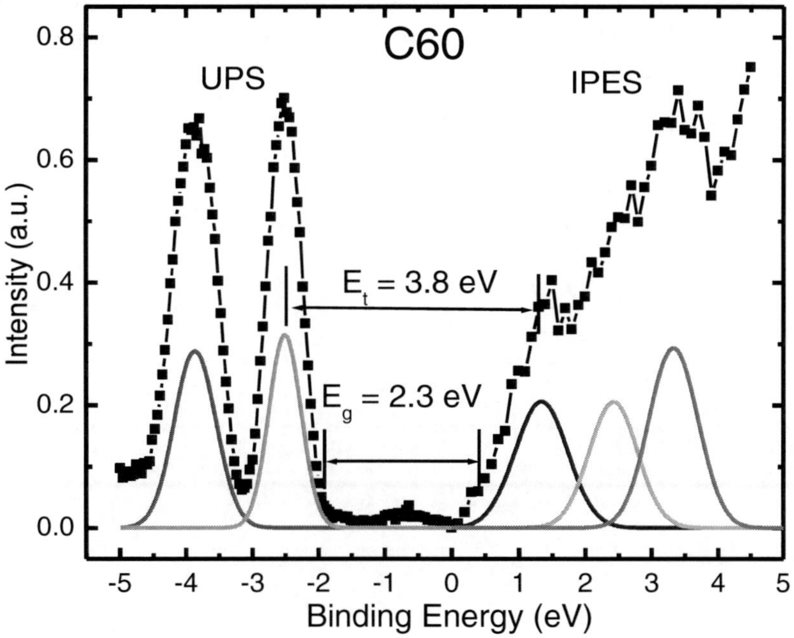

Figure 8. The occupied and unoccupied energy levels close to the Fermi level obtained by UPS and IPES measurement. The transport gap and the injection gap are noted.

2.4. Sample Preparation

Unlike the common practice in thermal deposition, alkali metals can only be evaporated from a SAES getter source, where pure alkali metal atoms can be released from their compounds as a heating current is applied. The pressure requirement for alkali metal evaporation is particularly critical, which is usually in 10^{-10} torr. In order to perform the surface analytical studies on the electronic structure modification induced by doping, it is essential to calibrate the doping densities (number ratios) between alkali metal atoms and organic molecules. Unlike the common practice in thermal deposition, the amount of deposited alkali metals cannot be monitored by micro balance, due to their extremely low sticking coefficient. Therefore, for most experiments discussed within this chapter, core level intensities obtained from XPS measurements were used to estimate the atomic ratio between alkali metal and carbon atoms with calibrated atomic sensitive factors, according to Equation 9. The doping densities can then be obtained, if one knows the total number of carbon atoms in a single organic molecule. In the experiments, the amount of deposition was monitored by the evaporation current and the sample exposure time.

A few other issues should be noted here, before we move on to the discussion of the photoemission results. First, since the alkali metals were deposited onto organic thin films, one may be concerned about the diffusion length of the metal atoms after they arrived at the organic surface. Previous studies have suggested that alkali metals can readily diffuse into organic thin films for as long as few hundred nanometers. [11] The depth profile with secondary ion mass spectroscopy (SIMS) shown in Figure 9 suggests that the diffusion length of alkali metals in organic matrix can be much longer than our photoemission probing depth. Therefore, we can safely conclude that the distribution of the alkali metal atoms inside organic films within the probing depth is reasonably even, and the doping densities obtained is thus reliable for the following discussions.

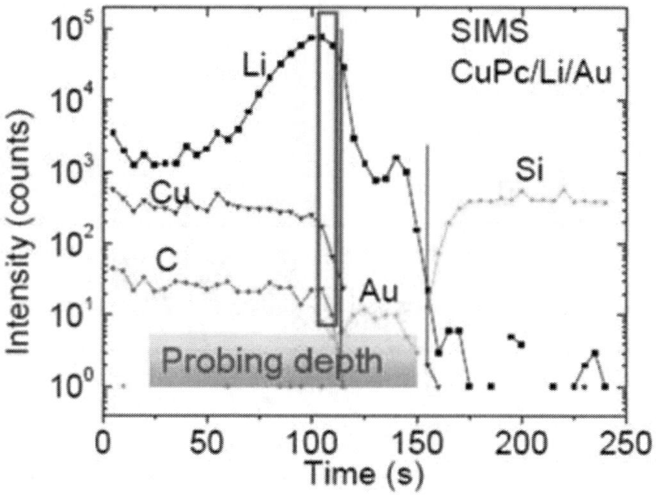

Figure 9. Depth profile of 100 nm CuPc/Li/Au/Si sample measured by secondary ion mass spectroscopy, indicating that the diffusion length of alkali metals are much longer than the probing depth of photoemission.

Another issue one should be aware of is the electronic structure modification (degradation) of the material induced by analytical probes, particularly for organic materials that are considerably soft. [45, 125-128] The damage always depends on the energy quanta and dosage, as well as the interaction of the probe with the organic materials. Generally, IPES is much more damaging than PES, as electrons interact strongly with a solid film. For PES, XPS is typically more detrimental than UPS if the same number of photons is involved, and a monochromatized XPS source is less damaging than a non-monochromatized one. The damage of the sample can be noticed by either the shift or the broadening of the measured features in the spectra, as the measurement progresses. The energy level shift induced by such a factor has to be excluded in the study of doping effect in organic materials. Discernable spectrum change can be induced by XPS on several organic materials, such as Alq. The UPS spectra of the cut-off and the HOMO of pristine Alq film before and after XPS measurement were presented in Figure 10 (a). It is clear that XPS measurement induces an energy level shift towards higher binding energy, which is most likely due to the generation of the trapped holes on the sample surface. Since the major damages of PES stem from the trapped charges, which modify the local electronic structure of the sample, metallic surfaces are typically safe from damage in PES measurements, even in the case of a small metal dosage on an organic layer. It is found that, when alkali metal atoms were deposited, such damage could be effectively limited, as shown in Figure 10 (b). This is attributed to the improvement of the charge transport ability by the metal dopants. For such reasons, great cautions should be taken when performing XPS measurements on pristine organic films, so that the influence from the measurement-induced error can be limited to the largest extent, leading to a more reliable interpretation of our photoemission data on the doping-induced electronic structure modification in organic semiconductors.

Finally, almost all experiments discussed in this chapter were carried out under UHV condition at room temperature. There are several reasons for the requirement of ultra-high vacuum, which usually refers to the pressure in range of 10^{-10} torr. First of all, sample cleanness is an exceedingly important issue in the investigation of the surface and interface. The interaction between the interested surface and contaminants will interfere with a clear characterization of surface properties. In a simple model, which describes the interaction between the surface and the surrounding air molecules, suggests that the rate k_P for the impact of atoms on a surface is given by [129]

$$k_p = \frac{P}{\sqrt{2\pi m k_B T}} \tag{12}$$

where P is the ambient pressure and m is the mass of the atoms. According to this equation, a monolayer of physisorption can be formed in the order of one second under the pressure of 10^{-6} torr, if the sticking coefficient is one. Therefore, detailed surface studies require a highly controlled environment with a pressure on the order of 10^{-10} torr, in order to ensure the surface cleanness throughout the reasonable measurement time (few hours) and exclude any possible interactions between the adsorbate and the sample surface. The second reason for the UHV requirement rises from the principle of the PES measurement, where the binding energy of the photoelectrons is derived from the measured kinetic energy the analyzer. It is thus necessary to ensure that the photoelectrons do not experience any energy losses from scattering in the travel towards the analyzer. In other words, the electron mean free path has to be much larger than the dimension of the measurement chamber. Last, but not least, the UHV condition is also critical during sample preparation, particularly for the evaporation of alkali metals. In reality, the UHV condition is usually achieved with multi-type pumping and long-term bake out to remove the water sticking to the chamber.

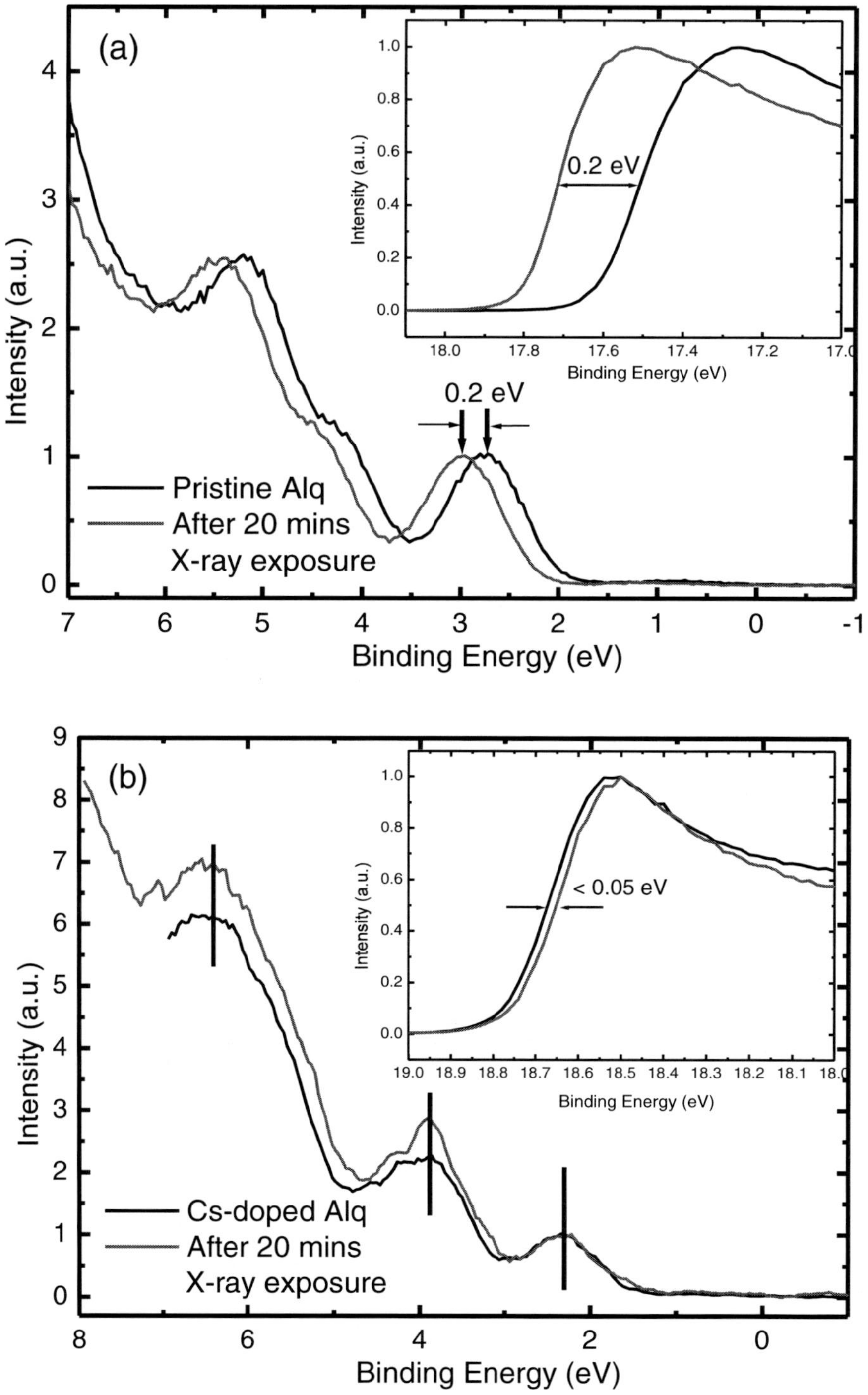

Figure 10. Difference in the UPS spectra before and after XPS measurement for (a) pristine Alq film and (b) Cs-doped Alq film.

3. Electronic Structure of Alkali Metal-Doped Organic Semiconductors

3.1. Electronic Structure Modification in Alkali Metal-Doped Alq

As one of the most important materials in the applications of organic electronics and prototypical for electronic and optoelectronic properties in organic semiconductors, Alq, whose molecular structure is shown in Figure 10, has been intensively studies for the investigation of doping-induced energy level shift. [13, 18, 130] Figure 11 shows the XPS spectra of C 1s and N 1s core level as a function of the doping concentration R_{Cs}, which is calculated from the core level intensity ratio. [13] Both C 1s and N 1s turn out to be a single peak at the beginning. This is quite consistent with the simple and unique chemical environment for these two elements in the Alq molecules. As R_{Cs} increases, both C 1s and N 1s peaks shift towards the high binding energy side. Particularly, for the first stage of $R_{Cs} < 0.64$, the change in peak position is very dramatic.An obvious shift of 0.3 eV can be observed even though R_{Cs} is as low as 0.01, and a total amount of 1.0 eV shift has occurred at $R_{Cs} = 0.64$. The shift then becomes saturated and there is a slight reverse shift for C 1s beyond $R_{Cs} = 1.07$. Another important feature is that the N 1s peak splits when R_{Cs} reaches 1.07. A new component emerges with binding energy about 2.2 eV lower than the main peak of N 1s, and the intensity of this new peak increases along with R_{Cs}. The behavior of N 1s core levels at higher doping concentrations indicates that there is clearly an interaction between the N and the Cs atoms at this stage. This assignment was supported by semiempirical (ZINDO) molecular orbital calculation by Burrows et al. [131] that showed the excess electronic charge in the Alq^- radical anion to be localized on the pyridyl side of one of the three quinolate rings.

In Figure 12, both the low energy secondary cut-off and the HOMO region of the UPS spectra of the same film were presented. [13] The Fermi level derived from the Au substrate is drawn as a dashed line and set as zero for the relative binding energy scale. It can be seen from the cut-off region that, similar to C 1s, the vacuum level shifts toward the high binding energy side as R_{Cs} increases. However, the significant changes only occur at $R_{Cs} < 0.38$. Above that, the shift is relatively small.

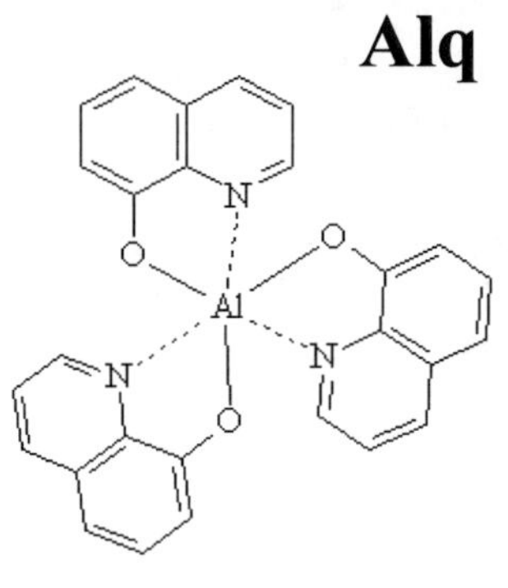

Figure 11. Molecular structure of Alq.

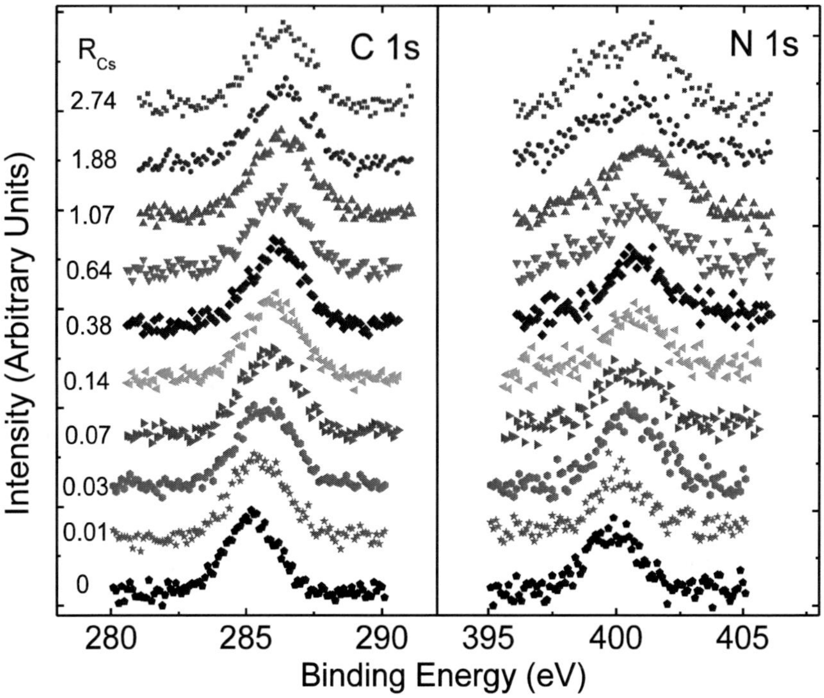

Figure 12. XPS spectra of C 1s and N 1s core level as a function of the doping concentration Rcs, the ratio of Cs atoms to Alq molecules, calculated from the XPS core level intensity.

Before the cut-off shift saturates, the total amount of the vacuum level shift is about 1.0 eV, which is quite close to the separation between the LUMO and the Fermi level in pristine Alq film. [130] The same conclusion can be drawn from the HOMO spectra (Figure 12 (b)). More importantly, a new gap state can be seen at $R_{Cs} = 0.64$, whose intensity increases with further increase of Cs doping concentration. A 1.6 eV separation between the original HOMO and the gap state can be observed. These spectral modifications were taken as evidence for formation of the Alq^- radical anion. The HOMO of the radical anion is closely identifiable with the LUMO orbital of the neutral molecule. These results indicate that for $R_{Cs} < 0.64$, the Fermi level of Alq can shift inside the gap toward the LUMO due to the charge transfer from Cs to Alq. While for the heavier doped cases, the saturated shift of the valence and core level characteristics in UPS and XPS spectra suggest a pinning effect of the Fermi level near the LUMO, which is also supported by the IPES data.

The evolution of the occupied and the unoccupied electronic states of Cs-doped Alq is shown in Figure 13 as a function of the Cs doping ratio R_{Cs}. [14] Unfortunately, pristine Alq thin film is particularly vulnerable towards electron beam exposure, [132] resulting in a broad and featureless spectrum. This suggests that further effort may be required to get a reliable IPES data for pristine Alq film. On the other hand, doping with alkali metal, even for a tiny amount, can significantly increase the stability of the film under the IPES scans. As one can tell from Figure 13, the shape of the IPES spectra remains relatively the same as Cs doping ratio increases, until the shift of the energy levels become saturated. The onset positions of the LUMO are noted by short bars. It can be seen that as R_{Cs} increases, the LUMO onset shifts towards, and eventually aligns with, the Fermi level. The total shift of the HOMO and the LUMO is almost identical, which agrees well with the rigid shift assumption. For high doping ratios above 0.8, a component at about –1.5 eV can be observed, which is assigned as the LUMO. It is very clear that its intensity decreases as the intensity of the gap state in the UPS spectra increases. The emergence of the gap state, at the expense of the LUMO intensity, and the new N 1s component in Figure 12 indicate that as the charge transfer progresses, correlation interaction becomes significant. The gap state was resulted from the occupation of what was one of the triplet of the empty LUMO orbitals in the neutral molecule.

The summary of both the valence and the core level shifts of Cs-doped Alq are presented in Figure 15. [13] The core level shifts were deduced by fitting the spectra with Gaussian curves. The vacuum level cutoff was obtained at the maximum of the differential curve of the UPS spectrum. Note that the doping ratio is in logarithmic scale to emphasize the initial shift. It is clear that, in both cases, the energy level shift can be seen as in two stages. For the low doping levels ($R_{Cs} < 0.64$), the shifts of all energy levels show a nearly linear dependence on the logarithm of the Cs concentration before they become saturated at about 1.0 eV. The slope of the line drawn from the first several data points is 0.250 eV, which is much larger than k_BT (0.028 eV). The detailed discussion about the origin of such a large slope will be provided in the section 3.5. At high doping ratios, the shifts of the energy levels become saturated, mainly because the Fermi level is pinned at the LUMO position.

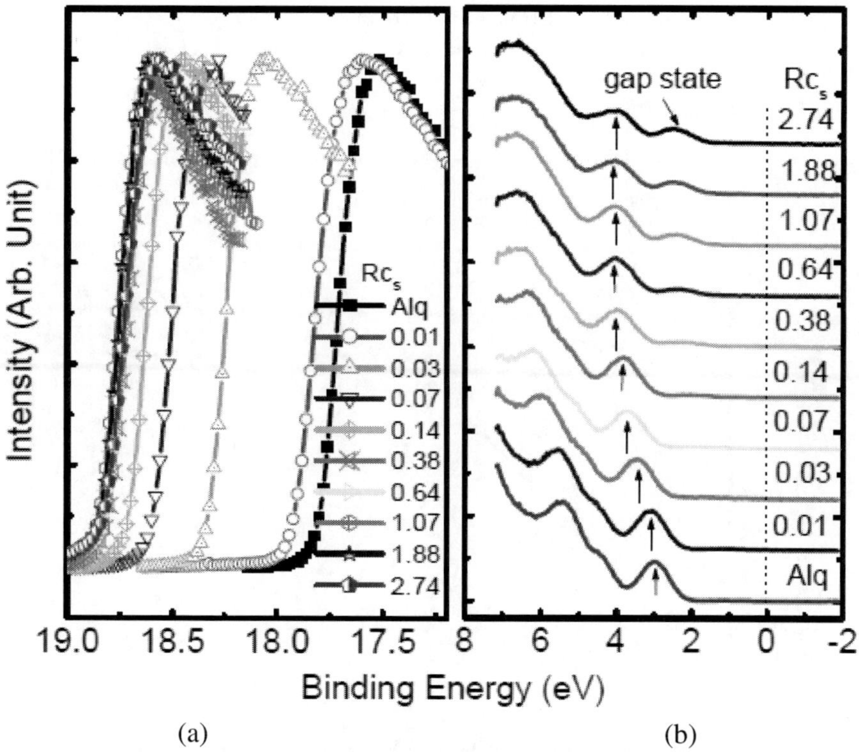

Figure 13. UPS spectra evolution of Cs-doped Alq film: (a) cut-off and (b) HOMO region. The Fermi level derived from the Au substrate is drawn as a dashed line and set as zero for the relative binding energy scale. A new gap state emerges and its intensity increases with Cs doping concentration.

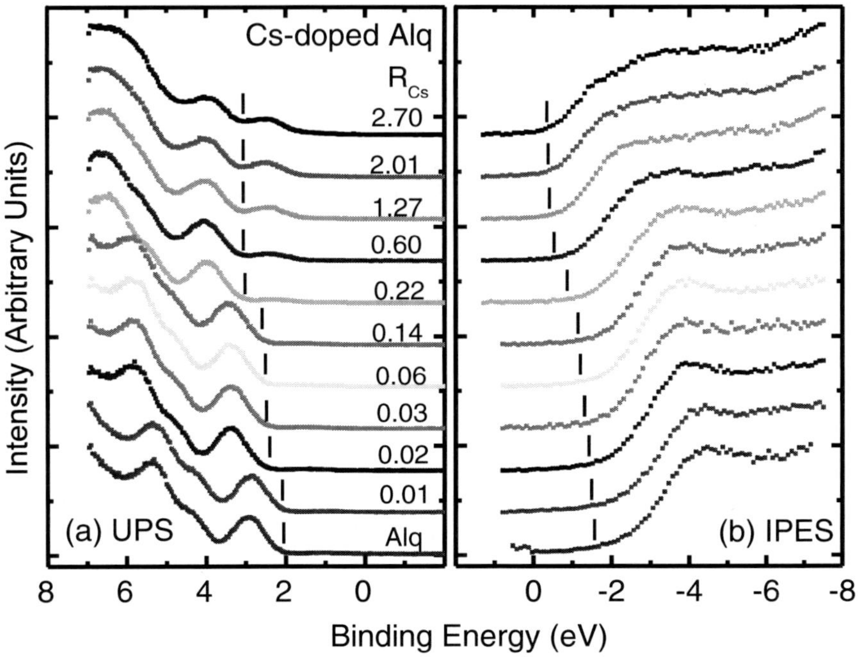

Figure 14. Evolution of the HOMO and LUMO region of Alq as a function of Cs doping ratio RCs. The spectra are plotted in binding energy relative to the Fermi level. The HOMO and LUMO onsets are denoted by short bars.

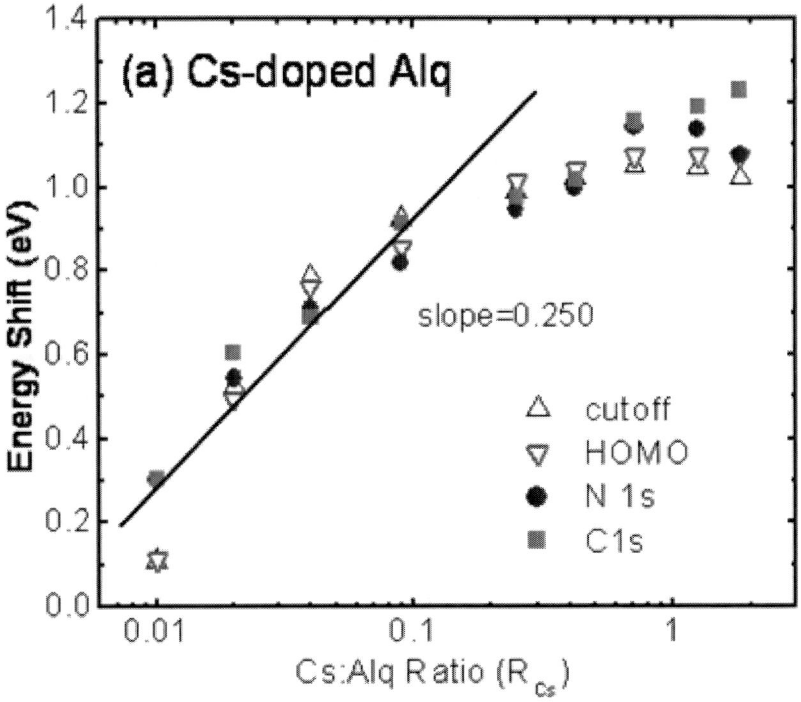

Figure 15. The energy level shifts as a function of doping concentration for Cs-doped Alq. Note that the doping density is in logarithmic scale to emphasize the initial shift.

Another possible explanation for the saturation is the increasing incompleteness of the ionization of Cs atoms. For R_{Cs} = 2.74, the intensity ratio of the gap state to the HOMO is 0.53. Since the HOMO of Alq is triply degenerated, a full ionization would result in a ratio of 2.74/6 = 0.46. So a complete ionization of Cs can be obtained. This indicates that incomplete ionization alone cannot explain the saturation of the energy level shift at higher doping concentrations. Instead, in the second doping stage, the change of the Alq electronic structure by the doping is obvious, manifested by the gap state, the new N 1s component, and different shift behavior of the valence and core features. At this stage, the doped electrons become more localized near the organic molecules, leading to a strong correlation effect, which explains the reverse shift for some of the core levels. Similar modifications can be observed in Na-doped Alq, where a slope of 0.220 eV can be deduced from the vacuum level shift. It should be noted that the slope in Na-doped Alq is noticeably smaller than that of Cs-doped Alq, which can be mostly attributed to the difference in the ionization potential of Na (0.55 eV) and Cs (0.47 eV). [133] In addition, incomplete ionization may also have contributed to the smaller slope.

The interpretation about doping induced electronic structure modification has been substantially strengthened by other works on Li, [134] Na, [14, 18] K, [134] Mg [18] and Ca [18] depositions on Alq. In Figure 16, the HOMO region of the UPS spectra for Alq films doped by all five of the metals was summarized by Mason *et al.* [18] The shape of the spectra in all cases are very similar, despite the small qualitative differences in intensity that result from variations in the metal coverage. In all five cases, the original HOMO level is shifted to higher binding energy and the gap state is formed in the energy gap at an energy of ~1.6 eV above the shifted HOMO level.

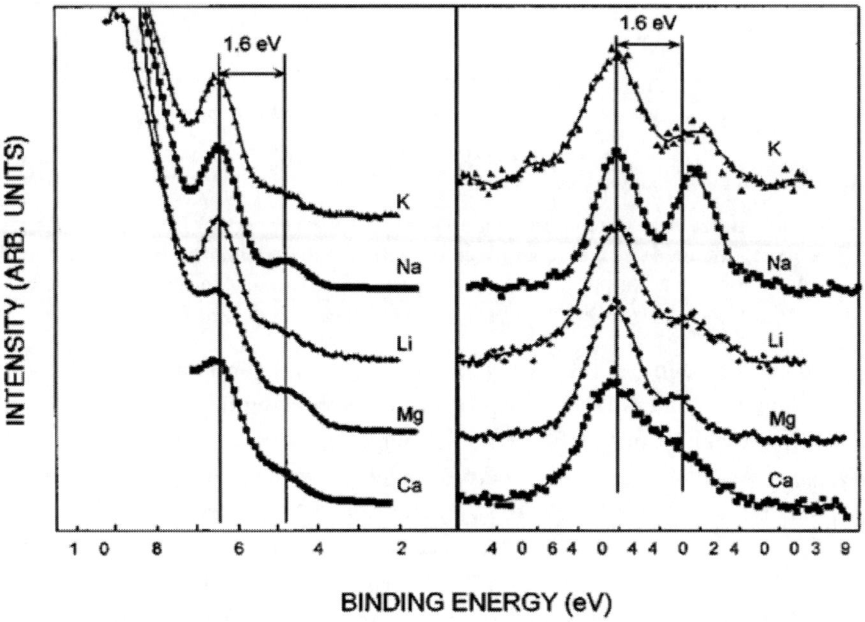

(a) (b)

Figure 16. The HOMO region of Alq after deposition of group I and II metals (a) and the N 1s region for the same series of metals (b). Binding energies are referenced to the vacuum level.

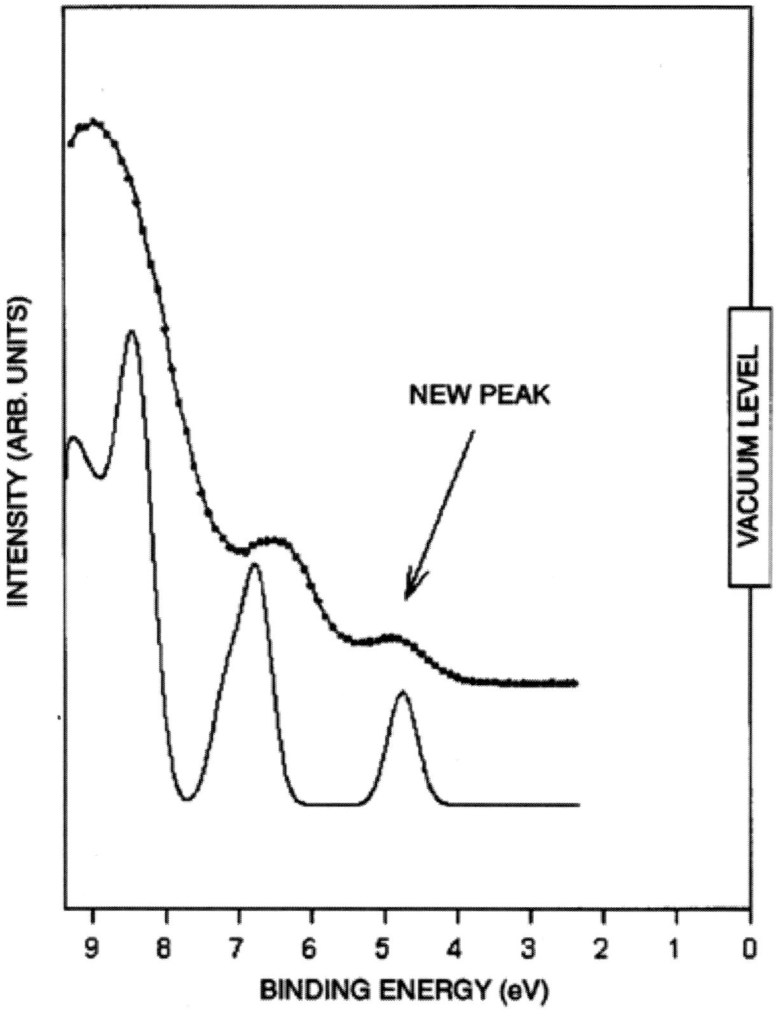

Figure 17. Comparison between the experimental He I spectrum (upper curve) and the theoretical DOVS (lower curve), obtained from a generalized transition state calculation on the basis of the DFT formalism, corresponding to a doping level of one potassium atom per Alq molecule. The calculated curve has been shifted to align with the experimental results.

A theoretical calculation of the UPS spectrum for the reaction product of K with Alq was obtained by using the generalized transition state method with the eigenenergies obtained from density functional theory (DFT) calculations and is presented in Figure 17. [18] The result shows an excellent agreement between the experimental spectrum and theoretical density-of-valence-states (DOVS) for the HOMO region at a doping level corresponding to one potassium atom per Alq molecule.The predicted separation between the gap state and the HOMO was, however, slightly larger (~ 1.9 eV). This can be attributed, at least partly, to the fact that the calculation is performed in the ''gas phase'' with no polarizing medium surrounding the Alq. Due to the presence of screening in a solid film, the repulsion is reduced, leading to a smaller energy separation between adjacent molecular orbitals. The DFT calculation, shown in Figure 18, [18] also indicates that the new peak corresponds to the

LUMO of pristine Alq, since it is clearly seen that the new HOMO of K-doped Alq is shown to be almost identical to the original LUMO of Alq.

Another interesting observation is that the basic structure of Alq is well preserved in the alkali metal doped state. No evidence of strong chemical reaction can be found from PES studies. The experimental evidence of this point is shown in Figure 19 where a much larger portion of the UPS spectra of Alq is shown for various doping densities of Na. [18] Besides the similar shift that can be observed in cases of other alkali metals, the molecular orbital structure of Alq appears to be preserved as the Na coverage increases. Such behavior is also observed for Li, K, Cs and Mg doped Alq film. It should be noted that, aside from some broadening, the energy separations between deeper Alq orbitals are not significantly affected by the doping process. It is thus strongly suggests that the basic Alq structure is preserved and the alkali metal (and some group II metal) doping will not introduce any significant modification of the electronic or chemical structure of Alq molecule.

The other spectral feature that is considered as the characteristic of the formation of the Alq^- radical anion is the splitting of the N 1s peak in the XPS spectra, as shown in Figure 12 and 16 (b) for a set of alkali and alkali earth metals doped Alq film. One can immediately notice in Figure 16 (b) that the separation between the doping induced peak and the original peak varies depending on the metal dopant.

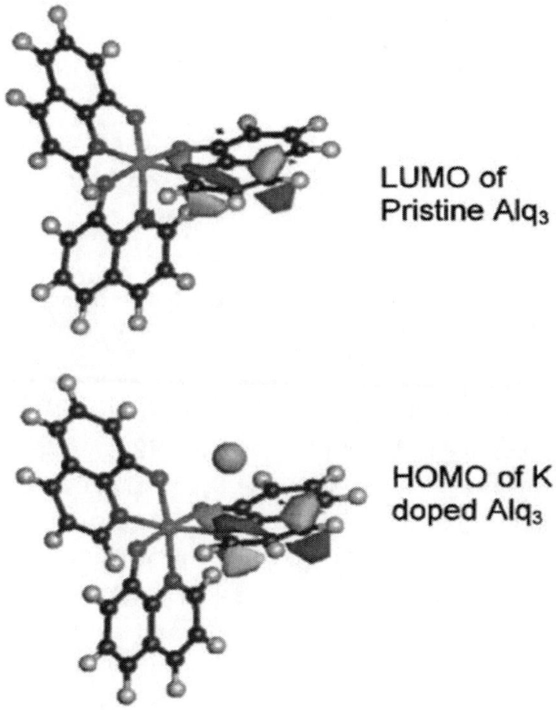

Figure 18. Sketch of the pristine Alq LUMO and the potassium doped HOMO as calculated at the gradient-corrected DFT level.

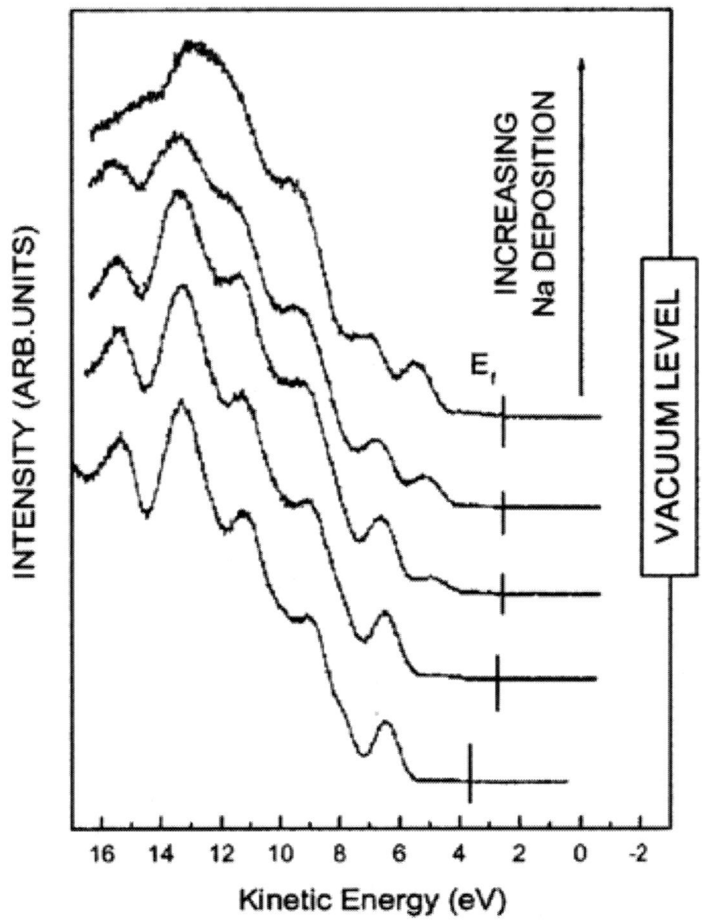

Figure 19. Evolution of the valence band spectrum with increasing Na doping. Note the appearance of the gap state on the high kinetic energy side of the HOMO and continued presence of the molecular orbital structure even at the highest level of metal doping.

Such dependence of the splitting is explained by Mason *et al.* as the difference in the ionicity of the "bond" between the metal and Alq. Atoms with low ionization potential tend to lose their out most electrons easily to Alq, resulting in the formation of a more ionic "bond" between Alq molecule and the dopant. Therefore, the charge distribution around the N atoms will vary for atoms with different ionization potential. And the lower the ionization potential of a dopant, the larger spectral splitting is expected, which agrees well with the experimental observations. Another interesting observation of the significant difference between the behaviors of the alkali metals (Li, Na, K and Cs) and alkali earth metals (Ca and Mg) is the intensity ratio of the doping induced peak in N 1s. Given a sufficient supply of metal atoms, alkali metals are able to donate up to three electrons to Alq, forming the Alq^{3-} radical anion, where each of the three different ligands has received one extra electron charge. [134] In this context, the intensity ratio of the doping induced peak can be higher than 50% of the original N 1s peak, as illustrated by Na-doped Alq shown in Figure 16 (b), where the doping density of Na is considerably higher than 1. It is possible that, if the alkali metal concentration is increased further, the low binding energy component will continue to increase in intensity to the point where it is the sole peak, which is suggested to be the signature for formation of the trivalent Alq^{3-} radical anion. [134] However, such situation does not happen for the alkali earth metals, such as Ca and Mg.

3.2. Electronic Structure Modification in Alkali Metal-Doped CuPc

CuPc is one of the most widely used molecules in both fundamental research and technological applications. [12, 135, 136] Its chemical structure is shown in Figure 20. It has an optical gap of 1.6 eV, [12] which is of great interest for visible absorption, and a transport gap of 2.3 eV, [137] which is very useful in electronic devices. The UPS and IPES spectrum of pristine CuPc film is presented in the bottom of Figure 21 (a) and (b), respectively. [15] The UPS spectrum shows the density of states in the valence region as a function of binding energy relative to the Fermi level, whereas, the IPES spectrum shows the unoccupied states above the Fermi level. The pronounced feature at about 1.2 eV in the UPS spectrum is the HOMO, whose position agrees well with the previous studies. [12] In the IPES spectrum, three main features can be observed, which are quite consistent with literature. [138] The first peak at about -1.9 eV is usually assigned as the LUMO, whose onset position, together with that of the HOMO, determines the energy gap of the CuPc molecule. The evolution of the electronic structure for both the HOMO and the LUMO region is plotted in Figure 21 as a function of Cs doping ratio R_{Cs}, which is the number ratio of Cs atoms to CuPc molecules, which is obtained from the XPS core level intensities. For R_{Cs} smaller than 0.4, a monotonic shift toward high binding energy can be observed for all the features in the UPS spectra, which is in good agreement with our previous studies of alkali metal doping in organic semiconductors. [12, 13] It is very interesting that the IPES spectra show a similar shift for the low doping ratios. The amount of the shift for the LUMO when R_{Cs} is equal to 0.4 is identical to that of the HOMO, which confirms the previous assumption of rigid shift.[12, 13] It suggests that the electronic structure of CuPc molecule is well preserved under low doping ratios, and that the energy level shift at this stage is mainly due to the Fermi level shift within the energy gap. Another issue worth mentioning is that, due to the downwards shift and the doping-induced broadening, the position of the LUMO onset in the IPES spectrum is very close to the Fermi level at this doping level.

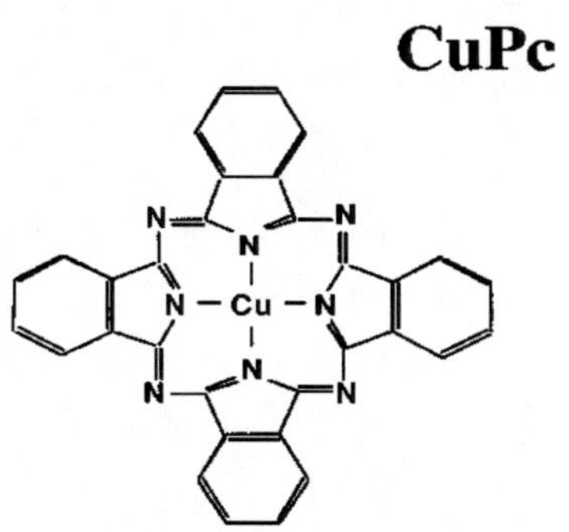

Figure 20. Molecular structure of CuPc.

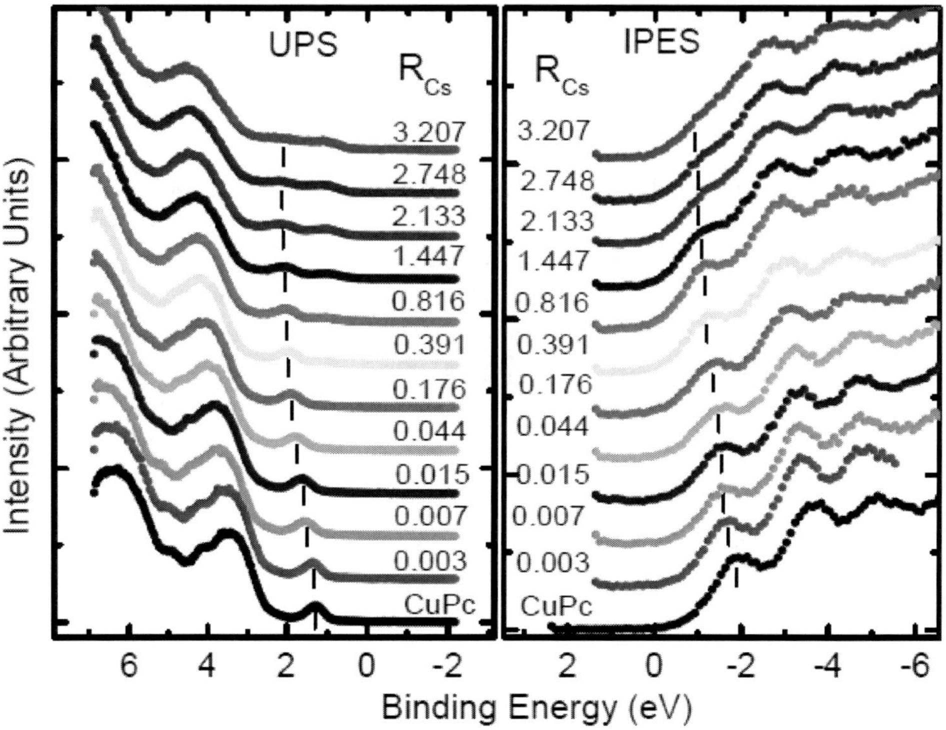

(a) (b)

Figure 21. Evolution of (a) the HOMO and (b) the LUMO region of CuPc as a function of Cs doping ratio RCs, derived from XPS core level intensities. The spectra are plotted in binding energy relative to the Fermi level. Negative binding energy values indicate energy levels above the Fermi level.

For more Cs doping, the shift of all energy levels becomes saturated. The saturation of the energy level shift provides direct evidence for Fermi level pinning in heavily doped system, as been proposed by Yan et al. [12] Although the energy level shift has saturated, the shape of both the UPS and IPES spectra are significantly modified at higher doping ratios. A new state emerges from the former energy gap in the UPS spectrum, and its intensity increases monotonically as R_{Cs} increases. The peak separation between the HOMO and the gap state is about 1.1 eV, which remains the same for all doping ratios. At the same time, the intensity of the CuPc LUMO in the IPES spectra starts to decrease as R_{Cs} increases. The increase of the intensity of the gap state, at the expense of that of the LUMO, suggests that the gap state induced by *n*-type doping is originated from the filling of the former LUMO in the CuPc molecules. [139] For high doping ratios, the spectral broadening becomes more and more pronounced in both the UPS and IPES. For R_{Cs} equal to 3.2, the HOMO and the gap state are so broad that they almost merge together. Similar broadening can be found in the IPES spectra, where the LUMO becomes too weak to be visually distinguishable. Finally, it should also be noted that no Fermi edge could be observed in either UPS or IPES spectra for doping ratios as high as 3.2, indicating a non-metallic surface.

In Figure 22, both the occupied and the unoccupied states near the energy gap are plotted for pristine and Cs-doped CuPc samples for comparison. [15] The experimental data were decomposed with Gaussian peaks after careful deduction of the background. The components presented in the figure are HOMO-1, HOMO, LUMO, and LUMO + 1, in the order of decreasing binding energy. The Fermi level is set as zero and located inside the energy gap. It is very clear that after doping the LUMO onset shifts downwards, reaching the Fermi level, and its intensity decreases substantially, compared to the pristine film. At the same time, the new state appears in the former gap, whose intensity increases gradually to be comparable to that of the HOMO. Therefore, the IPES measurements provide solid evidence for the origin of the gap state induced by *n*-type doping in organic semiconductors.

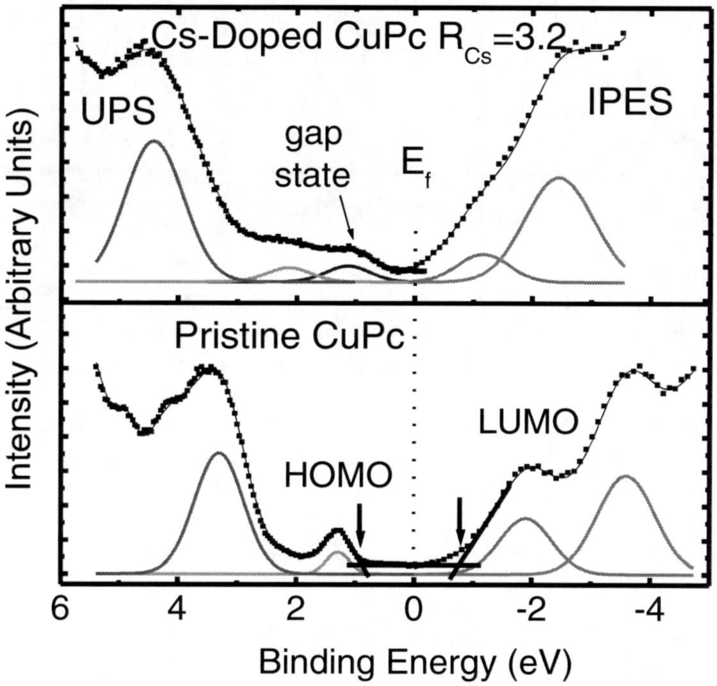

Figure 22. Comparison of the composite of the UPS and IPES spectra for pristine and Cs-doped CuPc films. After doping, the intensity of the LUMO is greatly reduced, while a gap state appears in the UPS spectrum.

In Figure 23, the XPS spectra of the C 1s, N 1s, and Cu $2p_{3/2}$ core level are presented as a function of R_{Cs}. [15] The C 1s core level consists of three visible peaks. The main peak (C 1s (A) as marked in Figure 23) can be easily associated with the aromatic rings of CuPc. The second peak (C 1s (B)) is due to the pyrrole carbon in contact with N, with some contribution of C 1s (A) satellite. The third (C 1s (C)) is a π-π^* satellite feature from the pyrrole carbon. [140] The first thing to notice is that all the three core levels shift toward higher binding energy as R_{Cs} increases. However, the shifts of Cu $2p_{3/2}$ and N 1s saturate or even reverse slightly as R_{Cs} increases beyond 0.8, whereas that of C 1s does not show any obvious reversion. Another important feature is that while the shapes of the Cu and N core levels remain nearly unchanged, the three components of C 1s at higher Cs doping ($R_{Cs} \geq 2.7$) merge into one broad and asymmetric peak. The different behaviors of the core levels at higher doping levels indicate that the electronic structure of CuPc is significantly modified at this stage.

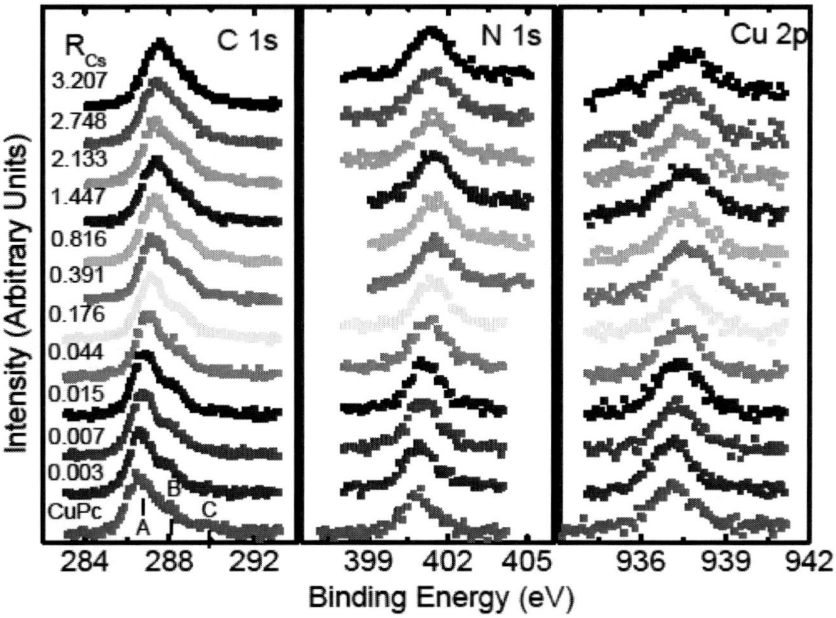

Figure 23. XPS spectra of C 1s, N 1s, and Cu 2p3/2 core level as a function of the doping concentration RCs, the Cs atom to CuPc molecular ratio, calculated from the XPS core level intensities. The three components of C 1s for pristine CuPc are noted by short bars.

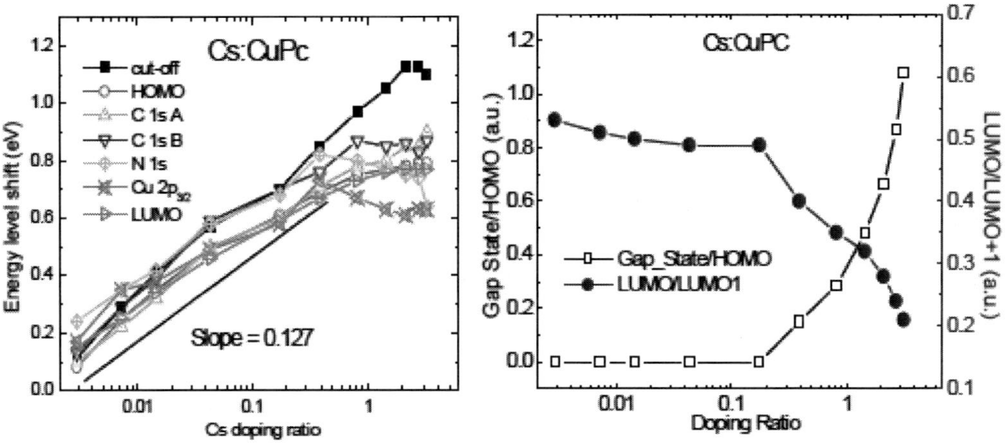

(a) (b)

Figure 24. (a) Summary of the energy level shifts for Cs-doped CuPc as a function of Rcs. (b) The evolution of the LUMO/LUMO+1 and the gap state/ HOMO intensity ratio as a function of Rcs for doped CuPc film. Note that in both figures, Rcs is in logarithmic scale to emphasize the initial modification. The dramatic change of the LUMO intensity occurs simultaneously with the saturation of the energy level shift.

The energy level shifts in Cs-doped CuPc are summarized in Figure 24 (a). [15] The vacuum level cutoff was obtained at the maximum of the differential curve of the UPS spectrum. HOMO and LUMO were determined by the extrapolation of the leading edge in the UPS and IPES spectra, respectively. The core level shifts were deduced by fitting the spectra with Gaussian curves for C 1s, N 1s, and Cu $2p_{3/2}$. Note that R_{Cs} is in logarithmic scale to emphasize the initial shift. The energy level shift induced by Cs doping can be seen in two stages, which is in consistent with the previous studies. [12] At the first stage where the Cs doping ratio is below 0.4, all energy levels shift simultaneously and linearly in the semi-log plot. The Cs doping experiment shows great reproducibility with a slope of 0.127 eV, which is in excellent agreement with the value obtained previously (0.132 +- 0.006 eV). [17] However, the slope is much larger than k_BT (0.026 eV), a value that is dictated by the classical theory for inorganic semiconductors. [102] Detailed discussions about the doping mechanism at low doping ratios can be found in section 3.5. One should also note here that at this stage, the amounts of the shift for all the energy levels are very close, especially for that of the HOMO and LUMO. This suggests that the rigid shift assumption is still valid under low doping ratios. As the doping ratio increases, the modification of the CuPc electronic structure moves into a second stage, characterized by the diverse behaviors for different energy levels. For the frontier orbital, the shift of the HOMO and the LUMO becomes saturated simultaneously, while the vacuum level shifts much further toward high binding energy. For the core levels, a reversal of the energy level shift is observed for N 1s, Cu $2p_{3/2}$, and C 1s(B), which can be attributed to the correlation effect as the donated electrons start to alter the electronic structure of CuPc. The reversal shift for the core levels occurs coincidentally with the appearance of the gap state in the occupied region and the reduction of the LUMO in the unoccupied region, which further corroborates the explanation given above.

The two-stage process induced by Cs doping is also suggested by the intensity evolution of the frontier orbital as shown in Figure 24 (b). [15] In this plot, Ding and Gao calibrated the intensity of the LUMO by that of the LUMO+1 of the same IPES spectrum to remove the uncertainty of the electron flux from measurement to measurement. Similar calibrations have been employed for the intensity ratio between the gap state and the HOMO in UPS. The assumption is that the intensities of the LUMO+1 and the HOMO remain relatively unchanged since Cs doping involves mainly the modification of the LUMO and the gap state. The intensity ratios of LUMO/LUMO +1 and gap state/HOMO can then be plotted as a function of the Cs doping ratio. One can easily see that at low doping ratios, there is no gap state and the change in the LUMO intensity is negligible. The dramatic change in the intensity occurs when R_{Cs} is larger than 0.4, when the LUMO intensity starts to decrease. At the same time, the gap state appears and its intensity increases continuously. It should be noted here that the appearance of the gap state in the occupied region and the reduction of the LUMO in the unoccupied region occurs coincidentally with the reversal shift for the core levels, which further corroborates the explanation given above.

A closer look at the intensity evolution of the LUMO provides important information that reveals the quantity of electrons involved in the charge transfer process in the guest-host system. For instance, the initial ratio between LUMO and LUMO +1 is about 0.5 for pristine CuPc film. When R_{Cs} is equal to 1.0, this ratio becomes 0.38, which means that the intensity of LUMO is reduced by about 24%. The theoretical studies show that the LUMO of CuPc has double degeneracy. [141] Thus, assuming that the LUMO could accommodate four electrons, the doped electron density can then be derived to be about 0.96 per molecule, which agrees very well with the doping ratio R_{Cs} derived from XPS. This suggests that the doped Cs atoms are almost fully ionized. The reduction of the LUMO continues beyond 1-to-1 doping. When R_{Cs} is equal to 3.2, the LUMO intensity has been reduced by 60% compared to the pristine film, which means that about 2.4 electrons have been transferred into a single CuPc molecule.

This doping ratio estimated from LUMO intensity is substantially lower than R_{Cs}. The discrepancy can be partly attributed to the attenuation of the C 1s core level intensity in the XPS spectra at high ratios, as well as possible incomplete ionization of the Cs atoms. The charge transfer amount per molecule can be estimated from the gap state/HOMO intensity ratio as well. According to the Gaussian fitting, at the final doping ratio of 3.2, the intensity of the gap state is about 1.1 times that of the HOMO. The XPS investigations suggested that there is no strong chemical interaction between Cs and CuPc at these doping ratios. [13] Therefore, the doped electron density can then be estimated to be 2.2, if one can assume that the HOMO remains unaffected after doping and consists of only one orbital. This number agrees very well with the earlier estimation from the IPES measurement, which suggests that the photocrossections for the gap state and the HOMO could be very similar. The results suggest that multiply charged CuPc species exist in the heavily doped film.

In another study conducted by Molodtsova et al., the possible lattice sites for the doped K ions in CuPc has been studied using core-level spectra obtained from synchrotron radiation. [142] They performed detailed decomposition of the core level spectra for C 1s and N 1s in K-doped CuPc, as shown in Figure 25 and 26, respectively. [142] Their study suggests that although doping induced broadening can be found in both aromatic and pyrrole carbon peaks (see C 1s (A) and (B) in Figure 23 or C-1 and C-2 in Figure 25), the pyrrole carbon, which is attributed to the carbon linked to nitrogen, spreads much more drastically. Such broadening in the core levels upon K doping was also reported in their previous studies. [143] It thus suggests that doped K ions most probably occupy equivalent positions near the pyrrole carbons of the CuPc molecules.

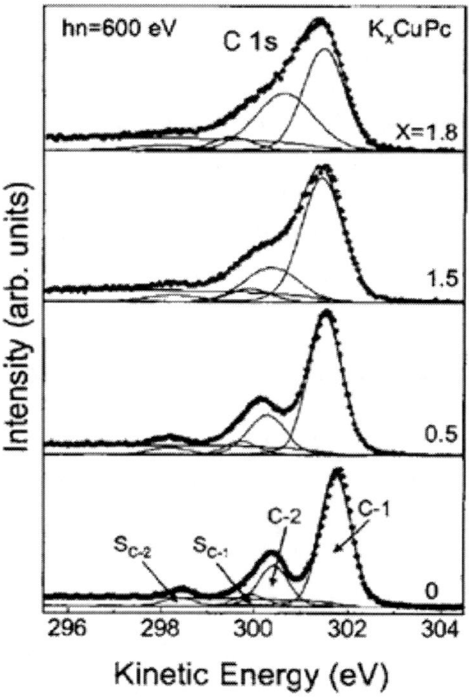

Figure 25. The results of the C 1s core-level decomposition for K-doped CuPc.

This conclusion is further confirmed with the detailed analysis of the variations of the N 1s core-level spectra upon K doping. The N 1s core-level spectrum of pristine CuPc shows a single component, though there are, in fact, two in equivalent sites for the eight N atoms in a CuPc molecule: four N atoms are connected to the central Cu; while the other four are linked to pyrrole carbon.However, the XPS measurements from different studies [12, 15, 27, 142, 143] seem to suggest that the binding energy difference between the N atoms in the two sites is very small, so that a single component fitting is good enough to describe the core level spectrum of pristine film. Upon K doping, N 1s peak becomes broadened. In order to analyze the N 1s spectroscopic information in more detail, Molodtsova et al. have carried out a numerical peak decomposition, as shown in Figure 26, [142] where the width of each component was taken from the N 1s core level spectrum of pristine CuPc and was kept unchanged at all K doping. Form the fitting, it is clear that aside of the shift of the main component A towards higher binding energy as a result of K interaction upon doping, a new component B appears at lower binding energy.

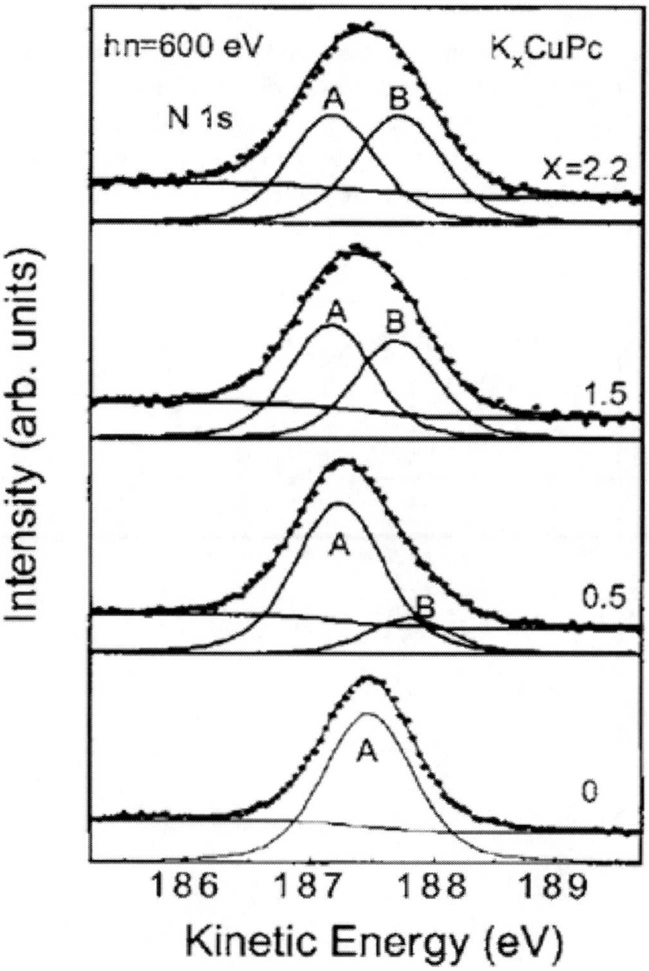

Figure 26. The results of the N 1s core-level decomposition for K-doped CuPc.

It is also interesting to note that the intensity of component B increases monotonically as the K doping ratio increases, while a rapid attenuation of component A is observed simultaneously. The evolution of the relative intensity for the two components of N 1s core level in K-doped CuPc is shown in Figure 27. [142] There is a linear increase for low doping ratios (< 1.5). However, the increase of the intensity ratio becomes saturated as the doping ratio approaches 2. At such doping level, the intensities of the two N components are about the same. It should also be pointed out that the total intensity of the N 1s core level, however, does not change. Thus, these results demonstrate different chemical surroundings for the two inequivalent N sites of the CuPc molecules. Component A is attributed to the N atoms connected to Cu and component B is attributed to the N atoms which are linked to pyrrole carbon, since its strong doping dependence is fully consistent with the significant broadening of the corresponding C 1s component as discussed above.

Based on the core level studies, it is suggested that the doped K ions are most likely located close to the pyrrole N atoms. Therefore, the electronic structure for the surrounding N and C atoms is significantly modified, and thus explains the changes observed in the core level spectra. Moreover, the fact that the two N 1s components have the same intensity at high doping ratios suggests that more than one electron can be transferred to a CuPc molecule, which seems to corroborate the discussion in Cs-doped CuPc. The observation of the multiply charged CuPc species is a very interesting phenomenon, since the theoretical calculations suggest that multiple charging is energetically unfavorable for isolated CuPc molecule. [135] It is thus interesting to study the electronic structure of CuPc under intensive doping levels, and to understand the lattice sites have to be occupied at such doping levels.

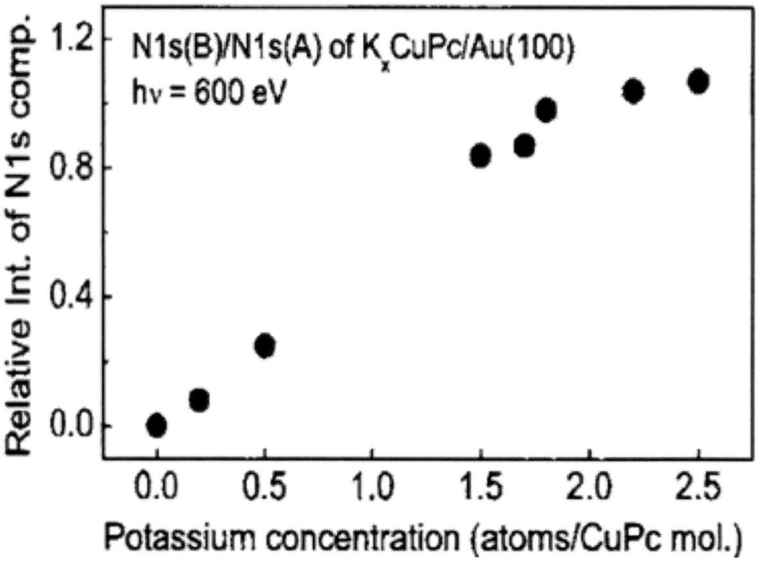

Figure 27. The evolution of the relative intensity of the B and A components of the N 1s core level (see Figure 26) with K doping.

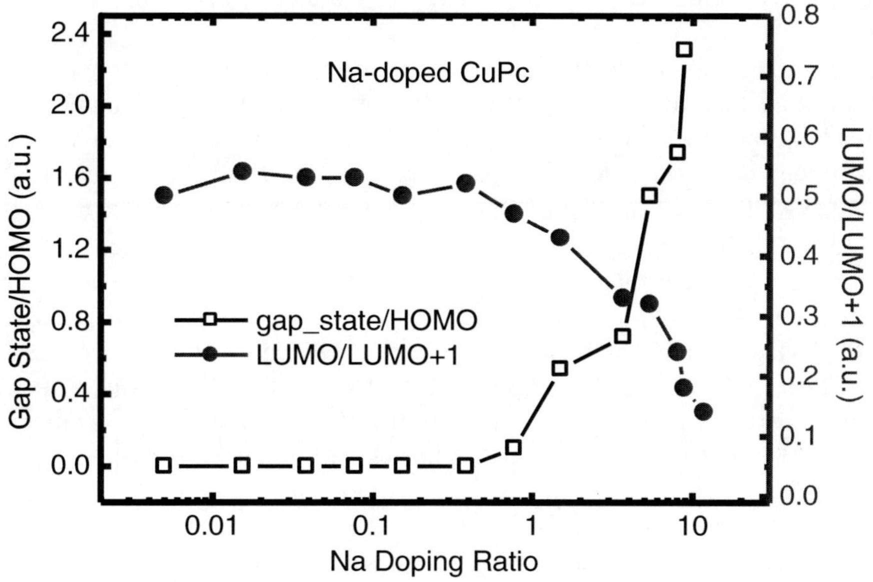

Figure 28. The evolution of the LUMO/LUMO+1 and the gap state/ HOMO intensity ratio as a function of RNa for doped CuPc film. For the heavily doped film, the reduction ratio of LUMO suggests the formation of multiply charged CuPc anions.

In the following experiment, [16] Na was doped into CuPc film with a doping ratio R_{Na}, which goes up all the way to nominally 22. However, it should be noted here that under such high ratios, the complete ionization is not expected. It is also observed that the intensity of the C 1s spectrum of CuPc film remains relatively constant until R_{Na} = 2.2, followed by a monotonic drop for doping ratios higher than that. When R_{Na} = 11.8, the XPS intensity of C 1s reduces to only about 85% of the pristine film. Therefore, R_{Na} thus calculated should be larger than the real doping ratio due to the attenuation of the C 1s.

The UPS and IPES spectra of Na-doped CuPc are similar to that of Cs-doped CuPc. The reduction of the LUMO starts at the same time the gap state is introduced in the UPS spectra. At high doping ratios, the original LUMO peak cannot be identified due to the spectral broadening and the intensity reduction. However, the intensity of the second main component in the IPES spectra is not reduced even at the final doping ratio. Under the previous assumption that the intensities of the HOMO and the LUMO + 1 do not change, the intensity evolution of the gap state and the LUMO is plotted in Figure 28. [16] The overall behavior is similar to that of Cs-doped CuPc, shown in Figure 24 (b). [15] In particular, for high doping ratios, the LUMO intensity can be reduced by more than 60%, which corresponds to more than 2.4 electrons per CuPc molecule. The results suggest that it is a general situation to form multiply charged species in alkali metal doped CuPc. More than one alkali metal atom can donate their outermost s electron to a single CuPc molecule. Although it requires much higher energy to form a multiply charged CuPc molecule in its isolated phase, the multiply changed anions may be stabilized by the nearby metal counter ions, possibly in the form of metal-organic complexes. Finally it should also be noted that the ionization of Na is harder than Cs, as the reduction of the LUMO intensity is much slower as a function of the doping ratio. This can be attributed to the higher ionization energy of Na atoms, as we have discussed in the previous section.

The detailed analysis of the core level spectra reveals the specific atomic sites that the doped Na ions are located. The most pronounced modification can be found in Cu $2p_{3/2}$, which only has a single component in the pristine CuPc. Broadening and shifts toward higher binding energy are observed as R_{Na} increases, until a reversed shift can be seen at $R_{Na} > 2.2$. More importantly, the spectrum splits into two peaks for $R_{Na} > 3.47$. A new component appears at lower binding energy and its intensity increases along with R_{Na}. High resolution XPS was carried out to investigate this important modification induced by heavy doping. The peak decomposition for Cu $2p_{3/2}$ at $R_{Na} = 6.19$ is shown in Figure 29 (a). [16] Gaussian peaks are used to fit the experimental data after careful deduction of the secondary electron background. Peak A in the Cu $2p_{3/2}$ represents the Cu^{II} ions in the pristine CuPc film, and peak B is the newly appeared component induced by heavy doping. The energy separation between these two peaks is about 2.0 eV, which is consistent with the energy separation between Cu^{II} and Cu^{I}. Therefore, it suggests that, at low doping levels, the doped alkali metal ions first interact with pyrrole N atoms in CuPc. However, as the doping ratios increases, strong chemical interactions take place due to the addition of the excess electrons into the CuPc molecular orbitals, resulting in a reduction from Cu^{II} to Cu^{I} by charge transfer. A similar behavior has been observed by Tang et al. for Mg deposition on CuPc. [144]

Another important observation is the evolution of the Na 1s peak, which can only be clearly distinguished at $R_{Na} > 0.48$. A dramatic change of the Na 1s spectra occurs at $R_{Na} = 2.2$, when a second component emerges at higher binding energy. The decomposition of the Na 1s at $R_{Na} = 2.2$ is presented in Figure 29 (b). [16] Peak A, which can also be observed at low deposition ratios, can be easily related to the ionized Na atoms due to the charge transfer towards CuPc. Unlike Cs-doped CuPc, where only single component can be observed for Cs $3d_{5/2}$ core level, another component, Peak B, at about 2.0 eV higher binding energy, appears at high Na doping ratios. One possible explanation is that at high doping ratios, Na forms complex with the presence of the organic molecules.

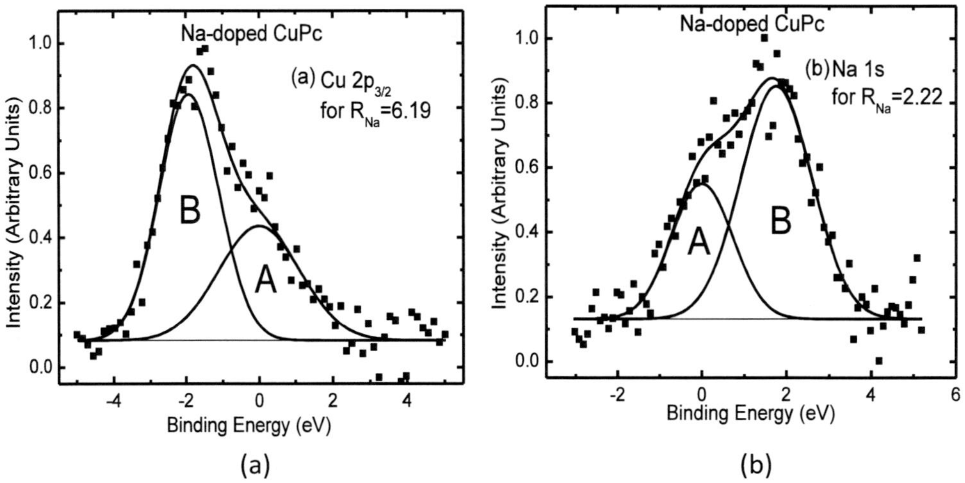

Figure 29. Deconvolution of the (a) Cu 2p3/2 and (b) Na 1s core levels at high Na deposition ratio.

The results from intensive doping studies indicate that at low doping ratios, electrons can be donated from Na 3s into the delocalized orbital of the CuPc molecules, causing the shift of the Fermi level within the energy gap. In the solid state process, the ionization potential of the atomic Na (IP = 5.14 eV) should have been reduced by the dielectric screening of the organic semiconductor for the charge transfer to occur. At high R_{Na}, the modification of the core level spectra for C 1s (B), N 1s and Cu $2p_{3/2}$ suggests a strong electron correlation effect and that the doped electrons are mainly located near the central part of the CuPc molecules. A recent theoretical calculation [135] pointed out that although the 3d-like orbital from the center Cu atom is half occupied and is located in the gap between the Pc HOMO and LUMO, the initial reduction steps correspond to deposition of electrons into the Pc LUMO, bypassing the lower orbital from Cu. It turns out that the Pc macrocycle plays the role of an electron buffer in the metal oxidation and reduction. However, as the doping ratio increases, the energy difference of adding electrons into Pc LUMO and Cu 3d for $[CuPc]^{x-}$ reduces dramatically. It is then expected that by intensive charge transfer, the center Cu atom can be finally affected and strong chemical reactions take place. In fact, the split of Cu $2p_{3/2}$ core level at high doping ratios supports the above predictions very well. Moreover, the intensity analysis shown in Figure 28 clearly suggests that more than one electron has been transferred onto the CuPc molecule. This contradicts the prediction that it is difficult to add electrons after the CuPc mono-anionic species are formed. [135] The discrepancy here may be due to the proximity of the Na^+ ions. Such multiple-ionic species have also been observed in other alkali doped organic materials, such as Na-doped Alq. [18]

The studies discussed so far focus on doping induced electronic structure modification in CuPc thin films. A different configuration, a single layer (SL) on Au (110), presenting an ordered (5 X 3) superstructure, as detected by the well-defined low energy electron diffraction (LEED) pattern, [145, 146] has been reported by Betti et al. [147] The high-resolution UPS results in the energy region close to the Fermi energy for CuPc-SL/Au (110) as a function of Cs deposition is reported in Figure 30. [147] The formation of a SL CuPc quenches the surface state on Au(110) substrate, and induces a new state at binding energy of about 1.22 eV, which can be related to the HOMO peak of the CuPc molecules. It is important to note that the density of states (DOS) is still present at the Fermi edge, suggesting the persistence of the metallic surface. This can be attributed to the formation of strong dipole as a result of the electron density redistribution due to the molecule-substrate interaction. Upon Cs doping, the HOMO shifts to higher binding energy and gradually diminished as the initial occupation of a first former LUMO starts. At higher deposition, a second former LUMO becomes occupied and the first former LUMO shifts toward higher binding energy.Different from what has been discussed in the thin film configuration, oscillating intensity of the former HOMO and LUMO when a new state becomes occupied is observed in SL doping. This can be attributed to the redistribution of the oscillator strength. At the same time, the DOS at the Fermi edge decreases exponentially, suggesting the transition from metallic to insulating surface.

The summary of the energy level shift in Cs-doped CuPc-SL/Au (110) is presented in Figure 31. [147] The work function of the system is significantly reduced upon doping, as a result of the decrease of the surface dipole due to charge transfer from Cs to CuPc. The most interesting observation here is that the evolution of the work function corresponds to the occupation of the former LUMO, as noted by the dashed lines in the top and bottom panels in Figure 31. At the first slope change, the first former LUMO is occupied, and in correspondence to the second slope change at about half of saturation, the second former

LUMO starts being occupied. The amount of charge transfer can be estimated by analyzing the former LUMO occupation as a function of deposition time. It is expected to have two additional electrons when the first former LUMO is filled. And the occupation of the second former LUMO and work function saturation correspond to four electrons per CuPc molecule.

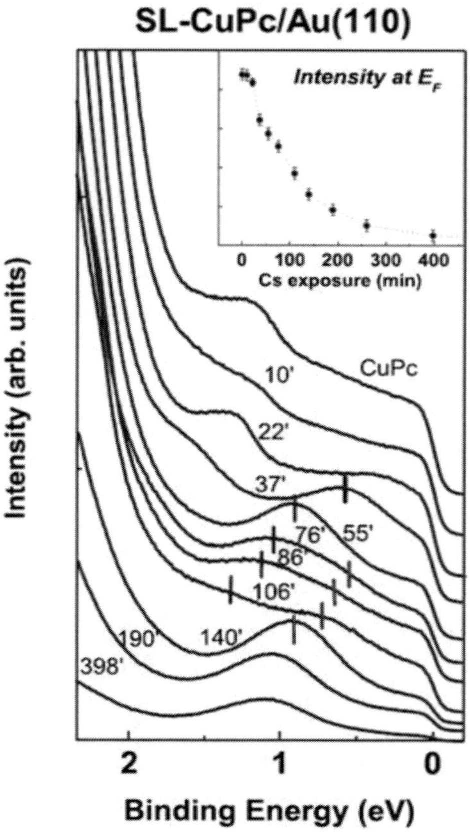

Figure 30. High-resolution UPS data of the Cs-doped CuPc-SL on Au(110) as a function of Cs deposition time(from top to bottom); two ticks mark the incoming occupation of the first and second former LUMOs. (Inset) Intensity of the electronic spectral density at the Fermi energy as a function of Cs deposition. Data were taken with 21.218 eV photon energy, at 20° out-of-normal emission, to enhance surface sensitivity.

Unlike Cs doping in its thin film configuration, two former LUMO state are observed in Cs-doped CuPc-SL/Au (110) interface. Although gas-phase and solid-phase CuPc molecules are characterized by a two-fold degenerate LUMO $2e_g$ state, [135] the symmetry of the former-degenerate $2e_g$ molecular states might be broken at the interface, due to the structure modification induced by the Au substrate in CuPc-SL. This explains the presence of the two former LUMO states at the interface. The effect of molecular distortion (Jahn-Teller (JT) effect) and different Cs chemisorptions sites may also contribute to the splitting of the former LUMO states. Another important observation is the exponential decay of the DOS at the Fermi edge as Cs doping level increases. Betti and the coworkers suggest several reasons for this issue. [147] On one side, alkali atoms may interact with the underlying metal through the CuPc-SL, as in the semiconducting CsAu alloy, [148] contributing to the non-conducting layer. On the other hand, the exponential decrease of the spectral density at the Fermi edge denotes an incoherent band sweeping, and strong correlation effects are the natural mechanism for the loss of coherence. They suggest that a repulsion energy U between molecules, which is larger than the single-particle bandwidth (0.1 eV) for the HOMO and former-LUMO states, can render CuPc a correlation-induced insulator: the introduction of electronic charge in the layer may activate the Coulomb repulsion between electrons trapped in the former-LUMO states. Such strong electron correlation effects are also confirmed by the insulating nature of the CuPc thin film at all Cs doping levels.

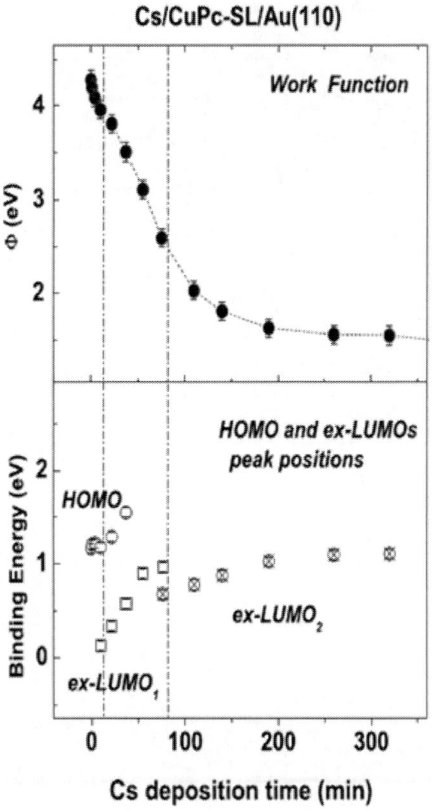

Figure 31. Top panel: Work function change of the CuPc-SL on Au(110) as a function of Cs deposition time. Bottom panel: HOMO and former-LUMO binding energies as a function of Cs deposition time.

3.3. Electronic Structure of Alkali Metal-Doped PTCDA

Perylene-3,4,9,10-tetracarboxylicdianhydride (PTCDA), shown in Figure 32, is another model system for molecular organic semiconducting materials. Its growth behavior [149-151] and its electronic, [152] vibrational, [149] and optical [153, 154] properties in a number of configurations are well studied by several groups. PTCDA has an extended π-conjugated system. At the same time, the oxygen atoms at the end of the molecule have a limited contribution to the formation of the HOMO and LUMO. Owing to such properties, PTCDA shows predominantly electron conduction, with a high electron affinity of about 3.6 eV, [155] which means that it can be easily doped by alkali metals.

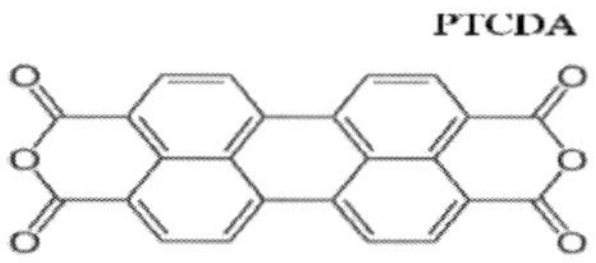

Figure 32. Molecular structure of PTCDA.

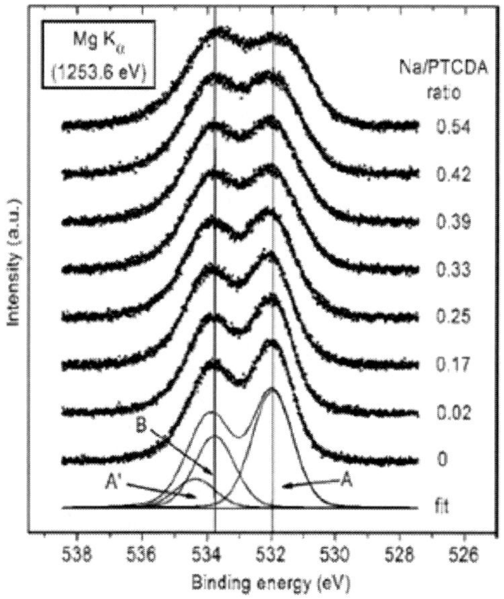

(a)

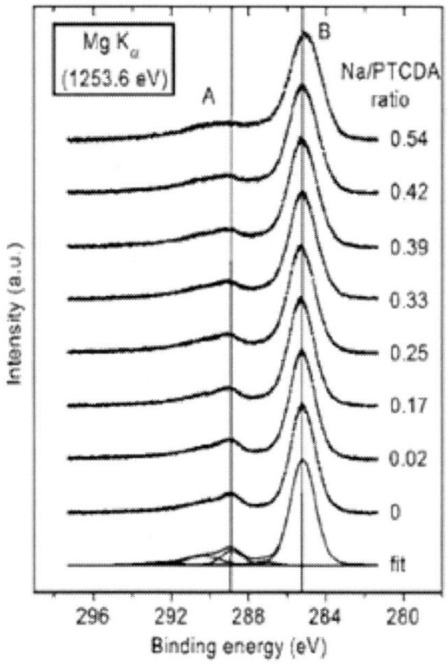

(b)

Figure 33. O 1s (a) and C 1s (b) peaks of pristine PTCDA and at rising dopant concentrations.

Alkali metal doping in PTCDA has been systematically studied by Ziegler's group. [149, 152, 155-157] This chapter will mainly cover their PES investigations about the electronic structure modification induced by Na, K and Cs doping in PTCDA thin films. The O 1s and C 1s core level spectra obtained from XPS measurement of PTCDA are presented in Figure 33, as a function of Na doping ratio. [152] The O 1s peak of pristine PTCDA has two main contributions which can be attributed to oxygen in two inequivalent positions in the PTCDA molecule: peak A at 532.0 eV binding energy can be assigned to the carbonyl oxygen and peak B at 533.7 eV to the bridging. The C 1s peak of the pristine PTCDA also has two main components: peak A at 288.9 eV binding energy has its origin in the carbonyl end groups of the molecule, and peak B at 285.2 eV is related to the perylene core. Besides the major components in O 1s and C 1s spectra, additional shake up structures can also be identified at higher binding energies.One of the most predominant shake up peak bears an energy loss of about 2.4 eV (A' in Figure 33), which is thereby assigned to a band-gap excitation. Na doping in PTCDA induces a primary shift towards higher binding energy and an overall broadening for all energy levels. In addition, a change in the intensity ratio of the two oxygen peaks, accompanied by a shift of peak B to lower binding energies is observed. This behavior suggests that the interaction of sodium with the PTCDA occurs at the carbonyl group of the PTCDA molecule. Such conclusion is corroborated by the modification of the C 1s core levels, where the intensity of peak A decreases while that of peak B remain fixed with rising dopant concentration.

To further understand the core level behaviors in doped system, detailed decomposition of the O 1s core level were performed for undoped and doped PTCDA, as shown in Figure 34. [152] For pristine PTCDA, the energy separation between the two components is estimated to be about 1.8 eV from the fitting of the experimental data, taking into account the stoichiometric ratio between the four carbonyl O atoms and the two bridging O atoms and the shake up contribution. The shake up of peak A (A') has an energy difference of 2.4 eV in the higher binding energy side. The shake up of peak B has to be assumed to have much less intensity with respect to its main peak to obtain satisfying fit results. This is attributed to the fact that the DOS in LOMO is nearly zero for the bridging O atoms in pristine PTCDA. Na-doped PTCDA was calculated assuming two Na atoms coupled to the PTCDA molecule in the arrangement shown in Figure 34 (b). [152] Three main components, which correspond to interacting carbonyl O (A2), non- interacting carbonyl O (A1), and bridging O (B), are used to fit the experimental data. The intensity ratio of the shake up contribution to the corresponding main component and the relative peak distances between A1 and B are kept the same as the pristine PTCDA. The intensity ratio among the three main components is constrained to the stoichiometric ratio of 1:1:1, including the shake ups. In comparison of the O 1s spectra of the pristine and doped PTCDA, one can notice that peak B, which corresponds to the bridging oxygen, remains single component after doping. It is thus suggested that the doped Na ions mainly interact with carbonyl oxygen atoms, leaving the bridging oxygen atoms intact.

The HOMO region of the He II UPS spectra for Na-doped PTCDA at different dopant concentrations is shown in Figure 35. [152] The peak of the HOMO of pristine PTCDA is found at 2.4 eV below the Fermi level. And the onset of the HOMO is about 1.8 eV below the Fermi level. The unoccupied molecular orbitals as measured by IPES are shown in Figure 36. [152]

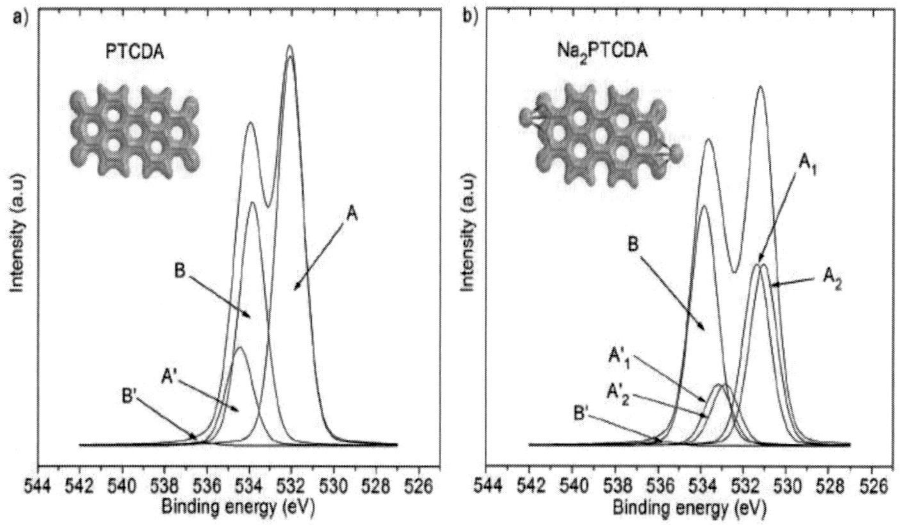

Figure 34. (Color online) Simulation of the O 1s peaks of (a) undoped and (b) sodium-doped PTCDA. In case (a) peak A is assigned to the four carbonyl oxygens and peak B to the bridging oxygen. The peaks A' and B' are shake ups of A and B. In (b) the peaks A1 and A2 are related to the two now inequivalent carbonyl oxygen species, peak B to the bridging oxygen. The primed letters indicate again the shake up peaks.

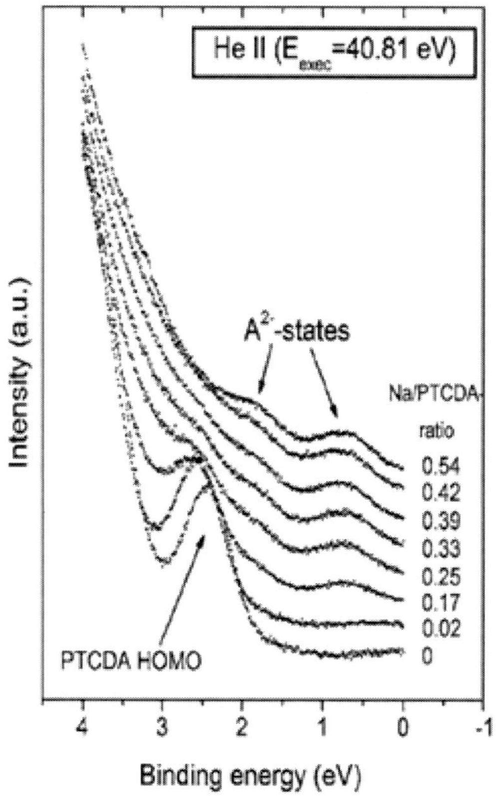

Figure 35. He II spectra of pristine PTCDA and at rising dopant concentrations.

The peak and the onset of the LUMO of pristine PTCDA are situated at -1.4 eV and -0.6 eV above the Fermi level, respectively. Combining the information from direct and inverse photoemission spectroscopy, the optical and transport energy gap of pristine PTCDA is derived to be about 2.4 eV and 3.8 eV, respectively. The width of the band gap supports the assignment of the most prominent shake up structure seen in XPS spectra to an energy loss of the photoemitted electrons by exciting a band-gap transition.

Doping the PTCDA thin film with Na results in Fermi level shift towards the LUMO onset. In addition two new states appear in the former energy gap at 0.7 and 1.8 eV below the Fermi level, accompanied by a decrease of intensities in the HOMO and the LUMO if compared to pristine PTCDA. The origin of the two new states are illustrated in Figure 37. [152] Doping of PTCDA with Na leads to negative charge density transfer from the Na's outer s orbital to the LUMO of pristine PTCDA, forming ionic coupling between the reaction partners, which is thereby lower the energy of the former LUMO. This results in the formation of the first new state at 0.7 eV below the Fermi energy (HOMO A^{2-} in Figure 37). Furthermore, the former HOMO of the neutral PTCDA is destabilized and split into the second new state in the gap (HOMO-1 A^{2-} in Figure 37).

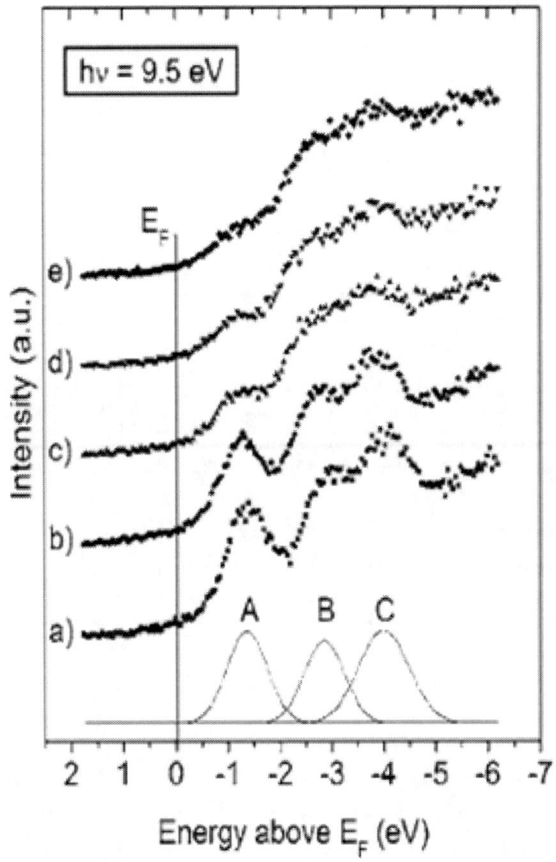

Figure 36. Inverse photoemission spectra of pristine PTCDA (a) and at rising dopant concentrations (b) – (e). The Gaussian peaks A, B, and C fit the three main contributions of the pristine PTCDA spectrum after background correction. Their maxima are found at −1.4 eV (A), −2.8 eV (B), and −4.0 eV (C) above the Fermi energy (E_f).

DFT calculations suggested that a maximum number of two Na ions can interact with a PTCDA molecule in a configuration with C_{2v} symmetry, as shown in Figure 38. [152] However, the quantitative XPS suggests a maximum doping level of about 1.4 Na atoms per PTCDA molecule at the saturation of the doping induced energy level shift. In order to clarify the mechanisms that determine the maximum charge transfer amount, it is necessary to investigate the doping process for other alkali metals besides Na. The normalized UPS spectra in the HOMO region are presented in Figure 38 for Na, K and Cs doped PTCDA. [155] It can be seen that, in the cases of all three alkali metals, doping induces two new gap states whose origin has been discussed in the previous paragraph. It is clear that the energetic position of these states is the same for all three dopants, which indicates that the changes in the density of occupied states is mainly related to a physical doping. The electronic structure of doped PTCDA is only modified due to the electrons donated from alkali ions, while the presence of the alkali metal counter ions has very little effect, since the latter would result in different energetic positions for the different alkali metals. Another important observation is that the intensity of the new gap states increases in order of Na, K and Cs doped systems. Since the photoemission cross sections are expected to be the same in all three cases, the discrepancy in the measured gap states intensities can be directly attributed to the amount of charge transferred onto PCTDA in average within the UPS probing depth.

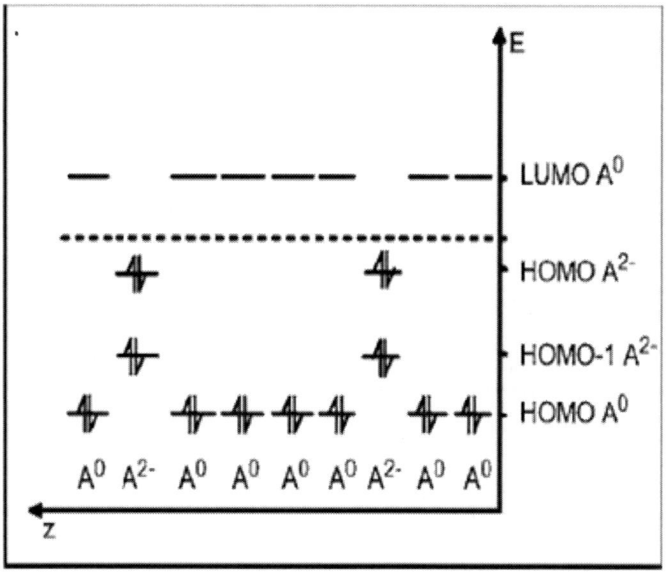

Figure 37. Model of the origin of the gap states below the Fermi energy.

A straight forward interpretation of the found differences can be given by taking into account the ionization energies of the alkali metals and the HOMO ionization energies of the alkali$_2$PTCDA complexes, as shown in Figure 39. [155] There are no significant differences in the binding energy of the HOMO in doped PTCDA, being about 5.0±0.1 eV, which is very close to the ionization of a Na atom (5.1 eV). It thus suggests that the remove of the doped electrons from PTCDA back to Na cost almost zero energy. However, as the ionization energy of the alkali metal atoms reduces, such as the case of K and Cs, the dedoping process becomes more and more energetically unfavorable. As a result, in the case of thermal equilibrium, Na_2PTCDA is the least stable, followed by K_2PTCDA and Cs_2PTCDA as the most stable complex. It is thus concluded that although a coordination of two alkali metal atoms with one PTCDA molecule is the preferred configuration for subsequently doped PTCDA thin films, Na and K are not able to form completely doped PTCDA thin films, due to their higher ionization energies. Whereas, when doped with Cs, a completely doped PTCDA layers at the surface can be established due to its lower ionization energy.

With the understanding of the electronic structure modification induced by alkali metal doping, Wüsten et al. performed further investigation on doped PTCDA film after annealing process. The UPS spectra of the HOMO region and IPES spectra of the LUMO region are presented before and after several doping and annealing steps in Figure 40 and 41, respectively. [152] It can be observed in the valence region that the intensities if the two gap states induced by alkali metal doping decreases after the annealing steps at about 380 K for 180 minutes. In addition, the formerly reduced HOMO increases again after annealing and shifts towards the Fermi level. A second doping step reestablishes the initial doping situation. Such reversal of the doping induced electronic structure modification after annealing can also be observed in the IPES spectra shown in Figure 41. [152] Alkali metal doping results in a reduction of the LUMO intensity and the Fermi level pinning at the LUMO onset position. The subsequent annealing shifts the LUMO away from the Fermi level and increases its intensity. A following doping step shows again the typical dopant influenced LUMO position. The presented spectra of the occupied and the unoccupied electronic states indicate that the doped electrons in the PTDCA molecule are partially removed after annealing.

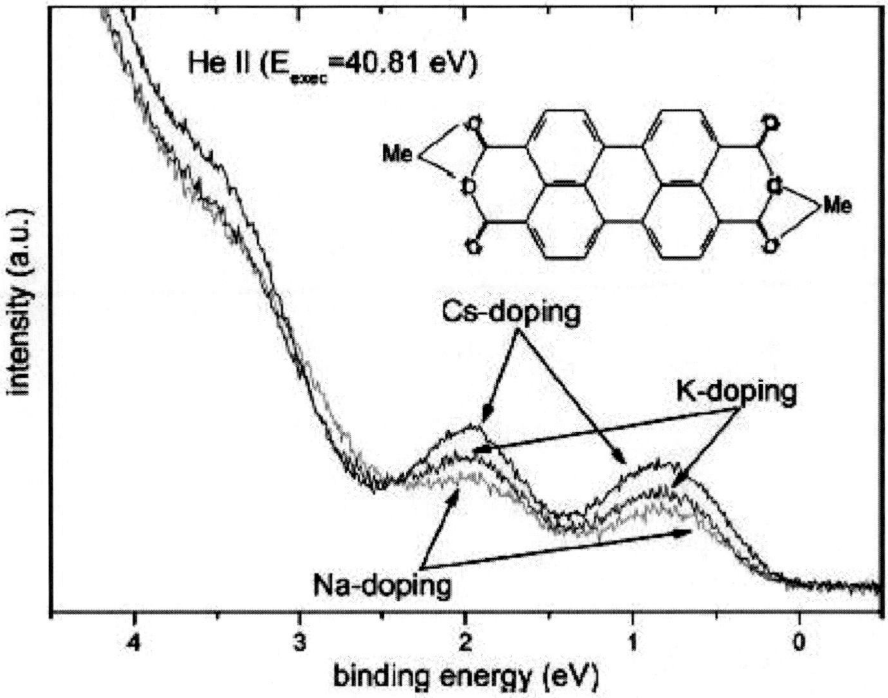

Figure 38. Comparison of the surface-near densities of occupied electronic states in Na, K and Cs doped PTCDA, each in state of alkali metal saturation.

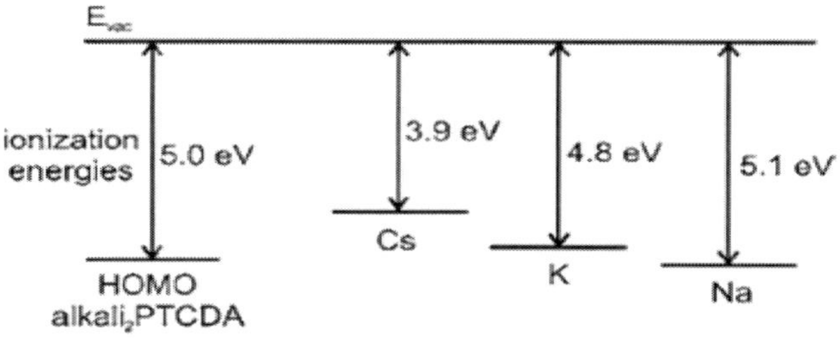

Figure 39. Ionization energies of Na, K, Cs and of the HOMO of the alkali-PTCDA complexes.

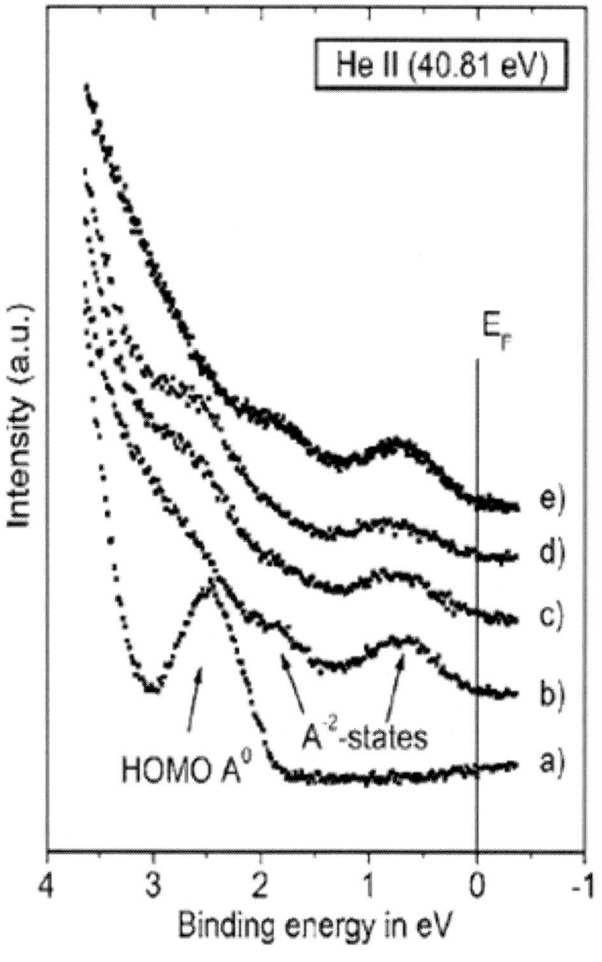

Figure 40. UPS spectra of pristine PTCDA (a), freshly Na-doped PTCDA (b), after 30 min. annealing at 380 K (c), after 180 min. annealing at 380 K (d), and after a second doping step (e).

Figure 42 and 43 show the behavior of the C 1s and O 1s core levels of Na-doped PTCDA after annealing at 440 K for 180 min. [152] Similar to what has been observed in the UPS and IPES spectra, the doping induced modifications are partially reversed upon annealing. The O and C atoms in the functional groups of the PTCDA molecule are no longer shifted by the presence of Na atoms. In order to study whether dissolved dopant accumulates on top of the PTCDA layer or leaves the sample by annealing, angle-resolved XPS measurements were performed. Figure 44 shows the ratio of the Na 1s to the C 1s peak area and Cs 4d to the C 1s peak area for Na-doped and Cs-doped PTCDA, respectively, as a function of the emission angle. [152] The probing depth in this study is reduced as the emission angle increases, resulting in an increase of the surface sensitivity.

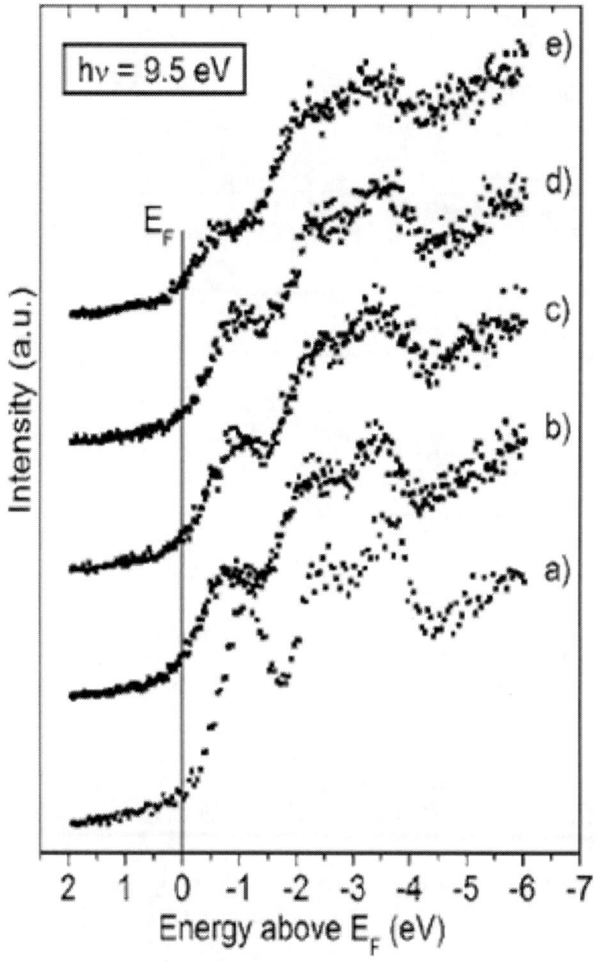

Figure 41. IPES spectra of pristine (a) and freshly Na-doped PTCDA (b), after two annealing cycles at 380 K (c) and (d), and after a second doping step (e).

The first observation is that the annealing process reduced the ratio between alkali metal atoms and the PTCDA molecules in both surface and bulk of the organic film. It is clear that some of the alkali metals atoms are evaporated from the organic film during annealing, which explains the reversal of the doping induced electronic structure modification shown in the valence and conduction region. Another important issue is that the distribution of Na in the PTCDA film shows clear depth dependence, i. e. there are less Na in the surface than in the bulk of PTCDA film. However, such depth dependence is not as clear in Cs-doped film. In summary, the annealing process above 390 K leads to the dedoping by sublimation of alkali metals out of the PTCDA film into UHV. The reversal of the doping effect after annealing is also confirmed by the electrical measurement, where the conductivity of the doped system decreases after annealing. [152]

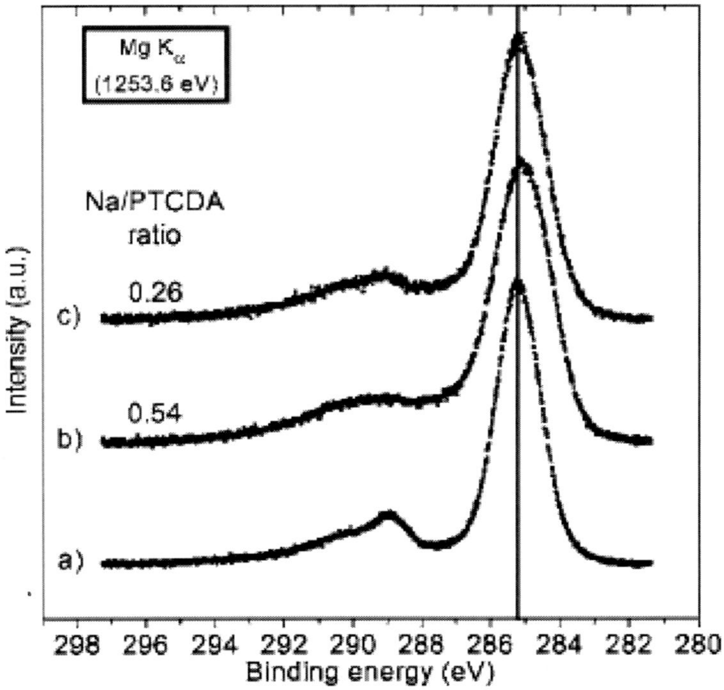

Figure 42. XPS spectra of the C 1s region of (a) pristine, (b) Na-doped, and (c) annealed (180 min at 440 K) PTCDA. The energetic position of the emission from the perylene core is 285.2 eV for pristine PTCDA (a), 285.1 eV for a dopant concentration of 0.54 Na/PTDCA (b) and 285.2 eV after annealing (c).

3.4. Electronic Structure of Alkali Metal-Doped Oligophenylene

Oligophenylene, such as para-quarterphenyl (4P) and para-sexiphenyl (6P), shown in Figure 45, is a model oligomer of poly(para-phenylene), a large band-gap organic semiconductor. This class of materials exhibit strong electroluminescence and has been intensively used in OLEDs. [158-160] Owing to its simple chemical structure and the ease to handle, oligophenylene is one of the best model systems for study the fundamental electronic properties of organic semiconductors. The electronic structure modification induced by alkali metal doping in vacuum has been extensively studied by Koch et al. [22, 23, 161-163]

Figure 46 shows the UPS spectra of the valence region for (a) a pristine, nominally 40 Å thick, 6P layer, and (b) to (d) the organic film after Cs doping. [23] Note that the binding energy used in this figure is calibrated relative to the vacuum level of the organic film. The UPS spectrum of the pristine 6P film exhibits good agreement with the result from the DFT calculations. [164] The HOMO and HOMO-1 states, which are not fully resolved due to the limited resolution in this experiment, are located at 6.5 and 7.0 eV, respectively. The HOMO onset is estimated to be about 6.1 eV, which agrees well with the ionization potential value reported in literature. [164] After Cs doping, several changes in the recorded UPS spectra can be identified. However, since the vacuum level is set as the zero for the binding energy scale used in this figure, the doping induced Fermi level shift cannot be directly observed. The energetic positions of the HOMO and other molecular orbitals of 6P remain almost intact with respect to the vacuum level, suggesting that, other than charge transfer, Cs doping does not introduced strong chemical interactions with the 6P molecules. It can also be noticed that the intensity of the former HOMO and HOMO-1 reduce, while new features appear first in the formerly empty band gap of the pure 6P material: two new electronic gap states appear at 5.8 and 3.6 eV, which can assigned to negative bipolaron states of n-doped 6P. [165] The intensity of the new gap states increases as the Cs doping increases. Finally, a new feature can be identified around 15 eV, which can be attributed, along with other lineshape modifications in 7 – 12 eV, to the contribution from the core level and valence structure of Cs. [23]

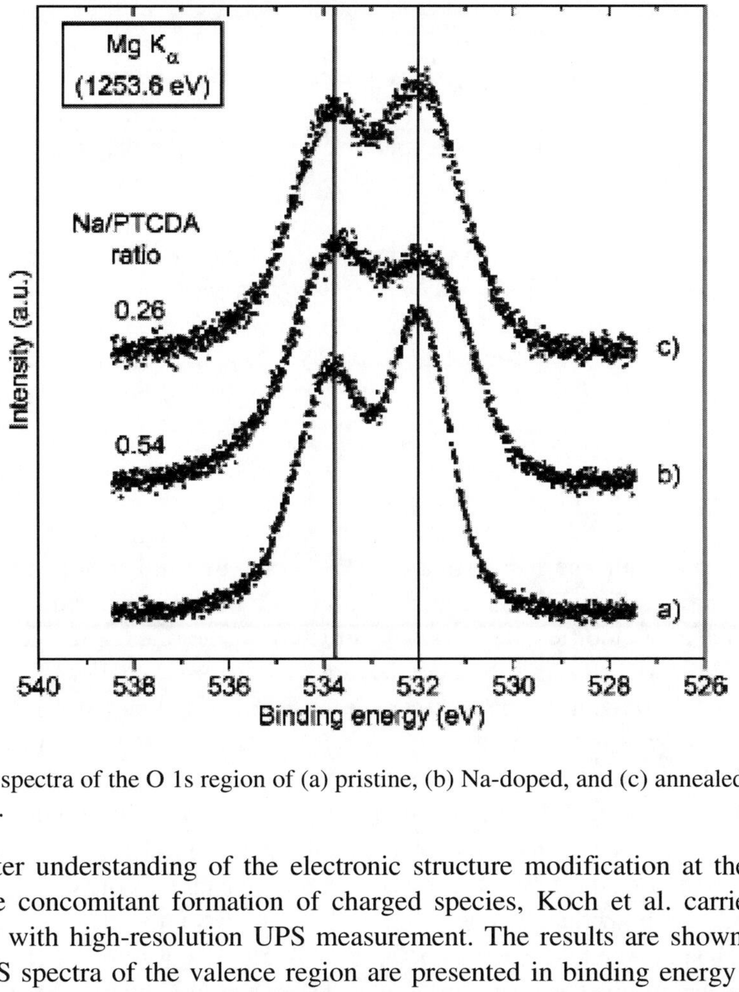

Figure 43. XPS spectra of the O 1s region of (a) pristine, (b) Na-doped, and (c) annealed (180 min at 440 K) PTCDA.

For a better understanding of the electronic structure modification at the early doping stages and the concomitant formation of charged species, Koch et al. carried out a more detailed study with high-resolution UPS measurement. The results are shown in Figure 47, where the UPS spectra of the valence region are presented in binding energy relative to the vacuum level. [23] This time the HOMO and HOMO-1 states can be clearly resolved due to the much better energy resolution used. The position of the Fermi level is noted by tick marks for each spectrum. It is clear that Cs doping first induces Fermi level shift towards lower binding energy, as expected for n-doping. However, such shift becomes saturated, as soon as the new gap states appear at 5.8 and 3.6 eV, at the expense of the delocalized p-states intensity. This doping behavior of 6P with Cs is ascribed to negative bipolaron states $6P^{2-}$.

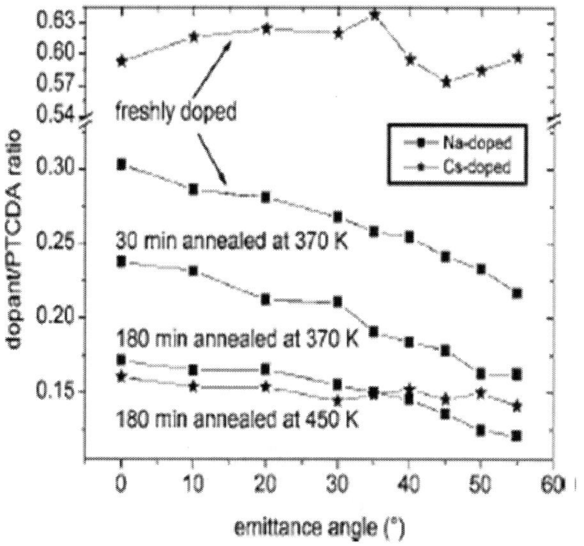

Figure 44. Angle-resolved Na-doped PTCDA and Cs-doped PTCDA ratios of freshly doped PTCDA layers and after two annealing steps for Na doping, respectively, one annealing step for Cs doping.

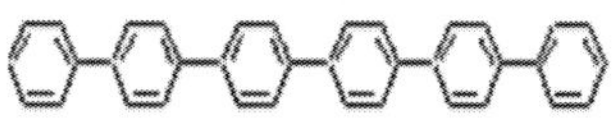

Figure 45. Molecular structure of para-sexiphenyl (6P).

The appearance of the new gap states in Cs-doped 6P film results from the charge transfer between the alkali metal atoms and the organic molecules. The addition of electrons to the organic molecule modifies the electronic structure of the pristine molecules, leading to destabilization of the HOMO and the stabilization of the former LUMO. The fact that no finite DOS can be observed at the Fermi edge seem to suggests that the formation of singly charged polarons is expected to precede the evolution of bipolarons. At this scenario, the Fermi level should be located midway between the fully occupied upper orbital and the empty new LUMO. The formation of the bipolaron state is also confirmed by Figure 48, [23] when the binding energy scale is adjusted so that the intense localized p-electron derived state at 8.5 eV with respect to the vacuum level is arbitrarily set at a constant energy (7 eV on the scale in Figure 48). One can immediately notice that the position of the new gap states remain unchanged, which suggest that there is no polaron to bipolaron transition.

Another interesting investigation carried out by Koch et al. is the study on the influence of the counter-ion on the electronic structure in alkali and alkali earth metal doped 6P thin films. [161] In this study, they have deposited K, Cs and Ca under UHV conditions on thin films of 6P and followed the changes in the valence electronic structure by UPS for each dopant used. The experimental results are shown in Figure 49, [161] where the UPS spectra of the two new gap states are aligned with respect to the one with higher intensity. Thus, it should be noted that the origin of the energy scale presented here is chosen arbitrarily. All three spectra correspond to complete doping level, when no obvious changes in the electronic structure can be found in the doped film.

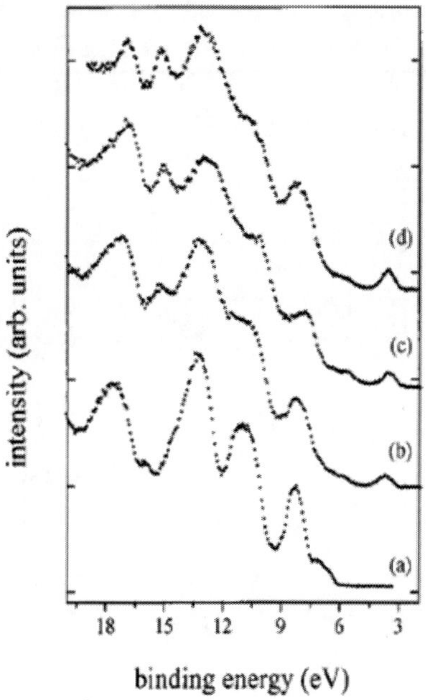

Figure 46. Valence photoemission spectra of the coarse doping study: (a) pristine 6P film, (b) ~ (d) increasing amounts of Cs on 6P; (d) saturation doping level, referred to as 100%.

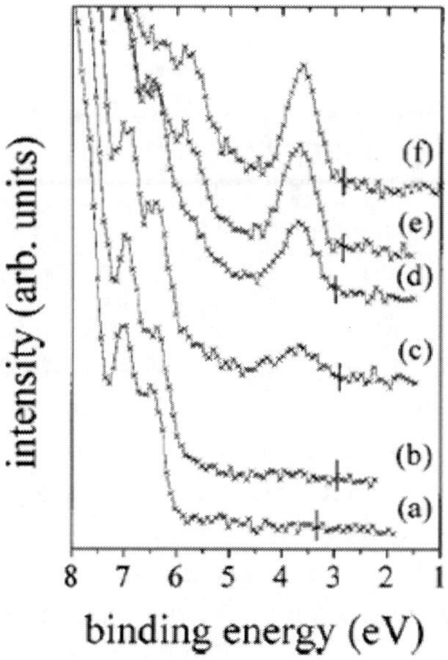

Figure 47. Close-up of the gap region of 6P for increasing doping with Cs: (a) pristine 6P film, (b) 2%, (c) 12%, (d) 25%, (e) 50%, and (f) 100% doping level.

As one compares the doping induced spectra modification with the presence of three different counter-ions, it is clear that the high intensity gap state in Cs-doped film has a very sharp and symmetric lineshape, while its feature is less defined for K and Ca doped films. A pronounced asymmetric broadening towards higher binding energy can be observed for the latter two cases. It can also be noticed that the other gap state is easily resolved at about 5.3 eV on this energy scale for Cs doping, while its separation to the maximum of the high intensity gap state is increased for K and Ca doping. Those differences suggest that the counter-ion plays an important role in determining the electronic structure of the charge-transfer complex formed.

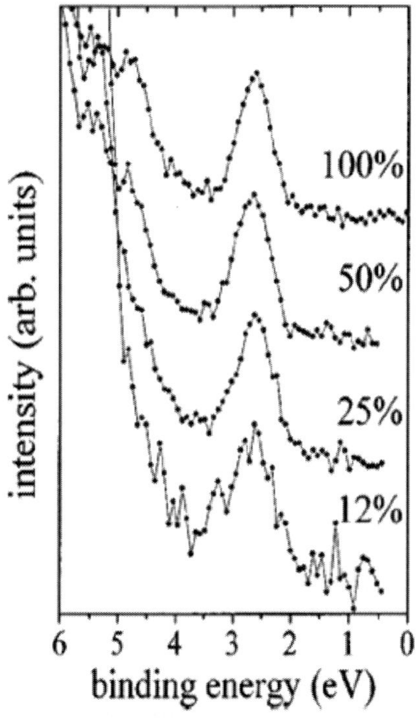

Figure 48. Close-up of the new low-binding energy emission. The binding energy scale is shifted for an alignment of this emission, and intensity normalized to it. The numbers indicate the doping level.

The 6P films used in this study is considered polycrystalline with many stacking faults and grain boundaries. [166] The alkali and alkali earth metals deposited on the organic film can diffuse on the surface and into the bulk of 6P, however, with different diffusion coefficient. Therefore it is reasonable to assume that there might exist a variety of different 6P/metal atom geometries in a sample. DFT calculations for various geometries was performed by Koch *et al.* to study the observed differences in the UPS spectra discussed in the previous paragraph. [161] The results suggest a strong influence of the position of the counter-ion on the energetic positions of the energy levels and on the overall electronic properties in doped organic systems. This explains the asymmetric broadening of the gap states observed in the K and Ca doped 6P films. It also implies that in a well ordered organic film with a homogenous distribution of the dopants, the asymmetry can be eliminated, as what has been observed in case of Cs doping. Moreover, the amount of charge transfer between the dopants and the organic molecules also differs among the studied atoms. It has been shown that Ca only donates partial charges to 6P molecule, while Cs could transfer a complete charge, due to the lower ionization potential, as we discussed in section 3.3. A complete charge transfer could mean that, for this system, the position of the counter-ions relative to the organic molecule is not as crucial as in the other systems.

Photoemission, along with metastable atom electron spectroscopy (MAES), studies were also performed for alkali metal doped 4P thin films. [22] Na, K, Rb and Cs were introduced as the dopants to the organic molecules. A similar result, as what is discussed for alkali metal doped 6P, has been reported. [22] The evolution of new electronic states in the former energy gap of the pristine 4P molecules is observed, and attributed to the formation of negative bipolarons.

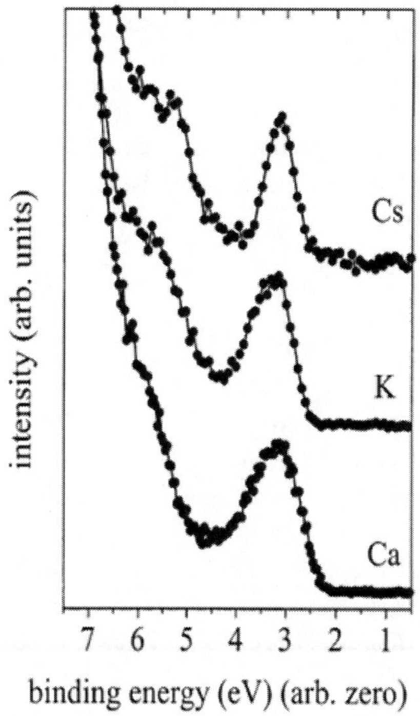

Figure 49. UPS spectra of the doping-induced gap states in 6P for various dopants as indicated.

3.5. Two-Stage Model for Doping-Induced Electronic Structure Modification

The experiments discussed in the previous sections about alkali metal doping in conjugated organic molecules all suggest that doping of alkali metals in organic semiconductors induces electronic structure modification, which can be seen in two stages. [13, 17] In the first stage, all energy levels of the organic molecule shift towards higher binding energy, in other words, the Fermi level shifts up towards the LUMO of the organic molecule, due to electron donation from the outmost s orbital of the alkali metal atom. At the second stage, the shift of the energy levels becomes saturated, followed by the appearance of the gap state in the valence region and the modification of the core levels, which are mainly induced by filling the unoccupied states of the molecules by doped electrons.

The energy level dependence on the doping density in organic semiconductor is fundamental for understanding the interaction between the dopant and the host system. The results presented in the previous sections clearly suggest that the doping-induced electronic structure modification in organic semiconductors can be seen in two stages, which are characterized respectively, by the rigid energy level shift and the charge transfer induced strong correlation effect. However, in the first stage, the slope derived from the semi-log plot is several times larger than what has been predicted with inorganic semiconductor theory. In an inorganic semiconductor, the electron concentration is given by Equation 1. For most crystalline inorganic semiconductors, the density of state is broad and smooth in comparison with k_BT, and in the non-degenerate limit, it leads to [102]

$$n = N_c \exp[-(E_c - E_f)/k_BT] \tag{13}$$

where E_c is the conduction band minimum. If the doping density N_D is much larger than the intrinsic carrier concentration and full ionization can be assumed, the Fermi level position can be deduced from

$$E_f = E_c - k_BT \ln\left(\frac{N_c}{N_D}\right) \tag{14}$$

Therefore, a linear dependence with slope of about k_BT is generally expected for non-degenerate inorganic semiconductors.

However, in most of the investigation of alkali metal-doped organic semiconductors, the slopes derived from the semi-log plot in the rigid shift region exhibit dramatically deviations from Equation 14. For instance, the slope of 0.250 eV and 0.127 eV can be deduced for Cs-doped Alq and CuPc from the linear fitting in Figure 15 and Figure 24 (a), respectively. Those values are much larger than k_BT (0.028 eV). Thus it clearly suggests that the classical result for crystalline inorganic semiconductors does not apply to organic semiconductors.

The fundamental difference between an organic and inorganic semiconductor is the bonding. The covalent bonding prevailing in inorganic semiconductors results in a large electronic wavefunction overlap, which is the basis for forming the Bloch waves and band structure. In organic semiconductors, however, the bonding is predominately van der Waals and the electronic structure remains largely molecular and localized. The localized nature of the electronic structure causes large correlation effects, either electron-electron or electron-phonon. In general, the bandwidth of intrinsic organic semiconductors is small, compared to k_BT, due to the weak interactions between the molecules. The simplest model to describe the molecular orbital in organic semiconductor is using an infinitely sharp energy level, which can be expressed by a δ-function

$$D(\varepsilon) = N_0\delta(\varepsilon - \varepsilon_0) \tag{15}$$

Equation 15 suggests that the organic energy level is fully isolated from the doping metal atom. In this case, Equation 13 will be replaced by

$$n \cong N_0 \exp[-(E_0 - E_f)/k_B T] \tag{16}$$

where N_0 and E_0 are the density and position of the molecular orbital, respectively. However, one would expect a same energy level dependence on the doping density as in Equation 2.21, i.e. logarithmic with slope of $k_B T$ in the semi-log plot.

In order to have a larger slope from Equation 13, one has to introduce certain broadening for the organic molecular orbital, which can be caused either by metal-organic or electron-phonon interactions. If the LUMO energy distribution is broad and exponential, it may result in a linear behavior in the semi-log plot with the slope given by the exponential width Γ that can be much larger than $k_B T$.

$$n \cong N_0 \exp[-(E_0 - E_f)/\Gamma] \tag{17}$$

In this context, the slope observed for alkali metal-doped organic semiconductors may be used as an estimation of the broadening of the LUMO due to the presence of the excess electrons or other impurities. Since electron-phonon interaction is intrinsic to the organic molecule, the fact that the slope for Na-doped Alq is noticeably smaller than that of Cs-doped Alq may favor metal-organic interaction as the more probable cause to the exponential broadening. However, incomplete ionization for Na-doped Alq may also be a contributing factor to the smaller slope.

Another surprise from the alkali metal doping experiments is that the slope cannot be explained by a simple Gaussian broadening of the LUMO. If the energy distribution is a broad Gaussian,

$$D(\varepsilon) \cong 2N_0 \exp[-(\varepsilon - E_0)^2/\sigma^2]/\sqrt{\pi}\sigma \tag{18}$$

with $\sigma >> k_B T$, the electron concentration will be dominated by the Gaussian and the charge density given by

$$n \cong N_0\{1 - erf[(E_0 - E_f)/\sigma]\} \tag{19}$$

where *erf(x)* is the error function. Experimentally, both UPS and IPES peaks appear to be Gaussian. However, the attempt of fitting using Equation 19 does not work well with the data obtained in Cs and Na doped Alq, indicating that Gaussian broadening cannot be applied, at least by itself, to explain the large slope of the doping induced energy level shift. The inapplicability of Gaussian distribution to the observed energy level shift may stem from the fact that the observed broadening is mainly from inhomogenuity of the molecular environment in a solid film. The localization of electronic wavefunctions in organic semiconductors may again dictate the doping and charge transfer predominately localized between the molecule and the dopant, and the Gaussian broadening resulted from the inhomogenuity of all the molecules may play a little role in the energy level shift.

The linear shift stage at low doping ratios can be understood as the Fermi level shifts within the energy gap, in order to reach equilibrium with the low work function metal dopant. At this point, the doped electrons are transferred into the tail of the broadened LUMO, leading to an increased charge density in organic semiconductors. However, at this stage, they can be shared by all organic molecules in the neighborhood. Since those electrons do not belong to any specific organic molecules, the interactions between the alkali metal cations and the organic molecules are still weak. Therefore, the electronic structure of organic molecule is well preserved, which can be concluded from the observation of the rigid shift in previous sections. However, as Fermi level approaches the LUMO of the organic molecule due to increasing doping, it becomes energetically favorable that the doped electrons fill the unoccupied states, leading to the formation of ionic band between the alkali metal and the organic ions. The most direct evidence is the appearance of the gap state and the reduction of the original LUMO. In this stage, the applicability of Equation 13, which is derived for the non-degenerate case, where the doping density is low and the overlap of wavefunctions between the dopants themselves is small, may be diminished. On the other hand, the correlation effect becomes important. As a result, the electronic structure of organic molecule is greatly modified, evidenced by the intensity evolution of the frontier orbital and the reverse shift of the core levels.

The evolution of the core levels provides further information about the modification of the electronic structure at the second stage. Generally, the shift of the core levels consists of a Fermi level shift when a rigid shift can be assumed, and a chemical shift from electron correlation effect as a result of the modification of the charge density environment. In the case of alkali metal doping, electrons are donated from the outermost s orbital to organic molecules, leading to a negative chemical shift towards lower binding energy for the core levels due to the Coulomb interaction. However, at the same time, the Fermi level of the system has been modified, leading to a positive shift towards higher binding energy for all the energy levels. These two shifts, which are actually of opposite directions, compensate each other and give the overall shift that can be observed in XPS. Even though a rigid shift may not be true for a charged system as we discussed above, the Fermi level shift is expected to affect all core levels the same way. Therefore, the difference in the energy level shift for the core levels can be mainly attributed to the chemical shift, which is directly related with the charge redistribution. In this context, the binding energy of the elements, at which the excess electrons are localized, will be lower than other elements. For example, the splitting of N 1s in lower binding energy for alkali metal doped Alq suggests that ionic interaction can be presented between alkali metal ion and one of the three N atoms in Alq molecule. This experimental observation has already been confirmed by theoretical calculations, as discussed in section 3.1. [18]

In the case of CuPc, the binding energy of N and Cu atoms in the center of the CuPc molecule is lower than the C atoms in the outer benzene rings. Thus the XPS results suggest that for a heavily doped CuPc, the excess electrons are mainly localized in the center part of the molecule, which involves the pyrrole and Cu, resulting in a reversed shift for C 1s(B), N 1s and Cu $2p_{3/2}$. Shown in Figure 50 is the DFT calculation of the frontier orbital for pristine and Na-doped CuPc molecule. [167] The calculation suggests that the electronic structure of pristine CuPc stems mostly from that of phthalocyanine. The contribution from the center Cu atom is very limited. The HOMO is non-degenerate and distributed over the whole Pc ring. On the other hand, the LUMO is doubly-degenerate, which is concentrated more at the center part of the Pc ring. When a single Na atom is placed close enough to the CuPc molecule, charge transfer process is allowed during the calculation. As a result, the electronic structure of the $Na^{+}CuPc^{-}$ complex shows significant difference from the pristine molecule. It can be seen that the shape of the original HOMO is not much affected; however, a new orbital is formed between the HOMO and the Fermi level. If one combines the experimental observation presented in section 3.2, this orbital can then be easily assigned as the gap state appearing at high doping ratios. The DFT calculation also suggests that the gap state orbital is mainly distributed at the center part of the CuPc molecule, which is consistent with observations of the reverse shift in some of the core levels for alkali metal doped CuPc. [15, 16, 140, 142] Another interesting result from the calculation is that the double degeneracy of the LUMO in the pristine CuPc is diminished in the doped system. Now, the new LUMO in Na-doped CuPc consists of only one orbital, whose distribution is different from that of the pristine molecule. The calculated evolution of the unoccupied orbital clearly suggests that the newly formed occupied orbital is originated from one of the original LUMO in pristine CuPc. In this context, the DFT calculation agrees very well with the photoemission studies about the second stage in alkali metal doped CuPc.

a)

b)

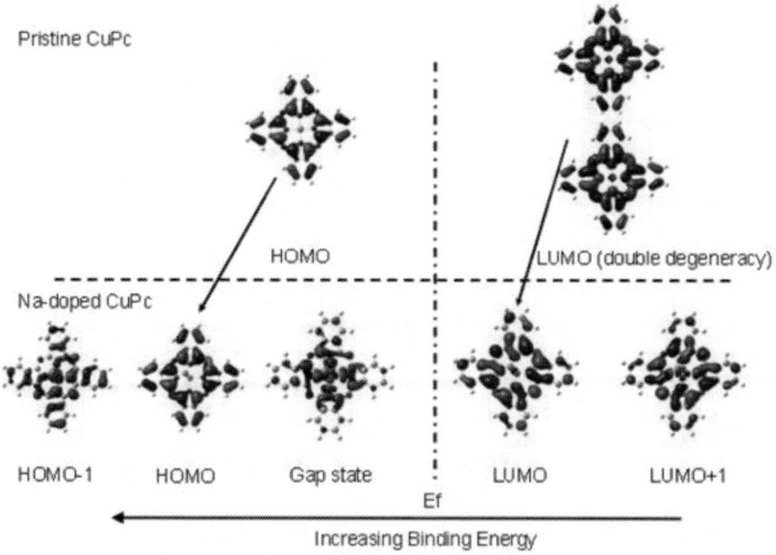

Figure 50. DFT calculation of the frontier orbital in (a) pristine CuPc and (b) Na-doped CuPc. The calculation suggests that the gap state induced by doped electron is located mainly in the center part of the molecule.

Besides the distribution of the molecular orbital, we can also estimate the energy of the frontier orbital and the gap state from the DFT calculation. In Figure 51, the DOS measured experimentally with photoemission and inverse photoemission is plotted in comparison with the calibrated value from the DFT calculations. [167] The calibration was carried out for pristine CuPc molecule in such a way that the energy separation between the calculated HOMO and LUMO is normalized relative to the measured peak-to-peak separation in the

UPS and IPES data. Using the same normalization factor for the doped molecule, the DFT calculation suggests that the gap state is 1.1 eV above the HOMO, which is excellent agreement with the experimental observations. One should also note that the energy separation between HOMO and LUMO in the doped system remains the same as that of the pristine molecule, which is consistent with the observation of rigid shift in the photoemission experiment. [15] Thus, the experimental investigations, as well as the theoretical calculations, clearly suggest that in the second stage of doping, the electronic structure of the organic molecule can be greatly modified through localized charge transfer. The doped electrons will fill in the unoccupied states of the organic molecules, leading to the formation of the gap state. Confined by such molecular orbital, those electrons do not act as free charges, as the case in the first stage. On the other hand, the distribution of their wavefunction is limited within certain region of the organic molecule, where strong correlation effect can be observed from the reverse shift of the core levels. Although charge sharing maybe still possible, a much larger barrier has to be overcome, comparing to the case in the first stage.

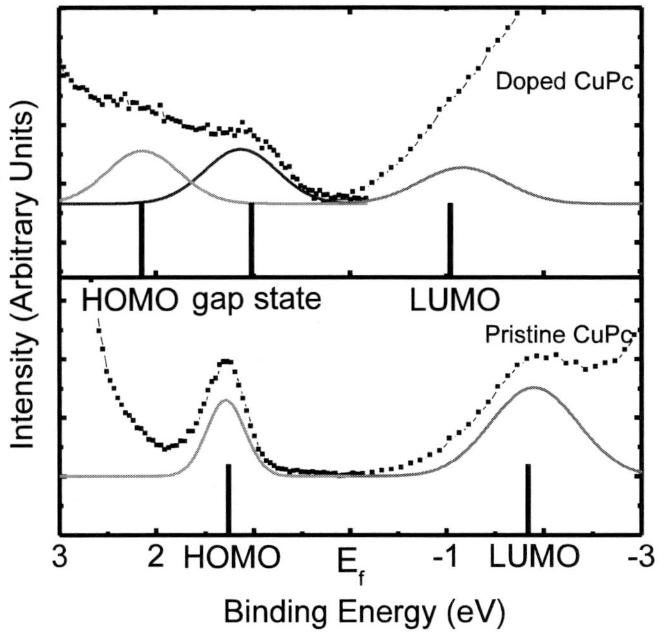

Figure 51. Energy comparison between the calculation results and the photoemission measurement for the frontier orbital in Na-doped CuPc. The calculated molecular orbital is shown as short bars at the bottom of the plot. The raw data from the UPS and IPES measurement and the Gaussian fittings are shown on the top.

3.6. Substrate Dependence of the Electronic Structure Evolution in Alkali Metal-Doped Organic Semiconductors

The two-stage model for alkali metal doping in organic semiconductor suggests that the energy level shift induced by doping cannot be explained by the equations derived for inorganic semiconductors. [13, 15] However, since most of the experiments discussed in previous sections were formed on organic films prepared on clean Au surface, there is also doubt if all the energy level shift is simply from the change of the substrate workfunction by the alkali metals diffusing through the organic film. To answer this question, Ding and Gao performed alkali metal doping experiments on metal substrates, other than Au, which show different workfunctions. [168] Thus the substrate dependence of the energy level shift can be studied.

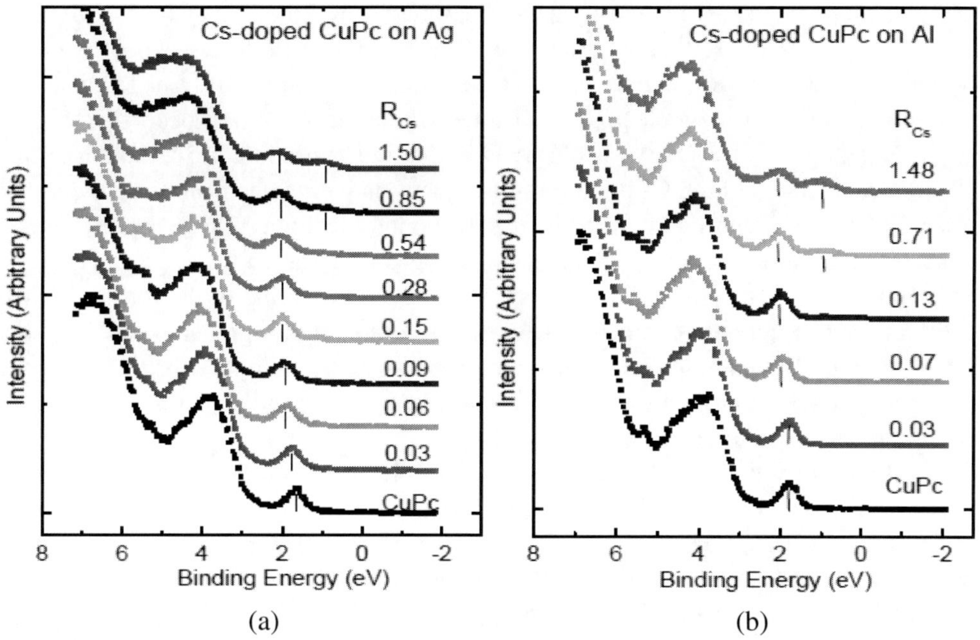

Figure 52. UPS spectra of the HOMO region of Cs-doped CuPc film prepared on (a) Ag and (b) Al substrates as a function of Cs doping ratio RCs, derived from XPS core level intensities. The spectra are plotted in binding energy relative to the Fermi level.

Figure 52 shows the evolution of the HOMO region of Cs-doped CuPc prepared on clean (a) Ag and (b) Al substrates as a function of R_{Cs}, the number ratio between the Cs atoms and the CuPc molecules. [168] The metal substrates were evaporated in UHV, right before the organic deposition. The formation of the metallic surface can be confirmed by the presence of a clear Fermi edge in the UPS spectra. The surface cleanness was also checked with XPS, where no discernable carbon or oxygen peaks can be observed. The workfunction of the metal substrate is measured with UPS (not shown), and is determined to be 4.32 eV and 4.09 eV for Ag and Al, respectively, which are consistent with the handbook values. [133]

The spectra of the pristine CuPc films on both metal substrates agree very well with the previous observations. [16, 27, 140, 142] For the film prepared on Ag substrate, the HOMO onset, which is usually determined by the linear extrapolation of the leading edge of the HOMO peak, is located about 1.2 eV below the Fermi level. Due to the lower workfunction of the Al substrate, the HOMO onset of the CuPc film prepared on Al shifts further downwards below the Fermi level, being about 1.4 eV. Upon Cs doping, energy levels of the CuPc films prepared on both Ag and Al substrates start to shift towards higher binding energy, in other words, the Fermi levels of the two samples shift upwards inside the energy gap of the CuPc. This is in good agreement with the earlier discussion of Cs-doped CuPc film prepared on Au substrate. [14, 15] As the Cs doping ratio increases, despite the broadening, the valence features of the CuPc molecules are well preserved, indicating that there is no destructive chemical interaction between the organic molecules and the doped alkali metal atoms. For both samples, the shift of the energy levels becomes saturated at higher doping ratios. However, the saturation occurs at different doping levels. For the sample prepared on Ag, the shift of the HOMO becomes saturated after R_{Cs} reaches 0.25, when the total shift of the HOMO is about 0.4 eV. On the other hand, the saturation of the energy level shift happens much sooner when Al is used as the substrate. The HOMO shifts for a total of 0.2 eV and then becomes saturated at R_{Cs} equals 0.13. Those values are significantly different from the previous doping experiment on clean Au substrate, where the saturation occurs at R_{Cs} equals 0.4, and a much larger shift (0.8 eV) is observed before the saturation. [15] Another issue one should note here is that, at high Cs doping ratios, a new state emerges from the former energy gap in the UPS spectra, and its intensity increases monotonically as R_{Cs} increases. Similar to the earlier observation for samples prepared on Au substrate, the gap state lies about 1.1 eV above the HOMO peak. It is believed to be originated from the filling of the former LUMO in the CuPc molecules due to charge transfer. [15] For samples prepared on both metal substrates, the gap state can only be clearly identified after the saturation of the energy level shift and when the Cs doping ratio is large enough (0.85 for Ag and 0.71 for Al). Before that the features are too weak to be seen.

The evolution of the C 1s, N1s and Cu $2p_{3/2}$ core levels of Cs-doped CuPc prepared on Ag substrate are shown in Figure 53. [168] The C 1s core level spectrum of pristine CuPc (see bottom spectrum in Figure 53 (a)) consists of two main components: the leading peak C 1s (A), which can be easily associated with the aromatic carbon of the benzene rings, and the second peak C 1s (B) at 1.4 eV higher binding energy, which is due to the pyrrole carbon in contact with N. [140] A satellite of C 1s (A) overlaps with C 1s (B), whose contribution must be accounted for to get the proper intensity ratio of these two peaks. Both components shift simultaneously towards higher binding energy for about 0.4 eV at initial doping ratios, which is quite consistent with what has been observed for the valence features. After that, they merge into one broad and asymmetry peak, as a result of the slightly reverse shift of C 1s (B) towards lower binding energy. Both N 1s and Cu $2p_{3/2}$ core levels appear to have a single component in the XPS spectra. They shift towards higher binding energy for similar amount as that of C 1s core level for the initial doping ratios. An obvious reverse shift can be identified for these two elements as the doping ratio is beyond 0.85, which coincides with the appearance of the gap state in the UPS spectra. This may be attributed to the correlation effect induced by the doped electrons, when they fill in the unoccupied sates of CuPc and become more localized at one molecular site. No new component can be observed in the core level spectra, which further corroborate the absence of any strong chemical interactions, except for a doping induced broadening.

The energy level shifts in Cs-doped CuPc film prepared on Ag and Al substrates are summarized in Figure 54 (a) and (b), respectively. [168] The vacuum level cutoff was obtained at the maximum of the differential curve of the UPS spectra. The core level shifts were deduced by fitting the spectra with Gaussian curves for C 1s, N 1s, and Cu $2p_{3/2}$ after careful deduction of the background. Note that R_{Cs} is in logarithmic scale to emphasize the initial shift. It has been reported that complete ionization of the doped Cs atoms can be easily achieved even at high doping ratios. [15] One can thus safely treat R_{Cs} simply as the electron doping density, as each Cs atom donates single electron to the organic molecules. The overall evolution of the electronic structure turns out to be similar to the case of the Cs-doped CuPc samples on Au substrate. [17]

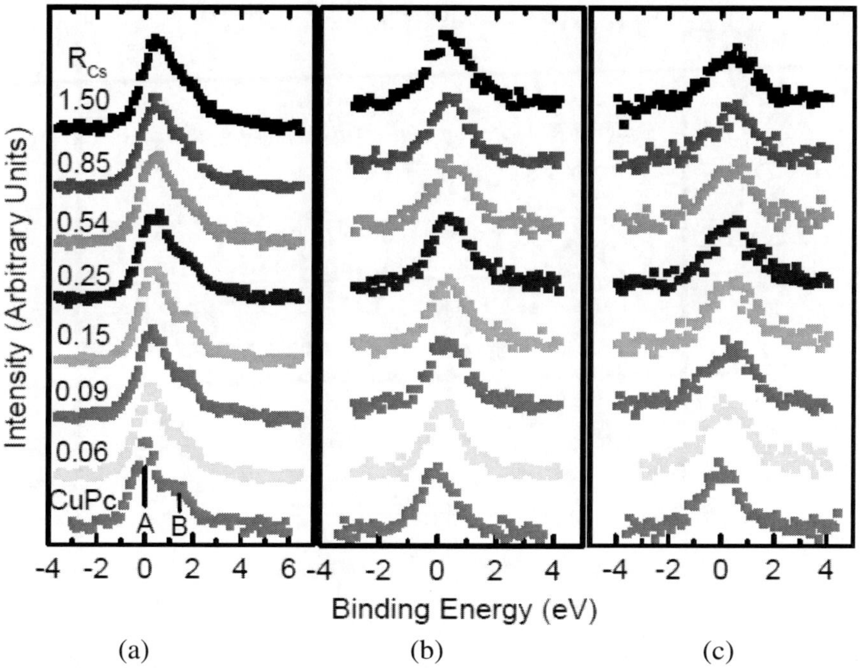

Figure 53. XPS spectra of C 1s (a), N 1s (b) and Cu 2p3/2 (c) core levels as a function of the Cs doping ratio (R_{Cs}) for CuPc thin film prepared on clean Ag substrate.

The energy level shift induced by Cs doping can be seen in two stages. At the first stage where the Cs doping ratio is low, all energy levels shift simultaneously and linearly towards higher binding energy in the semi-log plot. As the doping density increases, the shift of the energy levels becomes saturated, and some core levels, such as N 1s and Cu $2p_{3/2}$, start to shift back. However, the saturation turns out to be substrate-dependent, as it occurs at different doping levels with different amount of shift for samples prepared on Ag, Al and Au substrates. To understand the discrepancy here, one needs to first learn about the energy level alignment at the interfaces between the pristine CuPc film and the metal substrates with different workfunctions. The location of the Fermi level inside the band gap of CuPc can be estimated by the HOMO energy shown in Figure 52 and the gap energy of the pristine CuPc molecules, which is about 1.6 eV. [12] Thus, the energy separation between the Fermi level and the LUMO is derived to be 0.4 eV, 0.2 eV and 0.8 eV for Ag, Al and Au substrates, respectively, which agrees very well with the total energy shift before saturation for doped-samples prepared on each of the three substrates. One can then conclude that the amount of doping-induced energy level shift depends on the position of the Fermi level, and the saturation occurs as the Fermi level is aligned with the onset of the LUMO of the organic molecules. More importantly, despite of the workfunction difference of the metal substrates, the slopes derived from the linear shift region in the semi-log plot are in perfect agreement, being about 0.13 eV. [15] This is a clear indication that alkali metal doping induced energy level shift is determined primarily by metal-organic interaction, which is independent from the choice of the metal substrates.

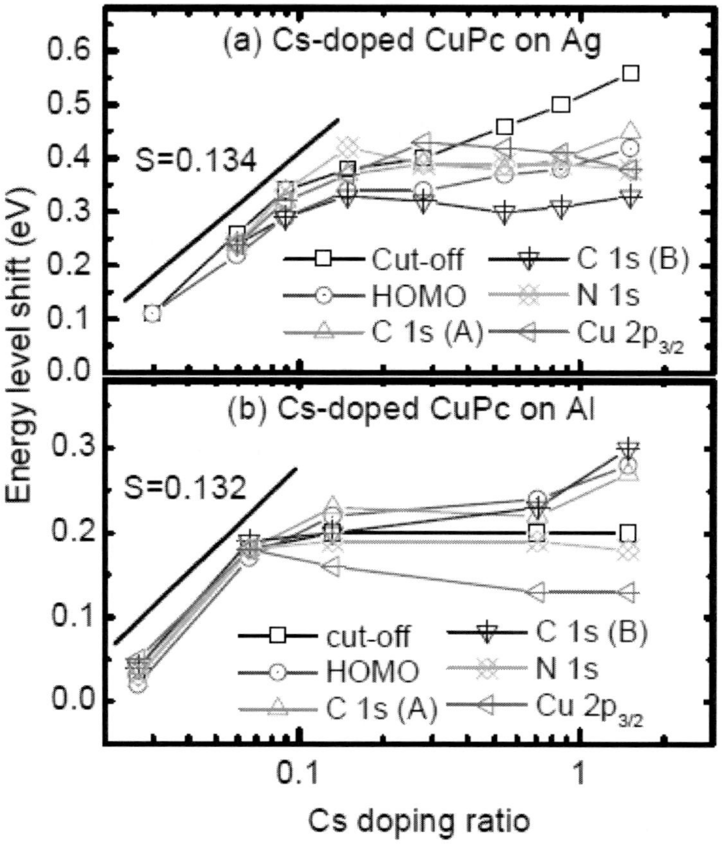

Figure 54. The summary of the doping-induced energy level shift for CuPc samples prepared on (a) Ag and (b) Al substrates. Note that Rc_s is in logarithmic scale to emphasize the initial shift. For both samples, the doping-induced energy level shift can be seen in two stages, and the slopes of the initial stage are almost identical.

The independence of the slope of the energy level shift in the semi-log plot on the substrate workfunction indicates that the shift is the result of dopant-organic interaction instead of the substrate workfunction change induced by the alkali metal. Further assessment of the statement can be obtained from the Debye length considerations. As a measure of the screening of an electric field by a semiconductor, the intrinsic Debye length is inversely proportional to the doping density. [102] The Debye length for 1% doping, assuming 100% ionization, is about 7 Å in the organics, supporting the notion that the substrate contribution to the measured energy level shift is insignificant given the probing depth (~20 Å) and the film thickness (200 Å).

As doping density increases, the energy level shift saturates because of the pinning of the Fermi level at the LUMO. [12] Further doping leads to the second stage of charge transfer, when the doped electron concentrates at the center of CuPc and a strong correlation of the electronic states takes place. This leads to the reversed shift of core levels of atoms at the center, including those of Cu and N, as well as the growth of the gap state. When low workfunction metal is used as the substrate, the energy level shift becomes saturated at much smaller doping ratio, comparing to the case of Au. In the UPS measurements, the gap state may not be clearly identified at this time, due to the weak signal intensity. However, as the LUMO now is aligned with the Fermi level, the doped electrons can easily fill in the unoccupied sates in organic molecules and stay localized at a single molecular site, leading to a strong correlation that can effectively modify the electronic structure of the doped molecule. The charge transfer at the second stage is therefore characterized by the formation of gap and strong correlation of the organic energy levels.

4. Reversal of the Doping-Induced Energy Level Shift in Organic Semiconductors

The investigations about the alkali metal doping in organic semiconductors indicate that the electronic structures of the organic semiconductors can be modified, manifested by the formation of the gap state. The DFT calculation also suggests that the new state can be formed due to localized charge transfer. However, fundamental issues still remain, such as whether such modification can be reversed. It is also important to investigate the nature of the interactions between the alkali metal cations and the organic molecules at high doping ratios to find out whether it is simply due to charge transfer, or more profound chemical reactions are responsible. Metal deposition and oxygen exposure induced reversal of the doping process in Alq and CuPc have been systematically studied by Ding and Gao with PES measurement. [139, 169]

4.1. Reversal of the Doping-Induced Energy Level Shift in Alq

In many applications, Au is commonly used as the electrode on top of doped organic semiconductors. Owing to its high work function, Au can be regarded as a p-type dopant in the organic matrix. It is thus interesting to see how the electronic structure can be modified by Au deposition in a heavily doped organic film. Figure 55 shows the XPS spectra of C 1s, Cs $3d_{5/2}$, and Au $4f_{7/2}$ core level in Cs-doped Alq film as a function of Au coverage. [169] As more Au is evaporated onto the film, both the C 1s and Cs $3d_{5/2}$ peaks become slightly broadened. The widths of the Gaussian function fitting of these peaks increase about 0.7 eV and 0.3 eV, respectively. The broadening is most likely due to increasing inhomogenuity by Au deposition. Both peaks shift gradually toward lower binding energy. The amounts of the total shift, however, are quite different, being 0.4 eV and 1.0 eV for C 1s and Cs $3d_{5/2}$, respectively. On the other hand, Au $4f_{7/2}$ undergoes little change compare to that of C 1s and Cs $3d_{5/2}$, either in the peak width or position. The lack of new component in C 1s, Cs $3d_{5/2}$, as well as Au $4f_{7/2}$, demonstrates that no chemical reaction has taken place.

Shown in Figure 56 are the low-energy secondary cutoff and the HOMO region of the UPS spectra for Cs-doped Alq as a function of Au coverage. [169] The Fermi level derived from the Au substrate is drawn as the dashed line. It can be seen that after Cs doping (R_{Cs} = 1.30), the cutoff shifts toward higher binding energy by about 1.0 eV, and a significant gap state 1.6 eV above the HOMO is introduced. Both of these are consistent with the previous studies about Cs and other alkali metal doping in Alq. [13, 18] It is very interesting that the initial deposition of 1 Å Au completely quenches the gap state while not inducing much shift in the cutoff and the HOMO. As the Au coverage increases, the cutoff shifts backward to the low binding energy side until it saturates after 30 Å of Au deposition. The total reversal is about 0.7 eV, corresponding to a work function of 3.4 eV that is substantially lower than that of pure Au (5.4 eV).Similar reversal of the Cs-induced shift can also be observed in the HOMO region. The Alq peaks become broadened and shift continuously toward the Fermi level as the Au coverage increases. However, an important difference from the cutoff shift is that the HOMO shifts faster and its saturation position actually turns out to be that of the pristine Alq film. Another interesting observation is that after the gap state is annihilated by 1 Å of Au deposition, the shape of the Alq HOMO peak does not change much by further Au depositions. The persistence of Alq HOMO feature even up to 60 Å of Au coverage further indicates that no significant chemical reaction has taken place, consistent with the results of the XPS spectra. Finally, it should be noticed that the position of the Fermi level, which can be seen beyond 15 Å of Au coverage, is the same as that of the Au substrate. Therefore, the charging effect during the measurements was insignificant at this stage.

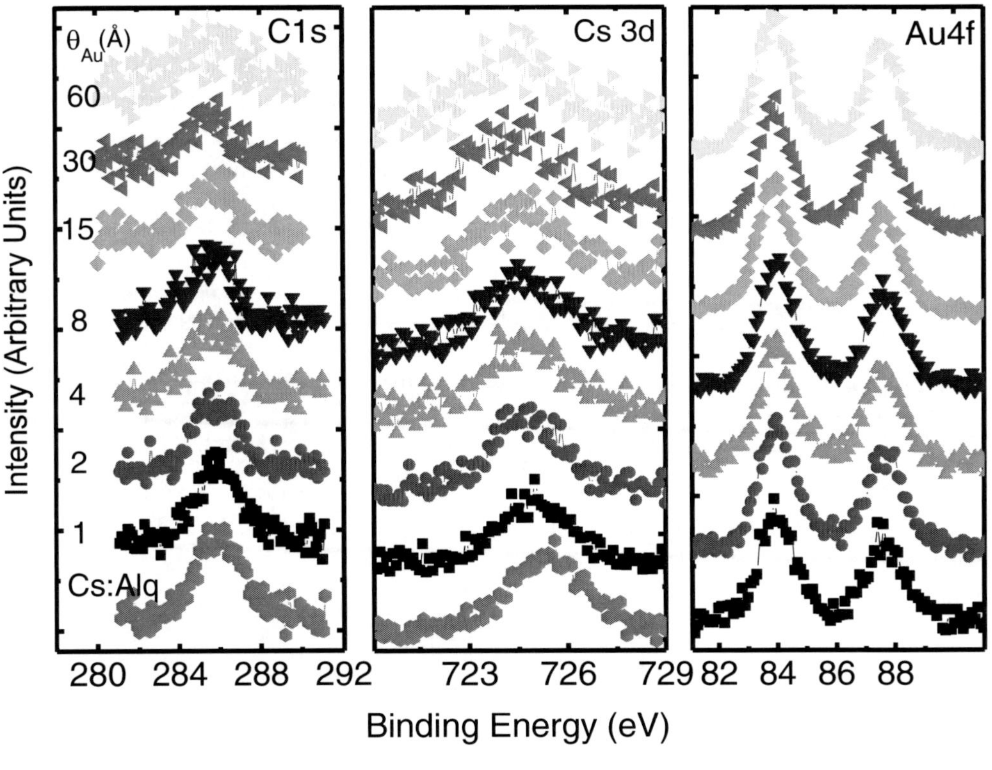

Figure 55. XPS spectra of C 1s, Cs $3d_{5/2}$, and Au $4f_{7/2}$ core level as a function of the Au thickness deposited on to the Cs-doped Alq film.

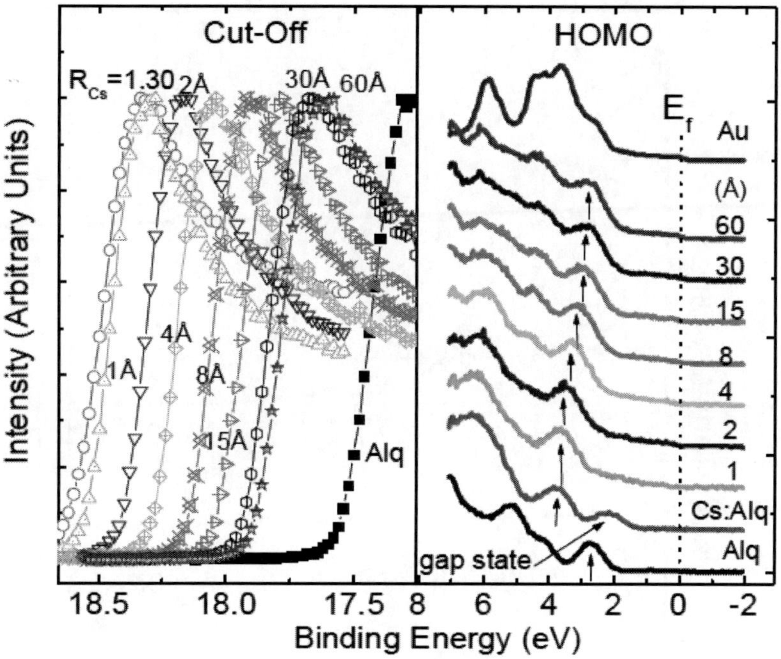

(a) (b)

Figure 56. The UPS spectra evolution as a function of Au thickness deposited on to the Cs-doped Alq film: (a) cut-off and (b) the HOMO region. The Fermi level derived from the Au substrate is drawn as a dashed line. It can be seen that the gap state caused by Cs doping disappeared as 1 Å Au is evaporated.

In order to study the effect of Au atoms on the doping-induced gap state, another set of experiment was carried out with intensive Cs doping and smaller Au deposition steps. The evolution of the occupied states is presented in Figure 57. [167] In this experiment, the Cs doping ratio, R_{Cs}, derived from XPS core level intensities is about 6.72. An obvious gap state formed above the HOMO. Fitted with Gaussian peak after careful deduction of the parabolic background, the gap state and the HOMO reveal an intensity ratio of about 1.05. If we can assume that the photon cross-sections for the two orbital are similar, which is quite reasonable according to the earlier discussion, the above mentioned intensity ratio suggests that 6.3 electrons has been transferred towards a single Alq molecule. That is to say that the electron doping density is about 6.3, which is comparable to R_{Cs} obtained from XPS measurement. Thus, it can be concluded that even under such an intensive doping level, most of the Cs atoms (94%) are completely ionized, donating their single outmost electron to Alq molecules. When Au is deposited onto the doped film, the intensity of the gap state starts to decrease. The position of the HOMO and the gap state does not change much within 0.5 Å of Au deposition. In the following steps, the HOMO starts to shift towards lower binding energy, which consistent with the observation in Figure 56. However, the gap state persists until 4 Å of Au deposition, above which the valence features from Au dominate the UPS spectra and that from Alq is greatly attenuated. The detailed analysis of the gap state intensity evolution will be discussed later in this section with the result from Cs-doped CuPc.

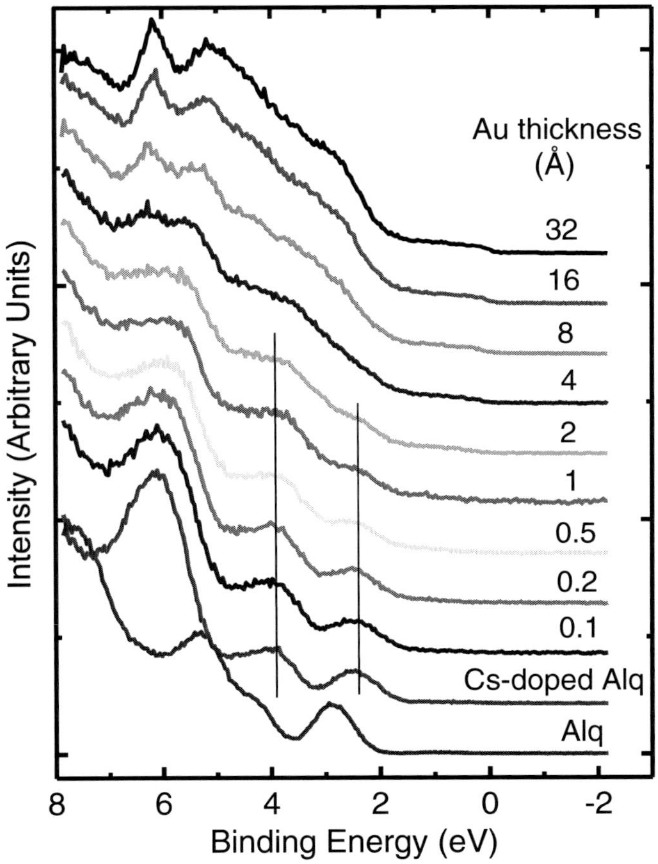

Figure 57. The evolution of the HOMO region as a function of Au thickness deposited on to the Cs-doped Alq film. It can be seen that the intensity of the gap state keep decreasing within the first few Å of Au deposition.

The simplest explanation of the reversal of the energy level shift is the compensation of the n-doping Cs by the p-doping Au as frequently observed in crystalline inorganic semiconductors. However, a more detailed analysis shows that the situation in an organic semiconductor can be more complicated. The energy level shifts are summarized in Figure 58. The core level shifts were deduced by fitting the spectra with Gaussian curves for C 1s, Cs $3d_{5/2}$, and Au $4f_{7/2}$. The vacuum level cutoff was obtained at the maximum of the differential curve of the UPS spectrum. The cutoff and HOMO positions of the pristine Alq film are also presented as the very first points to compare with the doped cases. It is clear that 86% Cs doping causes a Fermi level shift as large as 1.0 eV, which agrees with our previous data of Cs-doped Alq. [13] The deposition of Au induces a significant reverse shift of both the frontier orbital and the core levels. It is noticeable that even at very low Au coverage (1~2 Å), Cs $3d_{5/2}$ peak shifts toward the lower binding energy by amount of 0.6 eV, while C 1s peak shifts in the same direction for less than 0.1 eV at the same time. After that, both of them shift toward the lower binding energy for less than 0.4 eV. The abrupt shift of Cs $3d_{5/2}$ may stem from the removal of the donated electron from the N of Alq^-, as quantum chemistry calculations of alkali metal doped Alq have shown that upon doping, the excess electron is localized on the pyridyl side of one of the three quinolate rings. [18] The localization of the electron leads to the formation of the gap state and the split of the N 1s peak. This explanation is supported by the quenching of the gap state in the HOMO region, which is observed to be affiliated to the formation of Alq^- and the N 1s new component. Another important observation is that while the HOMO shift totally back to the position of the undoped Alq, the C 1s peak does not. It saturates after shifting about 0.4 eV, which is only halfway toward that of the pristine Alq film. Apparently, the modification of the Alq electron energy level by Cs doping cannot be fully compensated by simply removing the electrons from the host.

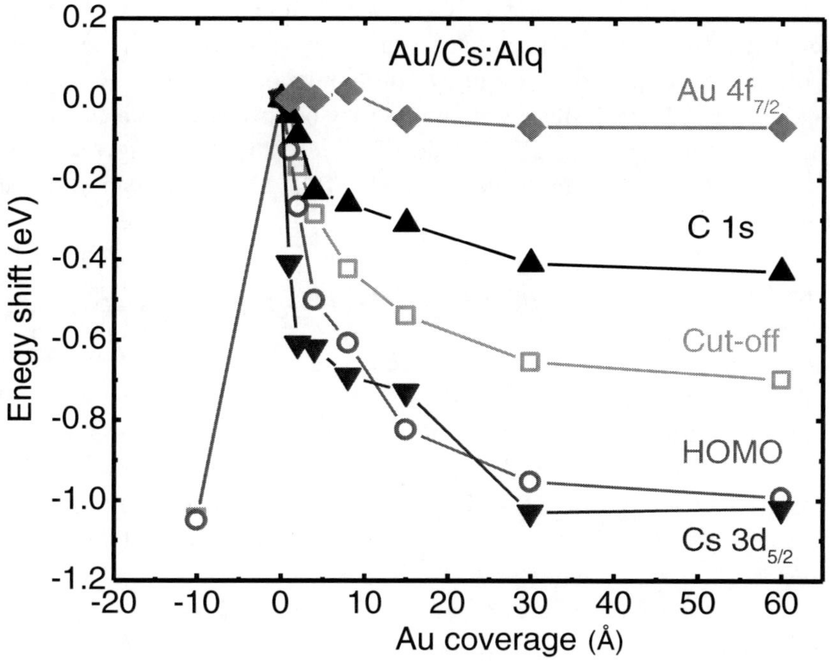

Figure 58. Summary of the energy level shifts as a function of Au coverage for Cs-doped Alq. The valence features as well as the C 1s of the undoped Alq film are shown as the first points of the curves.

The reverse shift of the Alq valence feature clearly shows that, opposite to the Cs doping case, the deposition of Au can be understood as the dispersion of the electron accepter into the organic matrix. Lee et al.[170] have reported that on Au/Alq interface, one can expect electron donation from the organic to metal layer and thus the formation of an interface dipole. The XPS data also show that there is no chemical reaction between Au and Alq, which is consistent with the other studies. [171] Thus, we can conclude that the electronic structure modification caused by doping results primarily from charge transfer, which can be reversed by charge compensation from the opposite dopant. We can also conclude that doping and un-doping are not related to the formation of new chemical bonds or chemical reactions. This result clearly shows the intriguing electronic correlation effects and the possibility of accurate control of electronic structure of organic semiconductors in the interface region.

The change of the C 1s binding energy with respect to the HOMO as Au deposition increases may reflect the change in the Cs^+Alq^- complex. As shown in previous studies, Cs^+ is close to the N atom where most of the doped electrons are located. [18] The Coulomb interaction may be still at play when electron transfer out of the complex takes place as Au is deposited, resulting in the reduction of valence electrons from the previous HOMO level that is more evenly distributed among the C 1s atoms. Such a reduction will cause the C 1s binding energy to increase as the screening by the valence electrons around the C nuclei is reduced. One may also consider the increased screening of holes left by photoelectrons as the interface region becomes more metallic as Au is deposited. However, such increase of screening will reduce the apparent binding energy of the measured photoelectrons, which is opposite to what we observed for C 1s.

It is understood that the Au-on-Alq interface is not abrupt and homogeneous. The interface is formed by deposition of the heavy metal atoms on the surface of a floppy molecular film. The diffusion of the metal atoms and a fixed interface can be expected. A. C. Durr et al. [172] have discussed that Au growth on diindenoperylene films does not proceed in a layer-by-layer mode. Instead small Au clusters may be formed, which can diffuse several nanometers into the organic film. The valence features shown in Figure 56 give direct evidence to the presence of cluster formation and diffusion, as one can still see the Alq features even after 60 Å Au deposition. It has also been reported that the Au cluster formed during deposition onto the organic film may cause a positive charging, especially for low Au coverages, since electron transfer from the neighbor organic molecules is difficult. However, in our experiments, charging induced shift, if exists, is very small. The Au $4f_{7/2}$ core level shift in Figure 58 shows that for the lower Au coverages, where the cluster-formed Au is predominate, the core level position is only slightly larger than the metallic Au by about 0.07 eV. This sets the upper limit of the charging in this experiment. For higher Au coverages, the Fermi level, which can be seen after 15 Å Au deposition, does not show any significant evidence of charging, as it aligned well with that of the Au substrate. The reason that there is no significant charging by the Au cluster is probably due to the fact that the conductivity of the highly doped Alq film is much higher than a pristine one.

The Au-induced reversal of the energy level shift in Cs-doped Alq indicates that, instead of forming new chemical bonds, the electronic structure modification induced by doping is primarily from charge transfer. The ionic interaction between the alkali metal cation and the organic molecule are weak and can be easily broken with the presence of another electron acceptor. To verify this result, another doping system has been investigated. [167, 173] This time the Alq film was doped with Na, which has relatively larger ionization potential (0.55eV) than that of Cs (0.47 eV). Figure 59 shows the evolution of both the cutoff and the HOMO spectra of Na doped Alq as a function of the Au coverage. [173] Similar to the case of Cs-doped Alq, the initial Au deposition does not induce significant shift in the vacuum level, while further Au deposition leads to a total shift of 0.7 eV. The final work function is determined to be 3.4 eV, which is the same as the Cs doping case. Another similarity is that the gap state is fully eliminated as 1 Å Au is deposited, while the other HOMO features persist up to 60 Å Au coverage.

In Figure 60 (a), the core levels of C 1s, N 1s, and O 1s in the Na doped Alq film are shown as a function of Au coverage. [173] It is clear that, as Au coverage increases, all three peaks shift toward lower binding energy. More importantly, for N 1s, the new component with lower binding energy, which is induced by electron transfer from Na to N, [18] is eliminated by the initial Au deposition (see Figure 60 (b)). [173] The quenching of both the gap state and the N 1s new component clear indicts that the excess electrons donated from Na have been removed by Au deposition.

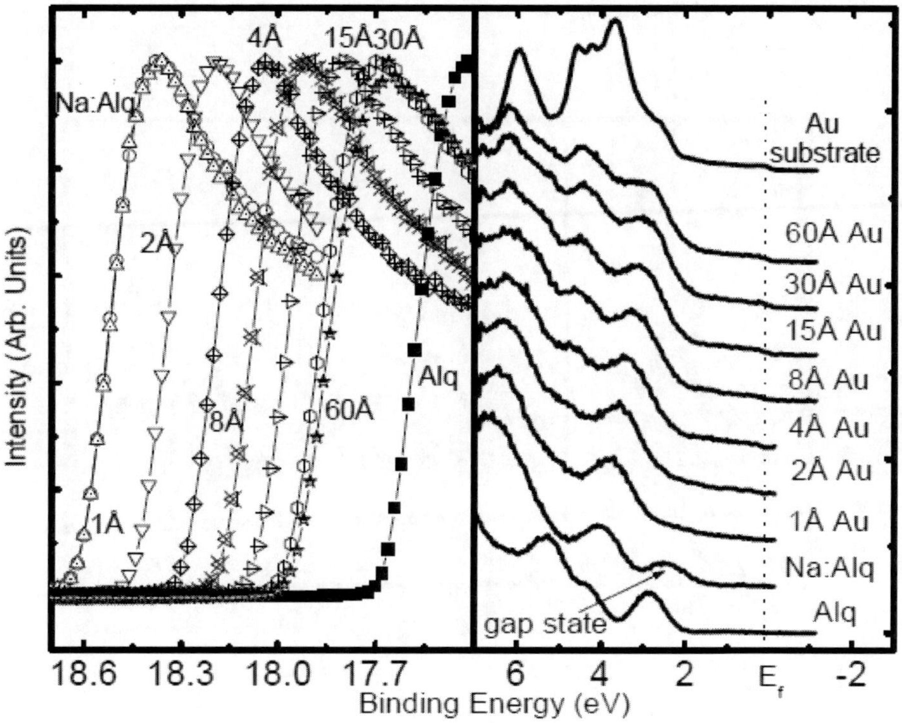

(a) (b)

Figure 59. The UPS spectra evolution as a function of Au thickness deposited on to the Na-doped Alq film: (a) cutoff and (b) the HOMO region.

The energy level shifts of Au on Na-doped Alq are summarized in Figure 61. [173] The energy level positions of pristine Alq are also drawn as the very first points. The most expressive thing is that although all energy levels show a reverse shift, the amounts of the shifts are different. The HOMO shifts back completely, while C 1s, O 1s and N 1s only shift halfway toward the position of the pristine film. The reversal of the energy level shifts presented here resembles that of the Cs-doped Alq after Au deposition. It is therefore confirms the earlier discussion that the electronic structure modification caused by doping results primarily from charge transfer, which can be reversed by charge compensation from the opposite dopant.

The reverse doping behavior of Au can be assumed from its high work function, it is thus interesting to see if other metals with lower work function lead to a similar evolution. Shown in Figure 62 is the evolution of (a) the cut-off and (b) the HOMO region for Cs-doped Alq as a function of Al coverage. [167] The work function of the pristine Alq film is reduced by about 1.13 eV after Cs doping. The Cs doping ratio R_{Cs} is derived to be 3.0 from the core level intensities. The Al deposition brings the work function back rapidly. The saturation of the reverse shift in cut-off occurs after about 16 Å of Al deposition, leading to a final work function of 4.25 eV, which agrees very well with the common value of 4.28 eV for metallic Al. Such observation is different from Au deposition, where the work function of Au was not reached even after 60 Å of Au coverage. It is thus suggested that the Al atom is less diffusive inside the Alq film. The surface can be completely covered by metallic Al within 15 Å of metal deposition. Such a conclusion is also supported by the XPS data, where the metallic feature of Al 2p core level becomes discernable at 8 Å, and dominates the whole spectra after 16 Å.

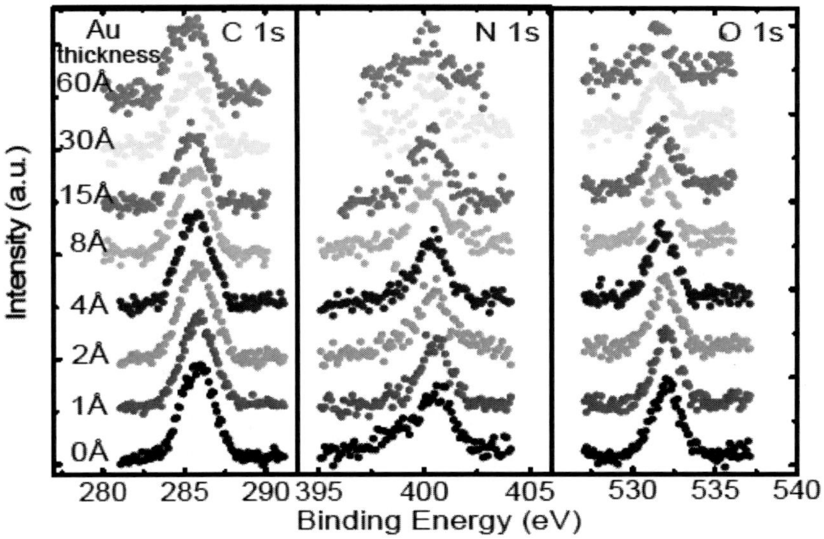

(a)

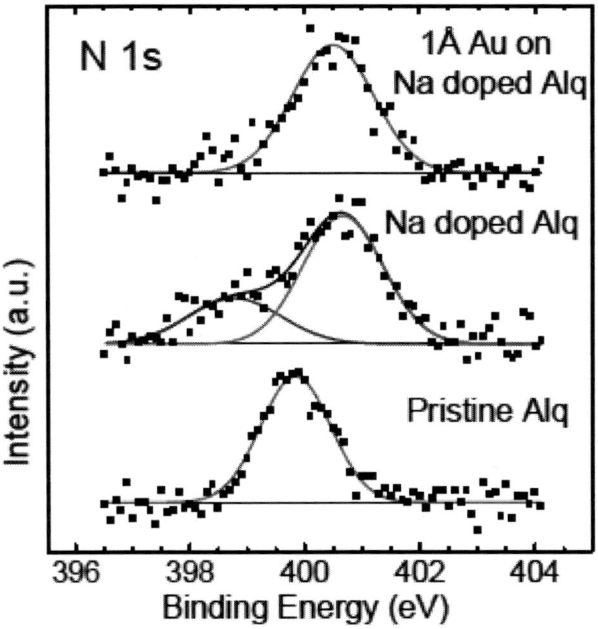

(b)

Figure 60. (a) XPS spectra of C 1s, N 1s, and O 1s core level of Na doped Alq film as a function of the Au thickness. (b) Fitting of N 1s of doped and undoped Alq.

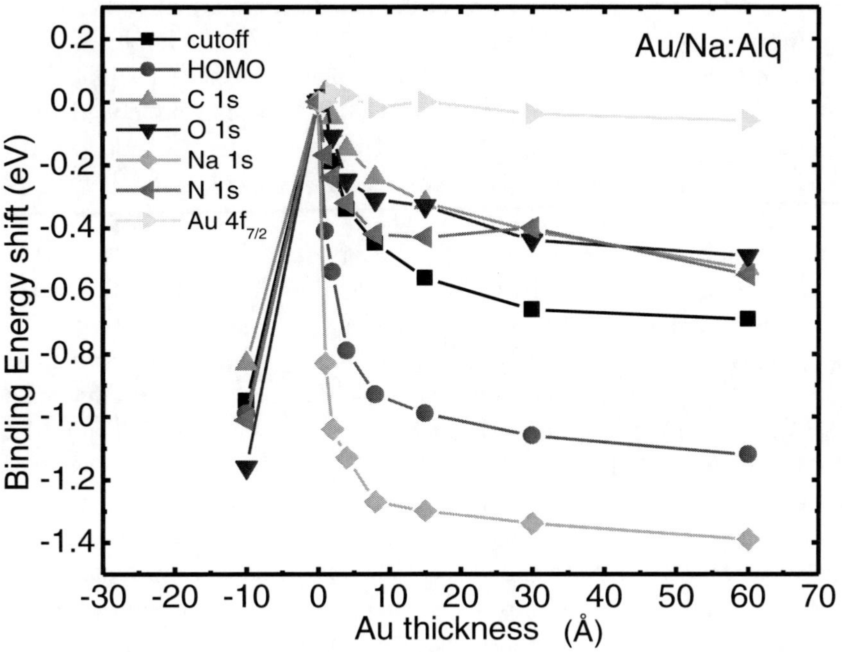

Figure 61. The summarization curves of energy level shifts for Au on Na doped Alq film. The energy level positions for pristine Alq are drawn as the first points.

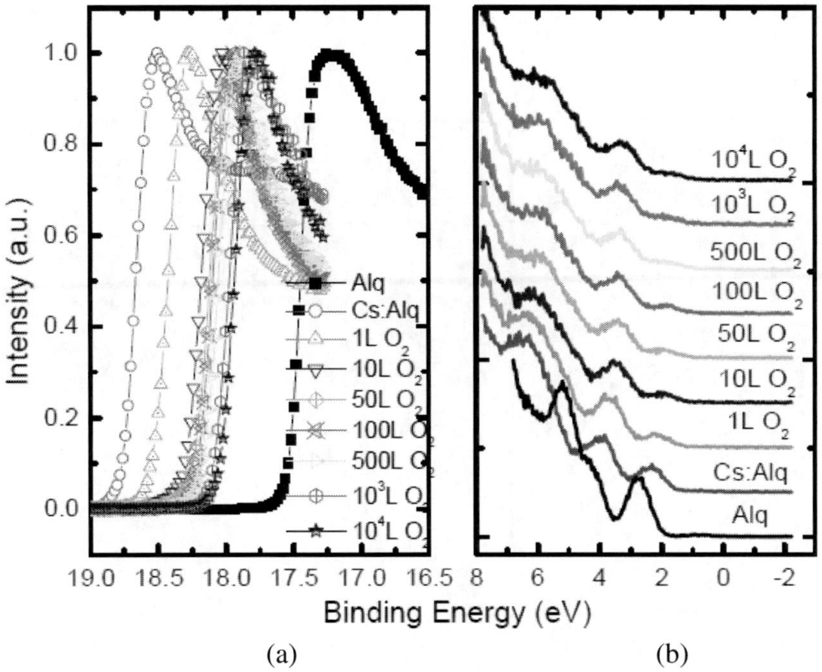

Figure 62. The UPS spectra evolution of both (a) cutoff and (b) HOMO region for Al deposition on Cs-doped Alq film.

Although the reverse shift is very clear in the cut-off region, the behavior in the HOMO region is significantly different, as shown in Figure 62 (b). [167] Cs doping induces a HOMO shift towards higher binding energy for 1.13 eV, which is consistent with the cut-off shift. The gap state has formed 1.6 eV above the HOMO. These results agree well with my earlier discussion in the previous section of this chapter. However, when Al is deposited, no reverse shift of the HOMO in Alq can be observed. Instead, the HOMO shifts continuously towards higher binding energy for about 0.1 eV within the first 1 Å of Al deposition. The gap state can still be seen until then, which shifts for about the same amount as the HOMO. A dramatic broadening can be noticed right after Al deposition, even for as little as 0.1 Å. Such broadening develops further with increasing Al coverage, and eventually eliminates all the Alq features after 8 Å of Al deposition. Previous studies have suggested that Al can react destructively with Alq. [18, 130] The current observation clearly suggests that the presence of Cs at the Al/Alq interface cannot effectively prevent such reactions. Finally, the Al Fermi level can be observed after 16 Å deposition, which further confirmed that the surface is completely metallic by then. The fact that the Al Fermi level aligns with that of the Au substrate indicates that there is not obvious charging effect during the measurement.

The XPS data of the C 1s, N 1s and O1s core levels are shown in Figure 63 for Cs-doped Alq as a function of Al deposition. [167] The shape of the C 1s spectra does not show too much of change, except the attenuation and the broadening induced by increasing Al coverage. However, a reverse shift (towards lower binding energy) of about 0.6 eV can be observed. The splitting of the N 1s is very clear for the Cs-doped Alq. The low binding energy component induced by Cs doping can still be clearly resolved until 1 Å of Al deposition, after which the spectra become too broad and scattered due to the poor counting rate at higher Al coverage.

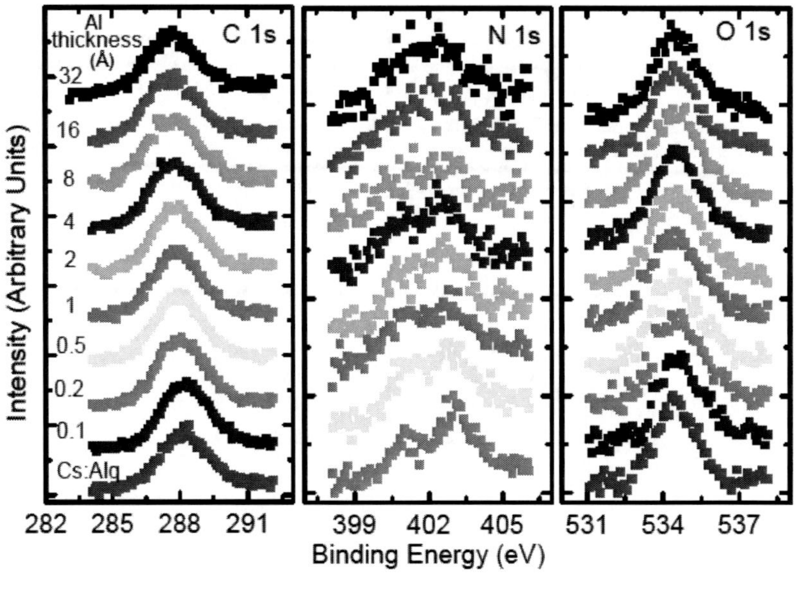

Figure 63. The XPS spectra of the C 1s, N 1s and O1s core levels for Cs-doped Alq as a function of Al deposition. The doping-induced splitting of N 1s core level is very clear, which persists even after Al deposition.

The earlier studies about Al/Alq interface (see detailed results in section 5.1) indicates that all core levels shift to higher binding energy with increasing Al deposition. [18] At the same time, O 1s developed a distinct asymmetry to the higher binding energy side from the initial Al deposition. [130] However, the current data suggests that with the presence of Cs, the position of the O 1s core level remains almost intact and the shoulder at the high binding energy side is not as obvious as the case in Al/Alq. Another interesting result is that the Cs doping ratio, R_{Cs}, has been greatly modified with Al deposition. It decreases from 3.0 to 2.0 as soon as 0.1 Å of Al has been deposited, and eventually saturated around 1.0 with thicker Al coverage. The reason for the modification of R_{Cs} by Al deposition is not clear at this time. According to the universal curve, the mean free path for Cs is slightly shorter than that of C, which will lead to a stronger attenuation for Cs signals as Al is deposited. However, such effect alone may not explain the dramatic change in the measured doping ratios. Another possible explanation is that the Cs has been driven away from the surface as more and more Al arrived. Thus the effective doping ratio within the photoemission probing depth is reduced.

Finally, since all organic electronics will be used under ambient condition in their real applications, it is important to learn the stability of such system after oxygen or air exposure. The effects induced by oxygen exposure on the electronic structure for Cs-doped Alq have been studied by Ding and Gao. [167] The UPS spectra evolution of both (a) cutoff and (b) HOMO region for Cs-doped Alq film are shown in Figure 64 as a function of oxygen exposure amount in unit of Langmuir (L), which corresponds to 1 second of exposure with pressure in 1×10^{-6} torr. The Cs doping ratio for this sample is calibrated to be 1.4. The reversal of the doping induced energy level shift can be noticed in both the cut-off and the HOMO region. The work function seems to become saturated after 10^3 L exposure. The reverse shift of the HOMO saturates even earlier (500 L exposure). The gap state induced by Cs doping can still be clearly observed even at the largest exposure. The shift of the gap state is identical to that of the HOMO, suggesting a rigid shift situation. A clear broadening in the HOMO features, as well as the core levels shown in Figure 65, [167] can be noted as the oxygen exposure amount increases. The core levels also show a trend of reverse shift towards lower binding energy. The splitting of N 1s is still noticeable at low exposures. However, it becomes less prominent as the exposure increases.

The summary of the energy level shift induced by oxygen exposure for Cs-doped Alq is presented in Figure 66. [167] It can be seen that oxygen exposure introduces an energy level shift to the reverse direction of that induced by Cs doping. The amount of the reverse shift is different for the core levels, as well as the HOMO, among which N 1s shows the largest reversal. The other thing one should note is that the oxygen induced reversal becomes saturated quickly within the first 10^3 L exposure. Further discussion about this issue will be provided in the next section.

In summary, a reversal of the doping-induced energy level shift occurs as Au is evaporated on the surface. The Cs induced gap state is quickly quenched by a very small Au deposition. As Au coverage increases, the HOMO shifts back to that of the pristine Alq film. However, the C 1s core level does not go that far, indicating that the interaction between Cs and Alq cannot be fully eliminated. Oxygen exposure of the Cs-doped Alq sample also shows a reverse shift, which is less pronounced than that of Au. On the other hand, when Al is deposited, the electronic structure of the organic film is further modified due to the interactions at the interface. As a result, the HOMO does not shift back. The doping and de-doping process suggests that rather than chemical reactions that form new bonds, the electronic structure modification caused by doping results primarily from charge transfer, which can be reversed by charge compensation from the opposite dopant.

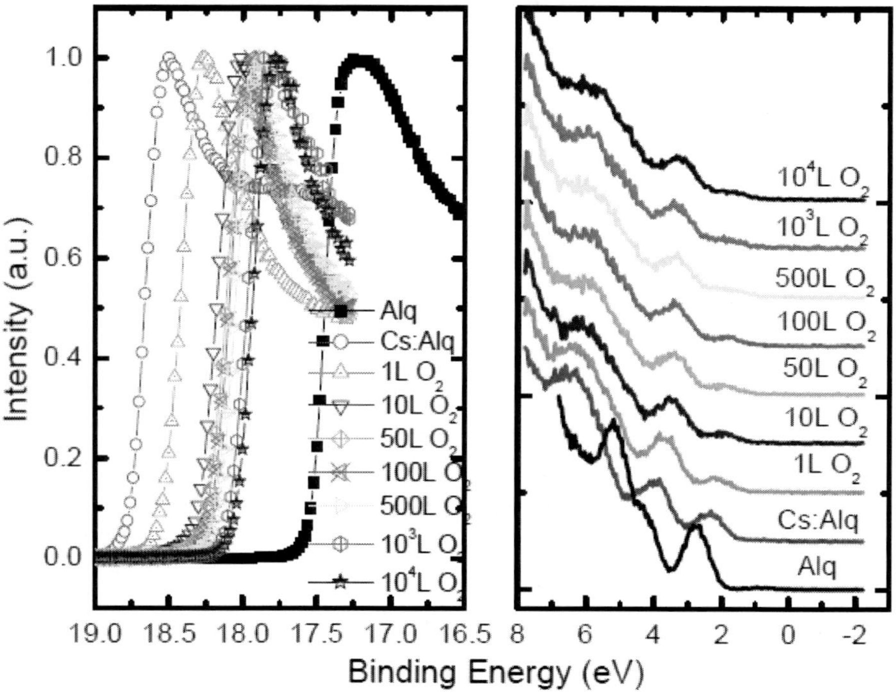

(a) (b)

Figure 64. The UPS spectra evolution of both (a) cutoff and (b) HOMO region for Cs-doped Alq film as a function of oxygen exposure.

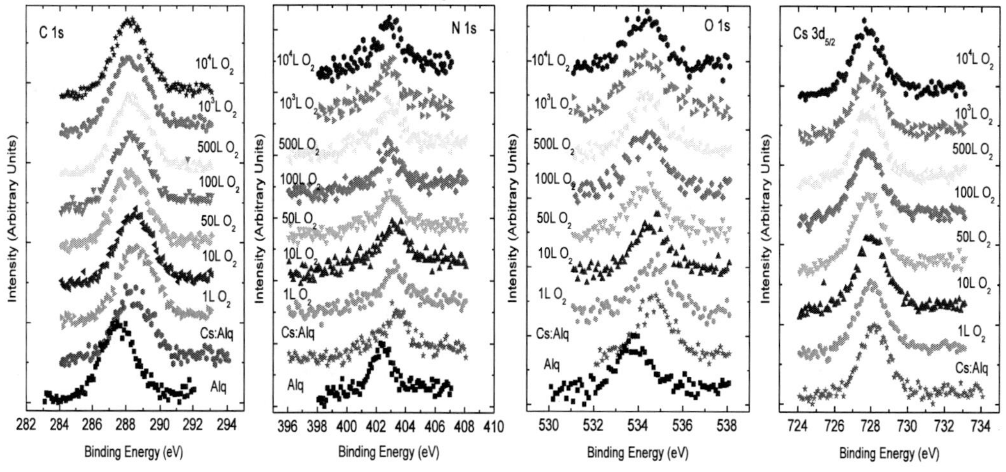

Figure 65. The XPS spectra of the C 1s, N 1s, O1s and Cs $3d_{5/2}$ core levels for Cs-doped Alq as a function of oxygen exposure.

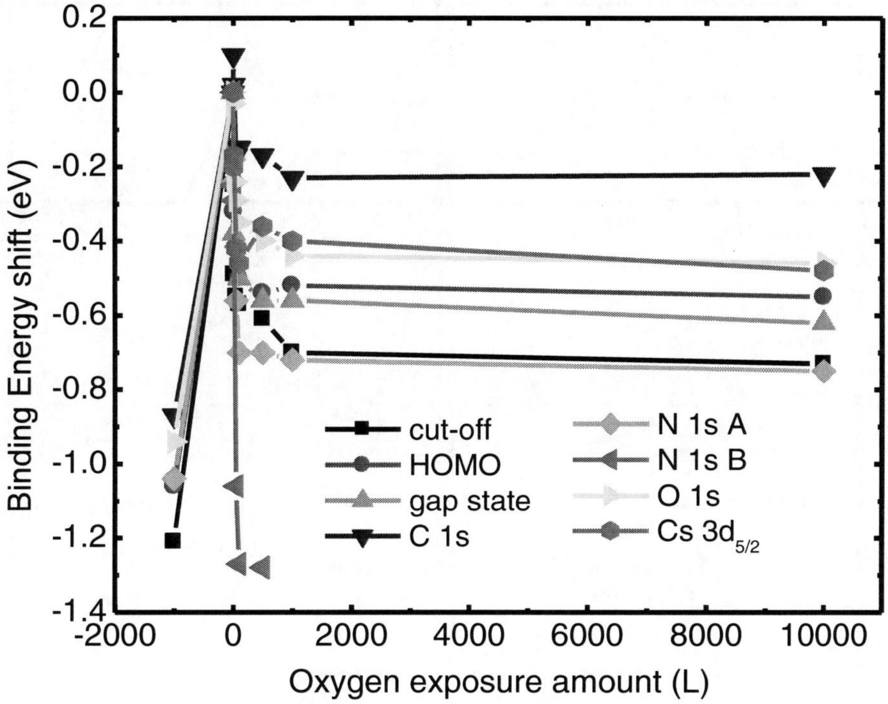

Figure 66. Summary of the energy level shift induced by oxygen exposure for Cs-doped Alq film.

4.2. Reversal of the Doping-Induced Energy Level Shift in CuPc

For further investigation, it remains to be seen if such reversal process can occur to other organic semiconductors. Shown in Figure 67 are the low-energy secondary cutoff and the HOMO region of the UPS spectra, measured with increasing oxygen exposure of Cs-doped CuPc. [139] One can easily notice that, upon Cs doping, the cutoff and the HOMO shift toward higher binding energy for 1.25 eV and 0.93 eV, respectively. At the same time, a gap state 1.1 eV above the HOMO is observed, which is consistent with our previous observations. No finite electronic density of state can be identified at the Fermi level, indicating that metallic Cs is not formed with significant quantity at the doping ratio R_{Cs} = 3.4. The subsequent oxygen exposures induce a gradual reversal shift of the cutoff toward the lower binding energy, leading to an increase of the work function. The shift almost saturates after 1000 L (1 Langmuir equals one second exposure at 1.0×10^{-6} torr) oxygen exposure, giving a final work function of 3.53 eV. The valence features show a similar trend, as a gradual shift toward the lower binding energy can be seen as the oxygen exposure increases. A more detailed look at the valence features shows that the amount of the shift for the HOMO, whose peak is around 2.0 eV in binding energy relative to the Fermi level of the Au substrate, is substantially different from the next strong feature, HOMO-1, which has a binding energy of about 4.5 eV. Continuous broadening of the HOMO features is observed upon oxygen exposure. The gap state is not removed by the initial oxygen exposures.Instead, it seems to persist until the broadening merges the HOMO and gap state into one feature at high oxygen exposures (> 500 L).

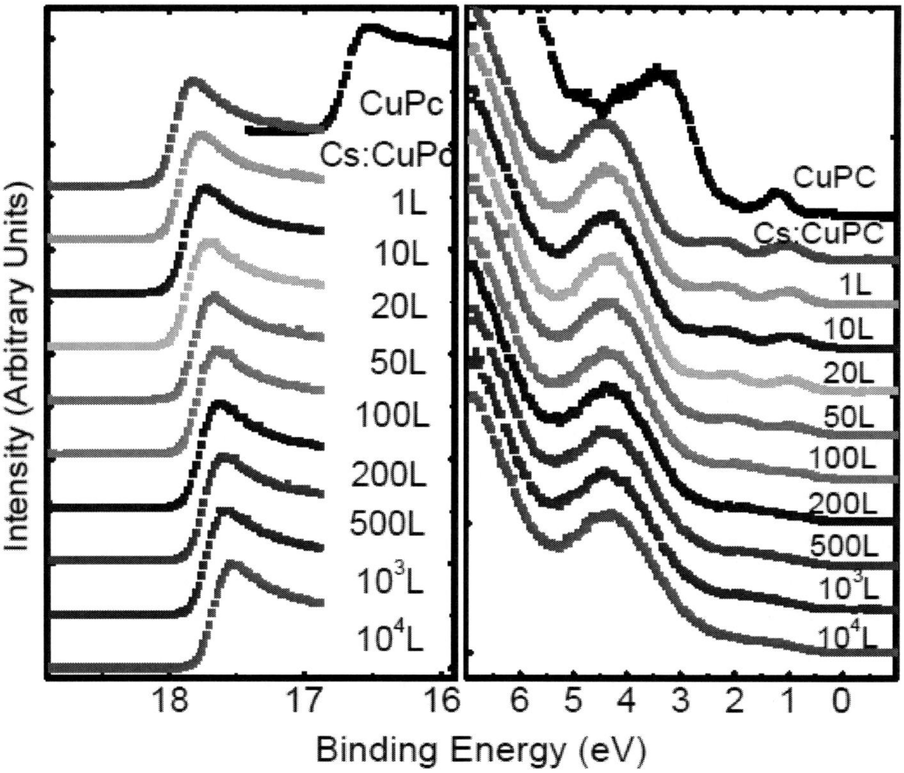

(a) (b)

Figure 67. The UPS spectra evolution of both (a) cutoff and (b) HOMO region for Cs-doped CuPc as a function of the oxygen exposure. All spectra are plotted in binding energy relative to Fermi level of the Au substrate.

In Figure 68, the UPS spectra of the low-energy secondary cutoff and the HOMO region of another Cs-doped CuPc sample are plotted as a function of the Au deposition. [139] Derived from the XPS core level intensities, the Cs doping ratio (R_{Cs}) is 3.1, which is slightly less than the oxygen-exposed sample. Consequently, the shift of the cutoff induced by Cs doping is slightly smaller than that of the oxygen-exposed sample, being 1.12 eV. The shift for the HOMO is 0.91 eV, which is about the same as that of the oxygen-exposed sample. After Au deposition, the cutoff shifts towards lower binding energy, reaching a final work function of 3.72 eV. The valence features shift in the same direction as plotted in Figure 68 (b). The CuPc HOMO becomes indiscernible as the Au coverage increases to beyond 15 Å. One interesting observation is that one can still observe the gap state even after 4 Å of Au deposition. The Fermi level of the top Au layer is observed after 8 Å of Au deposition, and is aligned well with that of the Au substrate, indicating that the charging effect can be neglected at this stage. Detailed data analysis of the HOMO features shows a larger reversal of the Cs-induced energy level shift by the Au deposition in comparison with that by the oxygen exposure.

Figure 69 shows the evolution of the gap state intensity during the modification process for the Cs-doped CuPc film by either the oxygen exposure or the Au deposition. [139] The intensity of the gap state is normalized to the intensity of the HOMO peak in the same spectrum. Given the fact that there is no degeneracy for the CuPc HOMO, the initial ratio of 1.56 for oxygen-exposed sample may indicates that about 3.2 electrons are transferred to CuPc from Cs, in good agreement with the core level intensity analysis that gives $R_{Cs} = 3.4$ if all Cs atoms are ionized. Assuming that the intensity of the HOMO does not change much during the modification processes, we thus observe gradual decreasing of the gap state intensity with either increasing oxygen exposure or Au deposition. At 500 L of oxygen exposure, the gap state intensity is reduced by ~ 65%, or only about 1.0 electron still remains on each CuPc molecule. After that, the peaks become too broad to identify the base line for peak fitting. However, it is still possible to judge the evolution of the gap state intensity from the trend indicated in this figure. For the Au-deposited sample, the initial ratio is slightly lower, which is also in good agreement with the lower Cs doping concentration for this sample. After 4 Å Au deposition, a reduction of 69% for the gap state is observed. Despite the slightly different initial Cs doping ratios, the evolution of the gap state during the modification process resembles each other in the two experiments, indicating a similar role for Au atoms and oxygen molecules as the *p*-dopant in CuPc. Given the nominal monolayer thickness of Au to be about 2.5 Å, it is interesting to notice that 1 ML of Au deposition is equivalent to about 300 L oxygen exposure in term of quenching of the gap state.

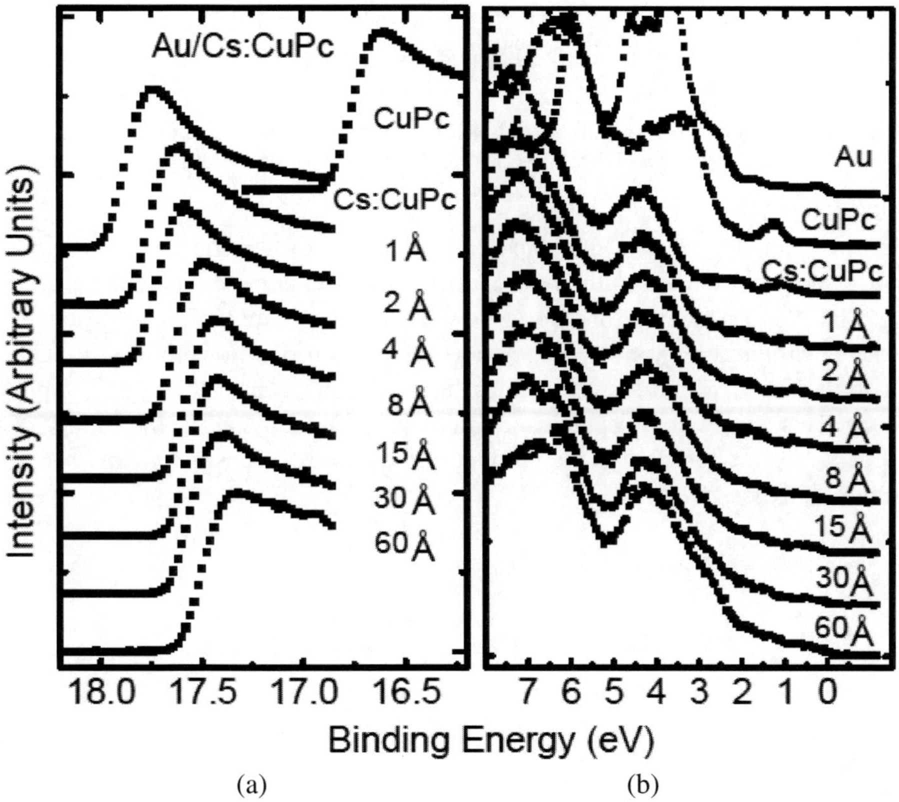

Figure 68. The UPS spectra evolution of both (a) cutoff and (b) HOMO region for Au deposition on Cs-doped CuPc film. All spectra are plotted in binding energy relative to Fermi level of the Au substrate.

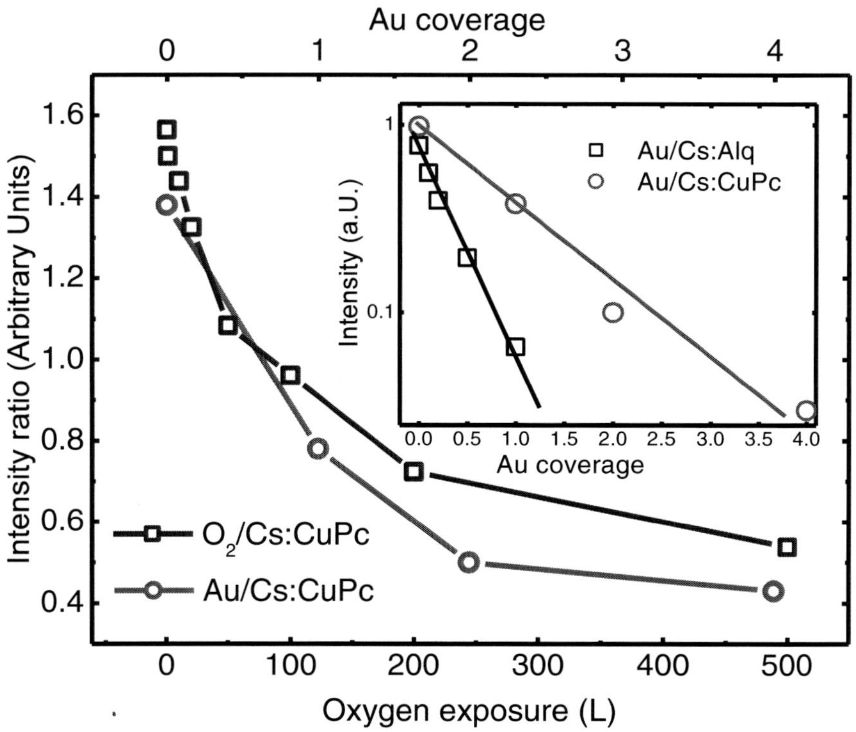

Figure 69. The evolution of the gap state over HOMO ratio in the UPS spectra for Cs-doped CuPc as a function of the oxygen exposure and the Au deposition. The inset figure shows the gap state intensity as a function of Au coverage in the logarithmic scale.

In the inset of Figure 69, the gap state intensity of Cs-doped CuPc is plotted in logarithmic scale as a function of the Au coverage. The plot shows a clear linear dependence, indicating that the quenching of the gap state by Au deposition is exponential. For comparison, a similar plot for Au deposition on Cs-doped Alq with $R_{Cs} = 6.72$ is also presented in the inset, which shows again a linear dependence in the semi-log plot with a steeper slope. This exponential dependence can be understand by taking the change of the gap state density N_{gap} as proportional to the product of the gap state density and the increase of the Au deposition thickness Θ_{Au}:

$$\delta N_{gap} = -(1/\Theta_0) N_{gap} \delta\Theta_{Au} \tag{20}$$

where Θ_0 can be taken as characteristic thickness that describes the electron removal ability of Au atoms in the organic material, and the minus sign indicates that it is a reduction. Integrating Equation 20 leads to an exponential relation between the gap state density and the Au deposition thickness as following,

$$N_{gap} = N_{gap}(0)\exp(-\Theta_{Au} / \Theta_0) \tag{21}$$

where $N_{gap}(0)$ is initial gap state density before the Au deposition. Equation 21 describes adequately the quenching process of the gap state by the Au deposition as presented in Figure 69. For oxygen induced quenching, the sticking coefficient is reduced at higher exposure, an effect not included in Equation 21. The effect of the reduction of sticking coefficient will be further discussed later in this section.

The evolution of the XPS spectra is shown in Figure 70 for C 1s, N 1s and Cu $2p_{3/2}$ core levels as a function of oxygen exposure. [139] Intensities of the spectra are normalized to the same height for visual clarity. Spectra of pristine CuPc are included for reference. As shown in Figure 71, [139] the C 1s core level spectrum of the pristine CuPc can be decomposed into two components: the main peak C 1s (A), which can be easily associated with the aromatic carbon of the benzene rings, and the second peak C 1s (B) at binding energy 1.4 eV higher, which is due to the pyrrole carbon in contact with N. [140] A small satellite can also be assigned at higher binding energy. Both the N 1s and Cu $2p_{3/2}$ core levels of pristine CuPc turn out to be single component. The intensive Cs doping of R_{Cs} = 3.4 induces not only a dramatic shift toward higher binding energy, but also a broadening to all three elements. Moreover, the two main components of C 1s spectrum merge into a large asymmetric peak, indicating that the shifts for C 1s (A) and (B) are not the same. During oxygen exposure, Cu $2p_{3/2}$ shifts monotonically toward lower binding energy. The total shift is about 0.3 eV for 10000 L oxygen exposure. However, no obvious shift can be observed for N 1s. The peak decomposition analysis of C 1s shows that the shift for C 1s (A) and (B) are not identical. C 1s (A) remains almost fixed during the oxygen exposure, while C 1s (B) behaves interestingly. At the beginning, it shifts toward higher binding energy for about 0.2 eV, introducing a clear broadening of the total C 1s spectra. After 50 L oxygen exposure, it then starts to shift backwards.

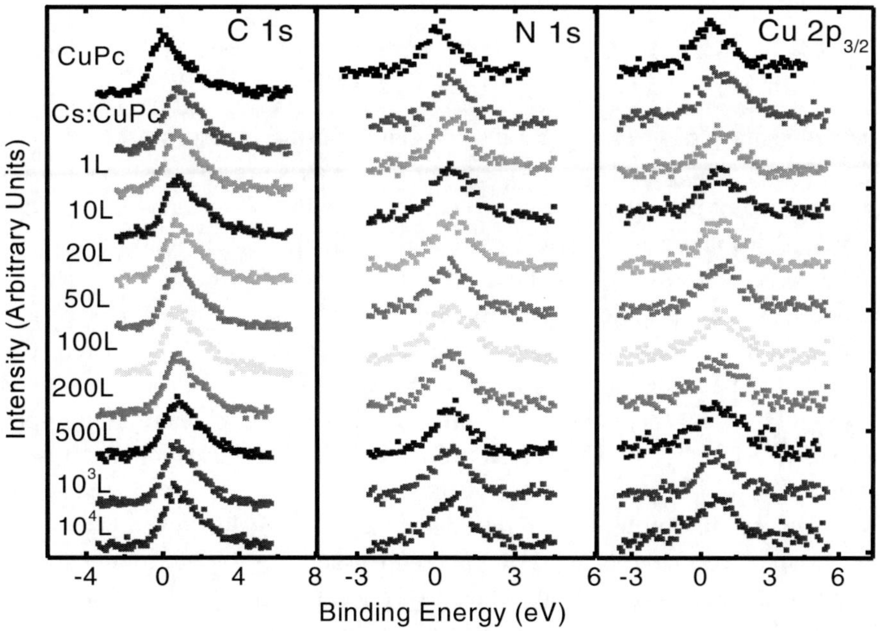

Figure 70. XPS spectra of C 1s, N 1s, and Cu $2p_{3/2}$ core levels of the Cs-doped CuPc film as a function of the gradual oxygen exposure.

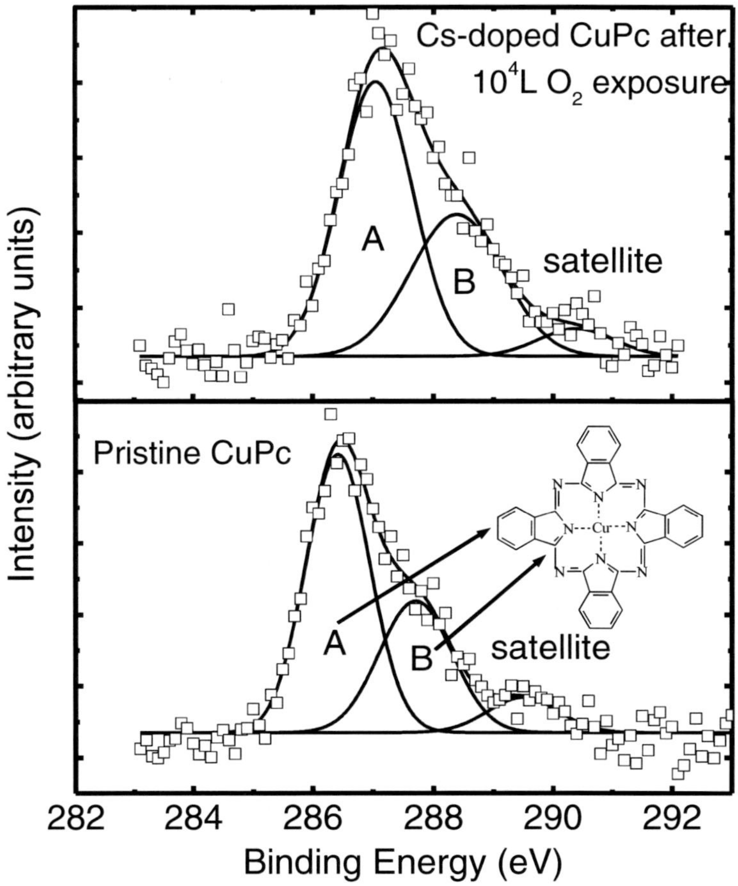

Figure 71. XPS decomposition of C 1s core level for pristine CuPc film and Cs-doped CuPc film after 10000 L oxygen exposure.

The decomposition of C 1s at the final oxygen exposure is also presented in Figure 71, where the peak broadening and the difference in the peak shift of C 1s (A) and (B) can be observed. The different behaviors of the two components in C 1s clearly indicate that there is charge carrier redistribution in the CuPc molecule. It should also be noted that no clear O 1s peak could be observed even at the highest oxygen exposure amount.

The modification of the electronic structure of both the oxygen exposure and the Au deposition are summarized in Figure 72 (a) and (b), respectively. [139] Here the binding energies of the Cs doped CuPc are taken as the origin, and the pristine CuPc positions are also presented for reference. The core level shifts were deduced by fitting the spectra with Gaussian curves for C 1s, N 1s, Cu $2p_{3/2}$ and Cs $3d_{5/2}$. The vacuum level cutoff was obtained at the maximum of the differential curve of the UPS spectrum. It can be seen that both the oxygen exposure and the Au deposition introduce energy level shift reversing to that induced by Cs doping, which can be explained by the compensation of the n-doping Cs by either the p-doping oxygen molecules or the Au atoms, as frequently observed in crystalline inorganic semiconductors. The recent studies about CuPc film preparation in oxygen atmosphere corroborate the effect of oxygen as a hole donor. [174] Therefore, it suggests that both the oxygen and the Au act partially as p-dopants, which increase the electron injection barrier for the n-type doped CuPc film by reducing the binding energy of the CuPc orbital. Similar to the quenching of the gap state shown in Figure 68, the effect on the energy level shift is also roughly 1 ML (2.5 Å) Au deposition equivalent to about 300 L oxygen exposure till approximately 10 Å of Au deposition. Further oxygen exposures show more pronounced saturation than Au, presumably due to the less sticking and penetration of oxygen at the higher exposure stages. Moreover, it can be seen that the total shifts for the core levels and the frontier orbital are not identical. The difference can, at least partly, be attributed to the correlation effect in heavily Cs-doped CuPc as the case for Alq. Another interesting observation is that, for the oxygen-exposed sample, the energy level shifts become saturated after 1000 L exposure. This indicates that the oxygen-induced reverse doping process is limited by the absorption and diffusion of the oxygen molecules in the organic film. Further exposure to oxygen may not enhance the reversal shift of the energy levels as effectively as the initial ones because of the reduction of the sticking coefficient.

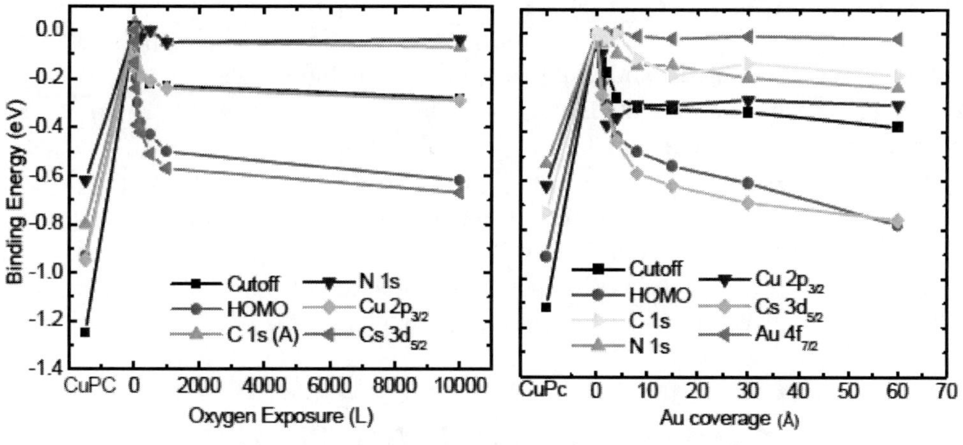

Figure 72. The summary of all the energy level shifts after reverse doping: (a) oxygen-exposed sample and (b) Au-deposited sample. The energy level positions of the pristine CuPc film are shown as the first points of the curves.

The discussion in section 3.5 suggests that n-type doping induces Fermi level pinning at the LUMO, leading to the appearance of the gap state, which is originated from the former LUMO. In the previous studies of the Au deposition on Cs-doped Alq, the gap state induced by Cs doping with R_{Cs} = 1.30 can be eliminated by 1 Å of Au deposition. For higher doping ratios (R_{Cs} = 6.72), the gap state may still persist at initial Au deposition with its intensity decreases exponentially as a function of Au coverage. As shown in the inset of Figure 69, [139] for the case of de-doping in Alq, the characteristic thickness Θ_0 is estimated to be about 0.33 Å, while in the case of Au deposition on Cs-doped CuPc, the same exponential fitting leads to a value of 1.0 Å, which is three times larger than that of Alq. Therefore, it suggests that the interaction between the doped alkali metal atoms and CuPc molecules is stronger than that of Alq, leading to an electronic structure modification that is more difficult to be eliminated by p-type doping. The other factor that one should also consider is the diffusion length of Cs atoms in organic materials. It has been reported that alkali metals, such as Li, diffuse much easier in CuPc than in Alq. [11] Thus the diffusion of the Cs atoms is also expected to be fast in CuPc, which may contribute to the persistence of the gap state.

Finally, it is also worth mentioning about the different behaviors for oxygen and Au as a p-dopant. For either case, most of the valence features of CuPc are preserved, which indicates that the mechanism of doping is dominated by charge transfer, and there are no strong chemical reaction during the reverse doping process. However, the HOMO recovers much faster after Au deposition than that of oxygen exposure. This is mainly due to the fact that all Au atoms stay in the organic and participate in the de-doping process, while only a small fraction of oxygen gets adsorbed. The lack of a clear O 1s peak in XPS suggests that there are not many oxygen molecules at the surface even after 10000 L exposure. Moreover, the hot Au atoms can diffuse deep into the organic film in its cluster forms, however, the saturation of the energy level shift at 1000 L oxygen exposure suggests that the de-doping process is limited by the inefficient diffusion of the oxygen molecules inside the film. Thirdly, the reaction between Cs and oxygen may lead to the formation of the oxygen anions, which further hinders the diffusion due to Coulomb interaction. Finally, the large Au clusters formed in the film may provide a better screening of the hole in the CuPc molecule after the photoemission process, which could lead to a larger shift in binding energy.

In conclusion, we have presented, in this section, the evolution of the electronic structure of oxygen exposed and Au deposited Cs-doped CuPc. The energy level shift induced by Cs doping is reversed by oxygen and Au, both acting to a certain degree as p-type dopants. However, the recovery of the electronic structures is incomplete and the gap state still persists, which suggests a strong interaction between the CuPc molecule and the Cs atom. Based on the experimentally observed exponential reduction of the gap state intensity by Au deposition, a formula was developed that adequately explained the process.

5. MECHANISM OF LIF INTERLAYER FOR ELECTRON INJECTION

Efficient charge injection into organic semiconductors is one of the most challenging issues in the research and development of organic electronic devices. In principle, low work function metals, like Mg or Ca, are preferable for electron injection. However, the reactive nature of such a low work function cathode is one of the limiting factors to the device lifetime. A more efficient approach is to insert a thin interlayer of ionic insulator such as LiF, [175-177] MgO, [175] MgF_2, [178] and Al_2O_3, [179] between the metal cathode and electron transport layer. For example, OLED devices with Al/LiF bilayer cathode show significant enhancement in the electron injection, as shown in Figure 73. [175] In addition, the interface stability has been great improved, which prolongs the device lifetime. Early suggestion by Hung et al. [175] was that the LiF layer reduced the electron-injection barrier height, but it is also found that it depends strongly on the metal deposited on top of the LiF. [180] Schlaf et al. measured the electronic structure of LiF on Al and Pt, and attributed the injection improvement to band bending in LiF caused by charged defect. [181] Zhao *et al.* suggested that tunneling in LiF and thus induced energy level realignment are responsible to the electron injection improvement. [182, 183] The mechanism of improvement induced by the LiF insertion layer remains controversial and to be fully resolved.

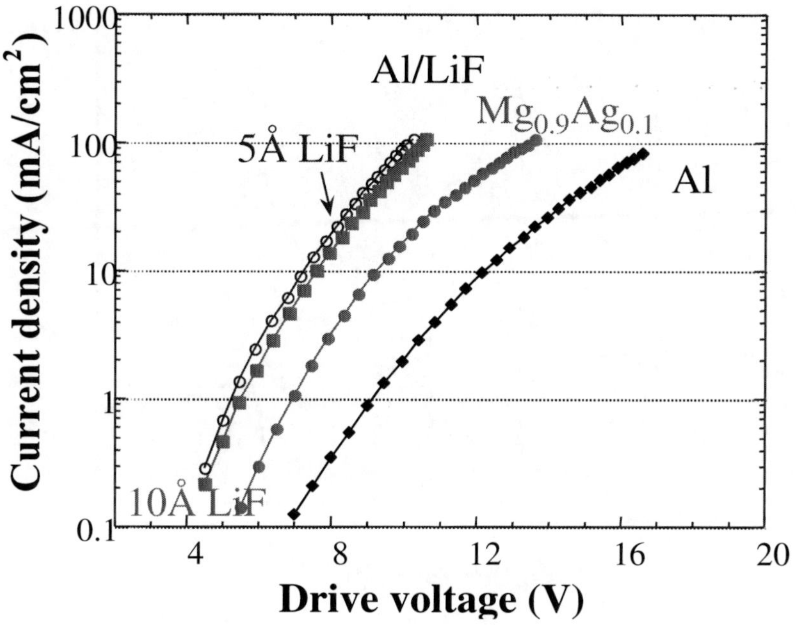

Figure 73. I-V measurements of OLED devices with different choices of cathodes.

5.1. Electronic Structure at Al/LiF/Alq Interface

In order to understand the electronic structure modifications induced by Al/LiF bilayer cathode on Alq, it is worth stepping back first to the simple situation and investigating the electronic structure at the Al/Alq interface. The evolution of the XPS O 1s, N 1s, Al 2p, and C 1s spectra at representative coverages of Al deposited on Alq film is shown in Figure 74. [99] The XPS spectra were normalized to constant intensity for comparison purposes. The bottom curve in each set represents pristine Alq, and can be fit very well by a single component, Gaussian line shape. All core levels showed significant broadening and shifting to higher binding energy with increasing Al deposition (Figure 75 [99]). However, the more detailed behaviors are very element specific. The C 1s exhibits only the earlier mentioned broadening and shifting as does the Al 2p up to a coverage of 4 Å where a low binding energy component begins to emerge.

The O 1s develops a distinct asymmetry to the high binding energy side while in the N 1s spectrum, a broad feature appears to evolve on the low binding energy side of the principal peak. This variety of element specific responses to the Al deposition is clear evidence that something much more complicated than simple band bending is occurring and implies that significant chemical interactions are taking place. In contrast to Ca/Alq, [184] where the nitrogen spectrum is singularly influenced by the metal interaction, it is the oxygen emission that is most significantly affected in the Al/Alq. DFT calculations predict that the incoming Al will bond directly to O atoms on two of the three quinolate rings in Alq molecule which explains the observed asymmetry and increased width of the O 1s line. [99] Above 4 Å of Al deposition, the intensity of the reacted O 1s component remains essentially constant.

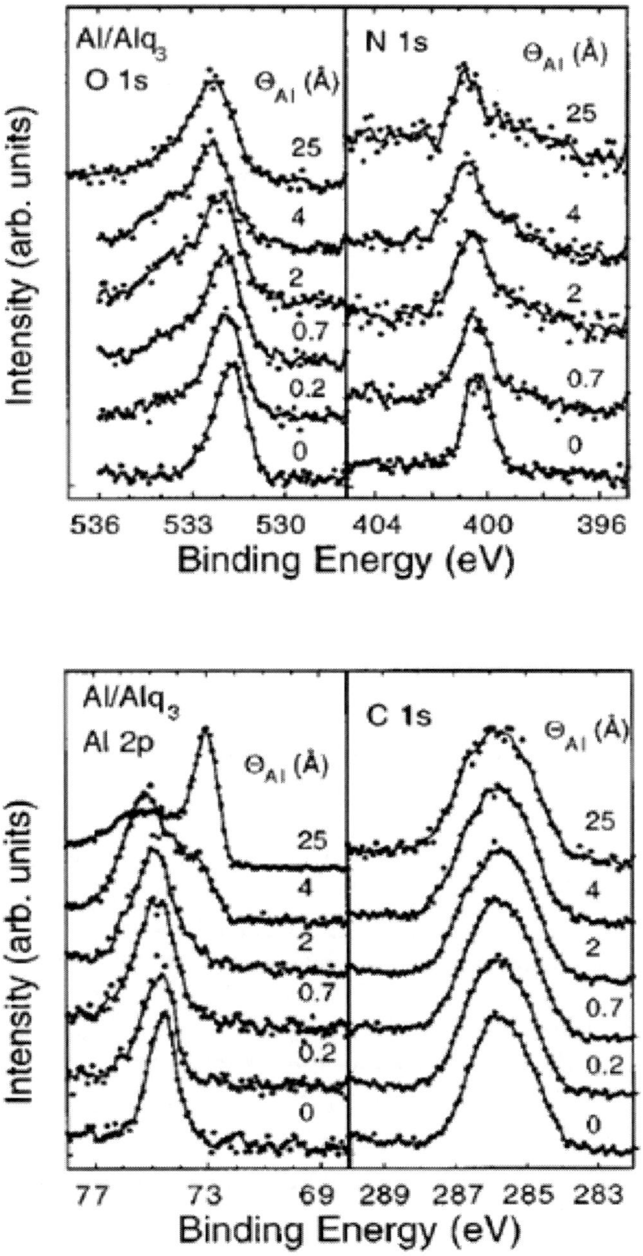

Figure 74. Evolution of XPS O 1*s*, N 1*s*, Al 2*p*, and C 1*s* spectra as a function of Al coverage observed at the Al/Alq interface.

The coverage dependence of the N 1s spectra can also be understood in terms of the theoretical model. Mulliken population analysis of the atomic charges shows that two of the quinolate nitrogens are very similar, while the third has a significantly higher electron density. Photoemission from the more highly charged nitrogen should occur at lower binding energy, as is observed in the measured spectra.

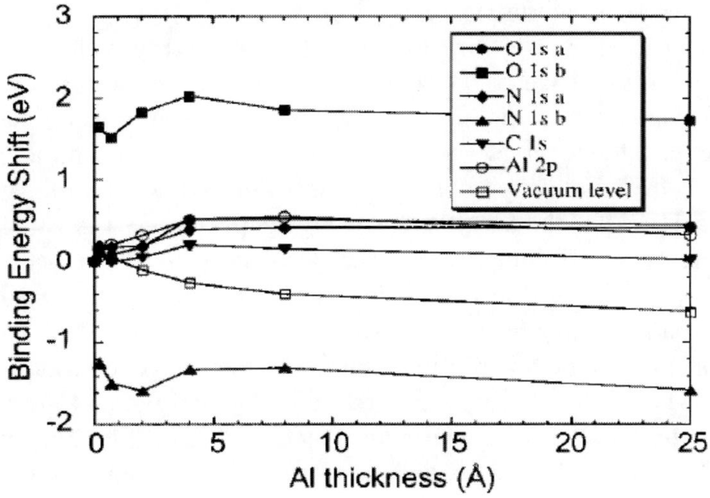

Figure 75. Binding energy shift of core-level spectra as a function of increasing Al thickness on an Alq substrate. (a) and (b) are referred to unreacted and reacted components of the core-level peaks, respectively.

The broadening of the Al and C spectra is simply due to the greater diversity in chemical states and environments of the atoms. While both the original Al peak from Alq molecule and the deposited Al are bonded to oxygen, both the coordination and charges are different and at least a modest increase in linewidth is to be expected. The C 1s linewidth is quite large, even for the pristine Alq. This results from the variations in bonding within the quinolate ring and the lack of symmetry in the Alq molecule that combine to make each carbon atom chemically unique. This effect is amplified by the reaction with deposited Al that produces an even less symmetric molecular structure. The low binding energy peak that first appears in the Al 2p spectra at 4 Å Al deposition is simply due to unreacted metallic Al. The intensity of this peak increases with coverage while that of all the other elements decreases as a true metallic film is formed. A clear demarcation is seen in the core level spectra for coverages above and below 4 Å of Al deposition. Above 4 Å, the binding energies and linewidths are virtually independent of coverage and all of the added Al is found in the metallic state. This clearly indicates that the reaction is limited within the first 4 Å of Al deposition. While a 4 Å layer of Al would be extremely thin, the modified layer which would result from a reaction between the deposited aluminum and the Alq film would be much thicker. Due to the extreme difference in molecular/atomic sizes, about 0.2 Å of Al corresponds to approximately one metal atom for every surface Alq molecule. If each deposited Al atom reacts with one Alq molecule and can freely diffuse into and react with the bulk, then the 4 Å Al deposition would produce a reacted layer of about 150 Å in the organic film. This also explains why such dramatic spectral changes occur when only 0.2 Å of metal is deposited.

The binding energy shifts of core-level spectra and the vacuum level as a function of Al thickness are summarized in Figure 75, in which (a) and (b) are referred to the "unreacted" and "reacted" components, respectively. [99] Unreacted is used in the context of a species that appears to remain very "Alq-like" after formation of the interface. These are the features which show the smallest energy shifts with metal deposition. Reacted implies that the species has been chemically changed in formation of the interface and differs significantly from its original state in Alq. These are the features which show the largest energy shifts with metal deposition. All core level signals shift to higher binding energy with the most significant shifts observed within the range of 0 – 4 Å Al depositions. Above this coverage, all levels gradually shift towards lower binding energy, as is expected for increased, final-state, screening as the metallic overlayer evolves. Among the constituent atoms of the Alq molecule, carbon is the least likely to be significantly affected by reaction with the deposited Al. The C 1s signal presents the least binding energy shift of only 0.15 eV. The overlapping of unreacted and reacted C 1s components makes the centroid of the peak differ from that of the initial, pristine Alq. The inability to distinguish a truly unreacted component makes any conclusions about band bending purely speculative.

In trying to understand the role of LiF in improving the charge injection characteristics of Al/Alq interface, XPS spectra of LiF deposited on to Alq and onto Al metal, as well as Al deposited onto LiF have been carefully examined by Gao's group. [99] The binding energy of Li 1s and F 1s, the F Auger parameter, and the Li and F atomic compositions as determined by their relative photoemission intensities have been carefully monitored. [99] No indication was found that the Li and F were in any state other than LiF. Unfortunately, Li has an extremely small photoionization cross section, which makes it almost impossible to completely rule out a reaction. However, these results suggest that it is necessary to study the complete ternary Al/LiF/Alq interface to elucidate the role of LiF.

In Figure 76, Le *et al.* show the O 1s, N 1s, Al 2p, and C 1s spectra after LiF deposition and with increasing Al coverage. [99] An unfortunate result of the LiF deposition is a significant increase in the observed linewidths. This probably results from the incomplete coverage of the surface that causes it to be very inhomogeneous. With the exception of the C 1s peak, all core-levels showed significant broadening with LiF and subsequent Al deposition. The increased FWHM with overlayer deposition makes it difficult to resolve multi-component lines. This effect is most relevant in the O 1s spectrum. Although this appears to remains a single component peak for all Al thicknesses and shows none of the asymmetry characteristics of the Al/Alq interface, some asymmetry may be hidden by the increased linewidth. Even though no splitting is observed, the increase in linewidth suggests a substantial alteration of the oxygen environment. In contrast to the O 1s spectrum, a new component in the N 1s spectrum is observable at 1.6 eV to lower binding energy, even at the lowest coverage of Al. The intensity of this reacted component is higher than that of Al on Alq in the absence of the LiF layer. The pyridyl species appears to be significantly modified by even a small amount of deposited Al. The evolution of the Al 2p signals is similar to that of the Al/Alq interface but the magnitude of the shift is significantly greater. Furthermore, the metallic component appears at a significantly lower coverage suggesting that LiF, at least, partially protects the Alq surface from reaction with the deposited Al. In the absence of LiF, the metallic signal is first noticeable at a coverage of 4 Å, whereas a similar relative contribution of the metal is observed at 2 Å when the LiF layer is present. It is a general characteristic of the spectra that the shifts are larger and occur at lower coverages of Al when LiF is present, compare Figures 75 and 77. [99]

In Figure 78, the UPS spectra are presented for Al depositions onto clean Alq and Alq with the 5 Å LiF buffer layer. The spectrum for clean Alq is comprised of several well-defined peaks and is consistent with those found in the literature. [13, 130, 171, 185-188] The position of the HOMO is found at 2.1 eV below the Fermi level, which is in good agreement with the values reported for Alq/Au.[13, 187, 188] Deposition of an Al layer as thin as 0.2 Å represents a significant change in the valence spectrum of Alq and induces the formation of new states which extend into the energy gap. The fingerprint of the Alq structure has virtually vanished after deposition of 0.7 Å Al, indicating the relatively strong reaction between the Al atoms and the Alq molecules. The appearance of the Al Fermi level for 4 Å of Al deposition is in good agreement with the XPS results, for which the Al metallic peak begins to grow after deposition of the same Al thickness. The spectra observed with Al deposition are consistent with an extensive chemical reaction forming a product significantly different from Alq, as predicted by the DFT calculations. The new states observed in the energy gap are a reflection of the molecular orbital structure of this reaction product.

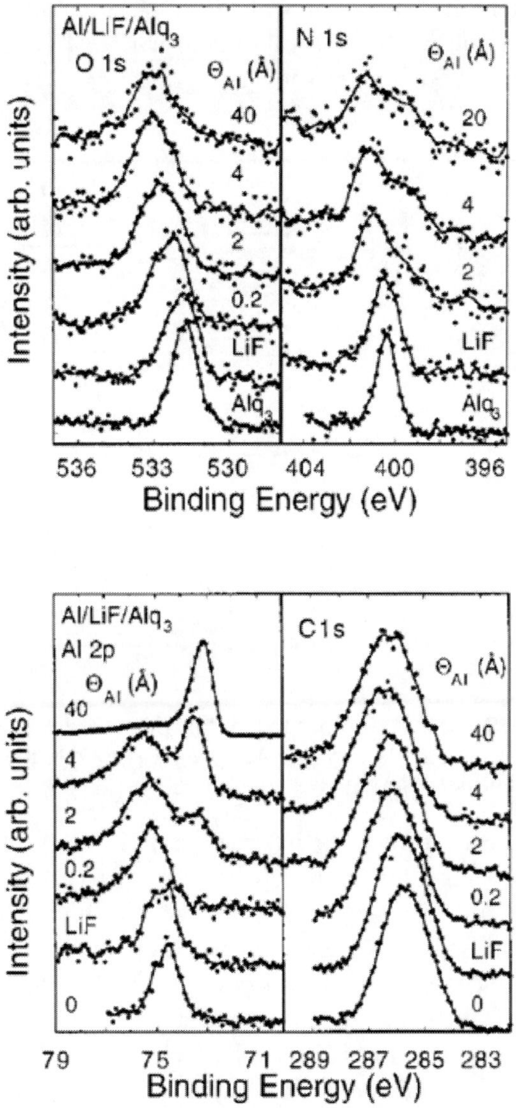

Figure 76. Evolution of XPS O 1*s*, N 1*s*, Al 2*p*, and C 1*s* spectra as a function of Al coverage observed at the Al/LiF/Alq interface.

As shown in Figure 78, [99] the presence of a thin LiF layer on the Alq surface does not substantially modify the HOMO position of Alq although it does induce significant broadening as in the XPS spectra. The deposition of Al induces new states above the HOMO in the energy gap, and results in a shift of the HOMO position to higher binding energy. These states are fundamentally different from those formed in the absence of LiF. The molecular orbital structure of Alq is preserved with LiF present. States are shifted to higher binding energy and the gap state that is formed is well defined. In the absence of LiF, the molecular orbital structure of Alq is broadened beyond recognition even at very low coverages of Al. The states that are formed in the gap appear to result simply from an extreme line broadening as a result of the Alq decomposition process. The presence of the LiF layer not only decreases the extent of the reaction with Al, it decreases the thickness of the reacted layer in Alq film. The Alq valence band is still identifiable at 4 Å of Al deposition, and the Fermi level is observed at an Al coverage as low as 2 Å.

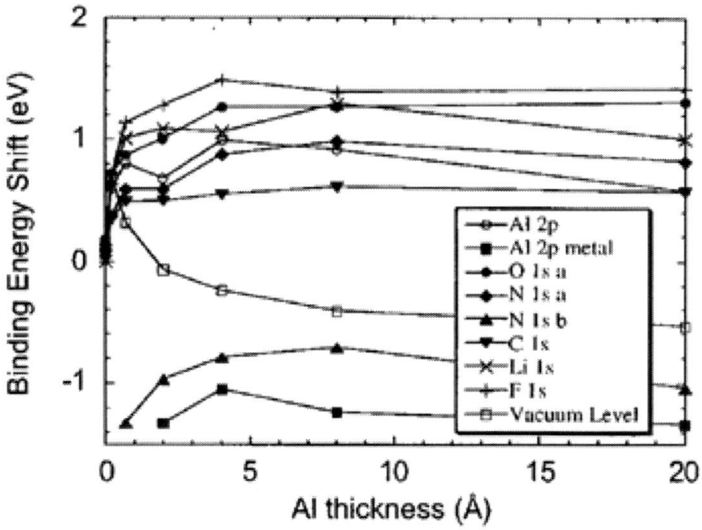

Figure 77. Binding energy shift of core-level spectra as a function of increasing Al thickness on a LiF/Alq substrate. The thickness of LiF layer was 5 Å; (a) and (b) are referred to unreacted and reacted components of the core-level peaks, respectively.

The XPS and UPS spectra of Al/Alq interface appear to be understandable simply in terms of a "destructive" chemical reaction between the organic and the reactive metal. To understand the role of LiF in modifying this interface, one should consider the discussions involved in the previous sections about the electronic structure modification in alkali metal doped Alq. In those cases the core levels shift to higher binding energy. The N 1s peak splits with a separation between 1.6 and 1.8 eV, and the O 1s broadens significantly but symmetrically. In the UPS spectra, the HOMO shifts to higher binding energy and a new state is formed in the previously existing energy gap and is separated from the original HOMO by 1.6 eV. All of these characteristics are observed in the case of Al/LiF/Alq interface. Therefore, it suggests that when Al, LiF, and Alq are all present, Li is liberated and reacts with Alq to form the radical anion.

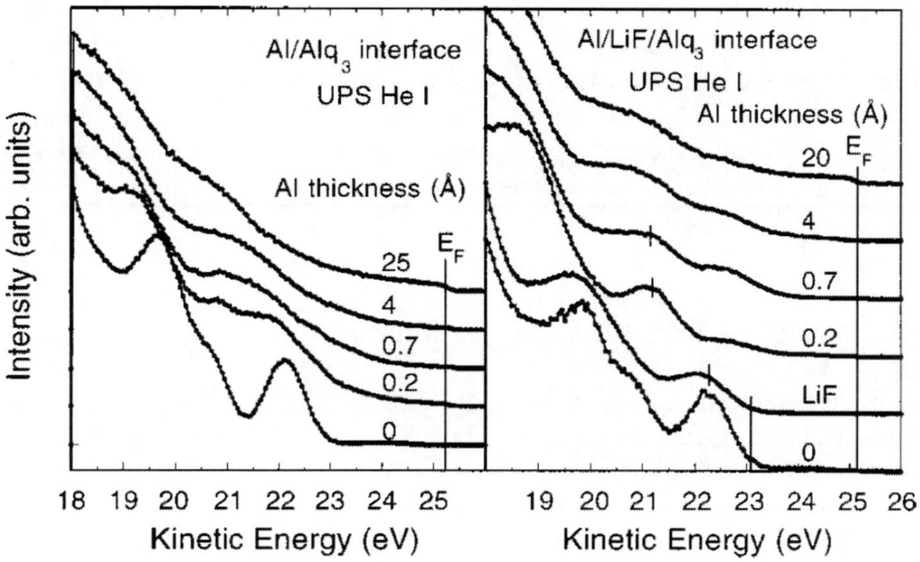

Figure 78. Evolution of the UPS spectra as a function of aluminum thickness for the Al/Alq and Al/LiF/Alq interfaces.

The above interpolation is addressed by considering both calculated and known bulk heats of formation in study from Mason et al. [18] As a very stable compound, the bond between Li and F atoms in LiF is one of the hardest to break by itself. The Gibbs free energy of formation (ΔG_f) for bulk LiF is reported to be 140.5 kcal/mole. [133] To break the bond between Li and F will require an energy as high as 138 kcal/mole. [133] However, it has been pointed out that there is a highly exothermic reaction between F and Al, as shown below,

$$3LiF \rightarrow 3Li + 3/2F_2 \qquad \Delta G_f \sim 421.5 \text{ kcal/mol} \tag{22}$$

$$Al + 3/2F2 \rightarrow AlF3 \qquad \Delta Gf \sim -340.6 \text{ kcal/mol} \tag{23}$$

$$3LiF + Al \rightarrow 3Li + AlF_3 \qquad \Delta G_f \sim 80.9 \text{ kcal/mol} \tag{24}$$

It is thus possible that the free energy required to form Li atoms can be reduced from 140.5 to about 26.9 kcal/mol, due to the exothermic nature of the AlF_3 formation. The fact that the overall reaction between LiF and Al themselves is still considerably endothermic is consistent with the experimental findings that no reaction occurs when LiF is evaporated on Al or Al onto LiF, without the presence of Alq. The detailed information about the heat of reaction between Li and Alq to form the radical anion is not known experimentally. However, Mason *et al.* have pointed that there is an exothermic reaction between K and Alq, which releases an energy of about 27 kcal/mol. [18] Because of the ionic nature of the alkali metals, it is believed that ΔG_f should not change much in case of Li. In this context, the complete reaction at Al/LiF/Alq interface may be written as

$$3LiF + Al + 3Alq \rightarrow AlF_3 + 3Li^+Alq^- \quad \Delta G_f \sim 0 \text{ kcal/mol} \tag{25}$$

The above equation suggests that the feasibility of the formation of the Alq^- radical anion due to doping process by released Li, when LiF, Al, and Alq are all present at the interface, is entirely reasonable. One should also note that the model is derived using thermodynamic properties of the bulk, which is a severe approximation of the actual situation. In the real case, only a few Å of LiF is evaporated, forming isolated clusters at the organic surface, and the Al arrives at the surface as thermally hot atoms. These facts may lead to much greater chemical reactivity than their bulk counterparts.

5.2. Electronic Structure at Au/LiF/Alq Interface

The results presented in Section 5.1 show that Al deposition on 5 Å LiF coated Alq induced substantial n-doping in organic material as well as suppression of the chemical reaction between Al and Alq, both important for efficient electron injection into the organic material. [18, 99] However, such performance improvement has been discovered to have a strong dependence on the selection of the metal cathode. [180] Moreover, in the fabrication of polymer-based devices, researchers have noticed that a much thicker LiF layer has to be used in order to have significant improvement. [189] It is thus interesting to explore the role of LiF at different interface, in order to have a better understanding about the cathode-dependent improvement and the thickness requirement for polymer devices.

In Figure 79, the UPS spectra of the low-energy secondary cutoff (a) and the HOMO (b) region are presented for Au/LiF/Alq interface, respectively. [83] All curves are plotted in the binding energy scale relative to the Fermi level of the system. The valence spectrum for clean Alq is comprised of several well-defined peaks that are consistent with those found in the literature. [190] The HOMO of Alq is attributed to the π-orbital located on the phenoxide side of the quinolate ligand. [190] The evolution of the HOMO is marked by arrows in the figure. The position of the HOMO peak is found at around 3 eV below the Fermi level, which is in good agreement with the values reported for Alq/Au. [188] The presence of a thin LiF layer on the Alq surface preserves most of the valence features, although it does induce a 0.14 eV shift toward higher binding energy and some broadening. The deposition of Au initially produces a slight shift of the HOMO position to higher binding energy, reaching a maximum of 0.26 eV at Au coverage θ_{Au} = 0.5 Å. Further Au depositions gradually reverse the shift toward lower binding energy. The final Alq HOMO position, however, cannot be observed from the valence spectra since the Au features dominate after θ_{Au} = 4 Å. In the low-energy secondary cutoff region (Figure 79 (a) [83]), the vacuum level shifts 0.15 eV toward lower binding energy, when LiF is deposited. The existence of an interface dipole can be ruled out since the HOMO shows the same shift at this stage, reflecting vacuum level alignment. [191] For the initial Au deposition ($\theta_{Au} \leq$ 0.5 Å), the vacuum level shift is consistent with that of the valence features, indicating the nature of rigid shift for the low coverages. As the Au overlayer grows thicker, the vacuum level shifts upward toward that of pure Au, but saturates at θ_{Au} = 8 Å at a position of 1.0 eV below that of pure Au, whose work function is 5.3 eV. This phenomenon has been observed in other Au deposition on organic systems, and can be attributed to Au intermixing with and clustering in the substrate material. [192]

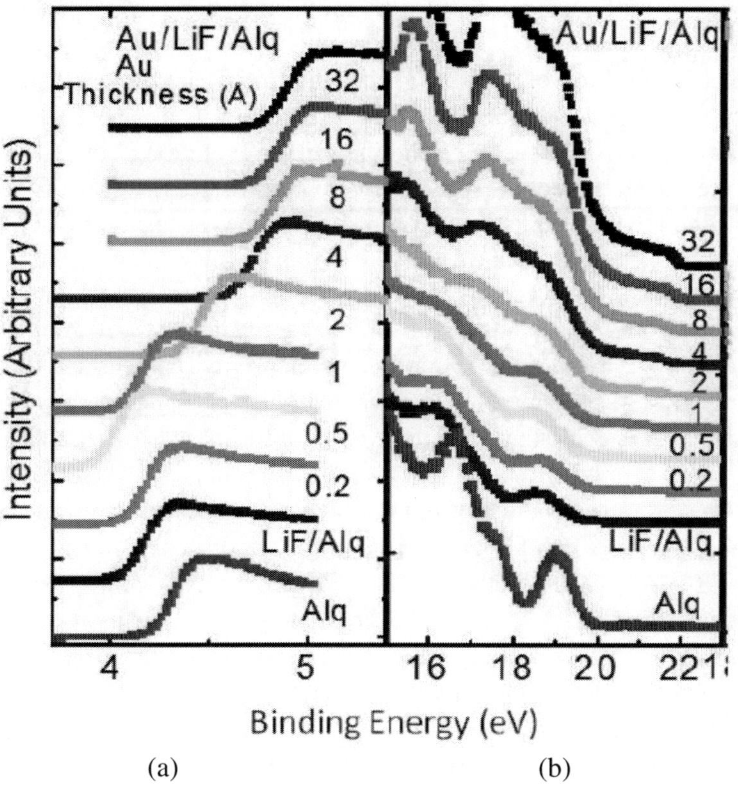

Figure 79. Evolution of the cutoff (a) and the HOMO region (b) of the UPS spectra as a function of Au thickness for the Au/LiF/Alq interfaces.

It is very intriguing to compare the valence energy level evolution of the Au/LiF/Alq interface with that of Al/LiF/Alq, presented in Figure 78. [18, 99] For both interfaces, the molecular orbital structure of Alq are preserved with LiF present. It is less of a surprise for the Au case than for Al, since the latter would induce significant chemical reaction with Alq if LiF was not present. [18] As seen in Figure 79, Au induces little energy level shift, leaving a "flat-band" in Alq in the interface region. As a result, the electron injection barrier is unaltered compared to that of pristine Alq. The situation is dramatically different in the case of Al/LiF/Alq, as can be seen in Figure 78, [99] where the deposition of Al on 5 Å LiF/Alq not only shifts the Alq HOMO to higher binding energy by as much as 1.0 eV, but introduces a new state in the former energy gap as well. Within the rigid shift assumption, the LUMO is also lowered by the same amount, resulting in a significant reduction of electron injection barrier that can lead to substantial improvement in electron injection as observed in devices. [175] Such electronic structure modification can be attributed to doping by Li released through the interface reaction by Al deposition, as we discussed in Section 5.1. [18, 99] The assessment of doping is also supported by the gap state in Figure 78 [99] and new N 1s component in Figure 76. [99] Those features have been observed for alkali metal doped Alq film with amazing similarity. [13, 18] More direct evidence of the existence of Li^+ in such a system has been demonstrated by Wu *et al.* [186] with synchrotron radiation photoemission spectroscopy, where the high photon flux allows direct observation of Li 1s. For the present case of Au/LiF/Alq, the HOMO region is convoluted by the features from both Alq and Au, however, no new features can be identified above the Alq HOMO before everything is attenuated by Au deposition.

It can be noticed in Figure 79 (b) [83] that the initial energy level shifts are gradually reversed by 0.5-2.0 Å Au depositions. For $\theta_{Au} > 2$ Å, the valence spectra become dominated by the contribution from Au 5d features, and the further evolution of the Alq valence features have to be inferred from the core level spectra from XPS. This is valid if there is no strong chemical interaction at the interface and a rigid shift can be assumed for all energy levels. The evolution of the O 1s, N 1s, C 1s and F 1s core levels of LiF-coated Alq as a function of Au overlayer thickness is shown in Figure 80. [83] The XPS spectra were normalized to constant intensity for visual clarity. The initial peaks can be fitted very well by a single Gaussian component, indicating a simple and uniform chemical environment for the elements. The single Gaussian feature is preserved as Au is deposited, suggesting no chemical reaction. The shift of C 1s core level is consistent with the UPS data for $\theta_{Au} \leq 2$ Å, reflecting a rigid shift in Alq. The F 1s shows a shift of 0.23 eV toward higher binding energy at the initial deposition of $\theta_{Au} = 0.2$ Å, much larger than that of C 1s (0.09 eV) at the same coverage, but the shifts of the two core levels are very similar for later stages. For higher Au coverages ($\theta_{Au} > 4$ Å), the energy levels shift gently back to the pristine Alq position, reflecting that there is a "flat-band" situation in Alq at this interface. The final energy level positions also verify that the change in cutoff after $\theta_{Au} = 1$ Å is mainly from the work function change by the Au overlayer, not the energy level shift in the substrate. The O 1s and N 1s core levels behave similarly to the C 1s.

Given the inertness of Au, it is not expected to react with LiF. However, it is also interesting to note that the initial Au deposition on LiF/Alq induces a small but discernable shift of the Alq features to the higher binding energy, and the shift is reversed to the pristine Alq position when further Au is deposited. This is somewhat surprising since one would not expect Au to donate electrons to Alq or LiF. Indeed the UPS study of Au/Alq by Lee *et al.* does not show such a downward shift. [170] This initial shift may be attributed to Li released by Au deposition. However, the reaction of Au with LiF is more endothermic than Al, and if any minute amount of Li is released, it is most likely due to defects in LiF and the thermal energy of the hot Au atoms. Comparing the core level shifts for Alq and LiF, one immediately noticed the interesting behavior of F 1s at initial Au deposition. A 0.23 eV shift toward higher binding energy is observed for F 1s at $\theta_{Au} = 0.2$ Å, while the energy level shift for the Alq substrate is relatively small at this point. The larger shift, which reaches a maximum of 0.27 eV at $\theta_{Au} = 0.5$ Å, may also be attributable to the decomposition of LiF by Au deposition. Another possible reason for this initial shift is the charging effect induced by Au cluster formation, as been reported for Au deposition on other organic films. [193] The reversal of the shift to the pristine Alq position by further Au deposition is easier to understand. After most of the Li atoms in the defect sites are released by the initial Au deposition, further Au atoms arriving to the interface region will act as p-dopants and compensate the n-doping effect by the Li. Such counter-doping and reversal of energy level shift have been demonstrated in the previous discussions of Au on Cs-doped Alq in Section 4.1. [169]

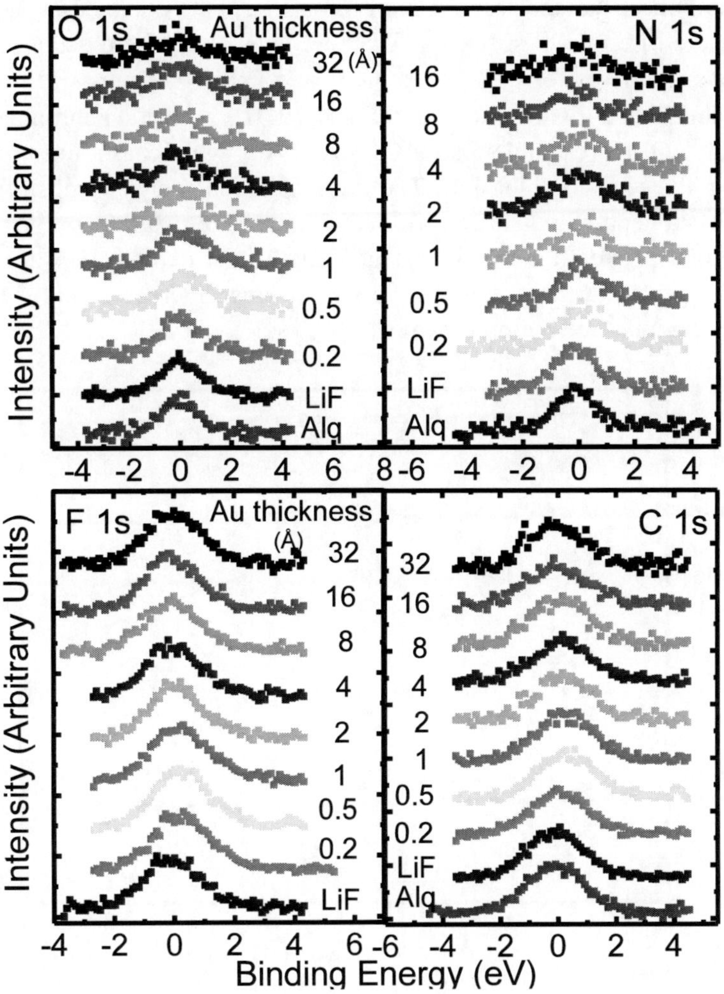

Figure 80. Evolution of XPS O 1s, N 1s, C 1s and F 1s as a function of Au coverage observed at the Au/LiF/Alq interface.

For further understanding of the interface formation, the attenuation curves of the core levels during the Au deposition are plotted in Figure 81. [83] The slight variation of the C 1s and F 1s core levels intensities at the initial Au coverages can be stemmed from the fluctuation of the x-ray photon flux. For further depositions, the attenuation curves turn out to be very typical, with exponential decay for the substrate C 1s and F 1s and corresponding growth of the Au 4f. The inelastic MFP of about 26 Å is found for C 1s, which is comparable to that for reactive metal depositions, such as Ca. [192] It suggests that the Au clustering effect is not too heavy in the interface region. It is also interesting that the effective MFP of F 1s, which is supposed to be shorter than that of C 1s due to its higher binding energy, is estimated to be 32 Å. The larger than expected MFP of F 1s may suggest the intermixing of the LiF and Au on the Alq surface.

The energy level shifts as a function of Au thickness are summarized in Figure 82. [83] The HOMO position was determined by extrapolation the leading edge of the UPS valence band spectra up to $\theta_{Au} = 2$ Å, above which no reliable value for Alq can be extracted because of the Au contribution to the spectrum. The vacuum level was obtained at the maximum of the differential curve of the UPS spectrum. The core level shifts were deduced by fitting the spectra with Gaussian curves for C 1s, N 1s, O 1s and F 1s. To emphasize the initial shifts, the coverage is plotted in a logarithmic scale. For $\theta_{Au} \leq 0.5$ Å, the synchronized shifts of the core levels and the HOMO resemble that of Cs-doped Alq, which has a linear dependence on the doping ratio in the semi-log plot. [13] The reversal of the energy level shift is introduced by the subsequent Au depositions. It is worth mentioning here that the core level positions of pristine Alq are fully recovered at the final stage. This further demonstrates the initial doping process followed by the later de-doping as the Au coverage increases, and the overall "flat-band" situation in the interface region is achieved.

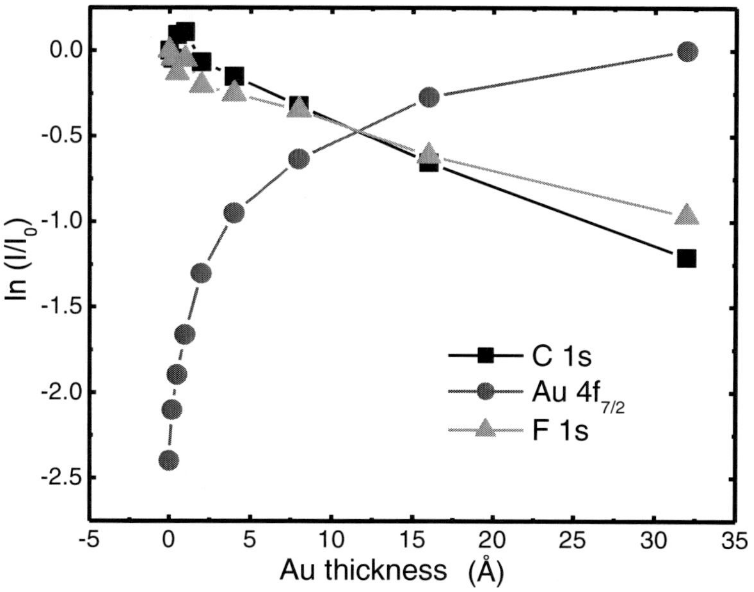

Figure 81. XPS attenuation curves for C 1s and F 1s in Au/LiF/Alq interface.

The observation of the overall flat-band condition in the interface region explains why for thin LiF interlayer, the metal overlayer material is critical for the improvement of the charge injection. As can be seen from Figure 83 (a), [83] the low reactivity of the deposited metal atoms, like Au, do not result in substantial n-doping of the Alq in the interface region, and the electron injection barrier remain unaltered. This is in sharp contrast to the case of more reactive metals like Al and Mg, where a downward shift of about 1.0 eV substantially reduces the barrier (see Figure 83 (b) [83]). Another contributing factor is the p-doping nature of Au, [139, 169] as the small amount of n-doping by LiF and by the initial Au deposition is completely eradicated by the further Au deposition. Similar situations can also be expected for Ag, which has been found ineffective in terms of improving electron injection when deposited on thin LiF covered Alq. [175] The p-compensation can also exist if there are other electron withdrawing impurities in the interface region, such as oxygen, as we have observed the reversal of Cs induced energy level shift in organic thin films by O_2 exposure. [139] For spin-cast polymer films, the surface is often covered by oxygen containing impurities, [184] and the compensation by the impurities may screen any improvement to electron injection by

Li doping. As a result, a flat-band situation may still be maintained and a much thicker LiF layer is needed for the optimum device performance even when non-p-doping metals are used. [189]

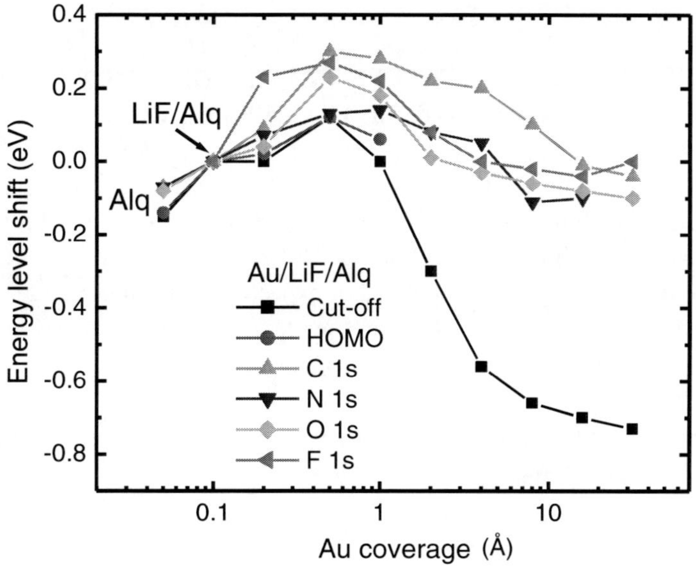

Figure 82. Summary of the energy level shift as a function of increasing Au thickness on LiF-coated Alq.

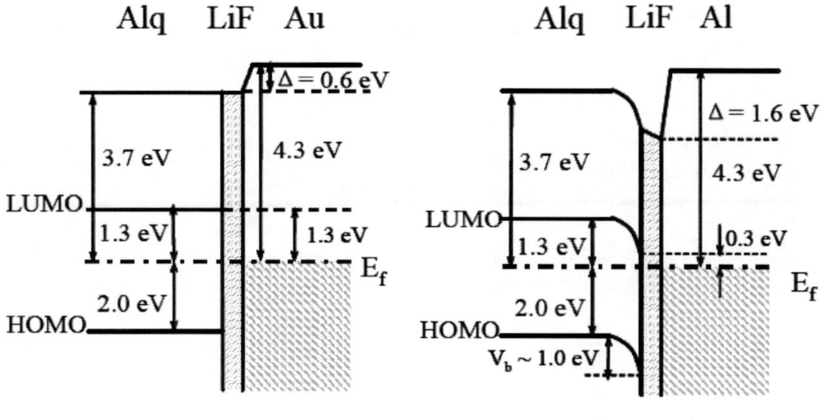

Figure 83. Energy level alignment at the (a) Au/LiF/Alq and (b) Al/LiF/Alq interfaces.

While n-doping of Al/LiF/Alq structure is clearly important for the reduction of electron injection barrier for thin LiF interlayer, the tunneling model [182, 183] should apply when such a doping is not taking place or compensated, as we see here for Au. The tunneling process should dominate when the LiF layer is thicker, for example in the case of Brown et al., [189] where several nanometers of LiF was found to be optimum for injection improvement. In that case, it is essential that the voltage drop across LiF is about the initial barrier height. In fact, for such a thickness, the Li doping mechanism that prevails for thin LiF and reactive metals may be completely blocked since the released Li may not be able to reach the organic material.

5.3. Electronic Structure and Interactions of LiF-Doped Alq

Besides the layered structure, LiF has also been reported useful as a dopant in organic films to improve the device efficiency and reliability. [194] A recent investigations show four orders of magnitude improvement of current density by co-evaporated LiF. [195] Optimal LiF doping not only leads to enhanced current transport, but also improves electron injection without using a low work function cathode. There have been few studies on the interactions between LiF dopant and the host organic semiconductors. [196] However, as one of the most strongly bonded ionic molecules, LiF itself is not expected to be an effective electron donor. It is thus also interesting to investigate the electronic structure of the co-evaporated material and its interaction with the metallic cathode in order to understand the role of LiF doping.

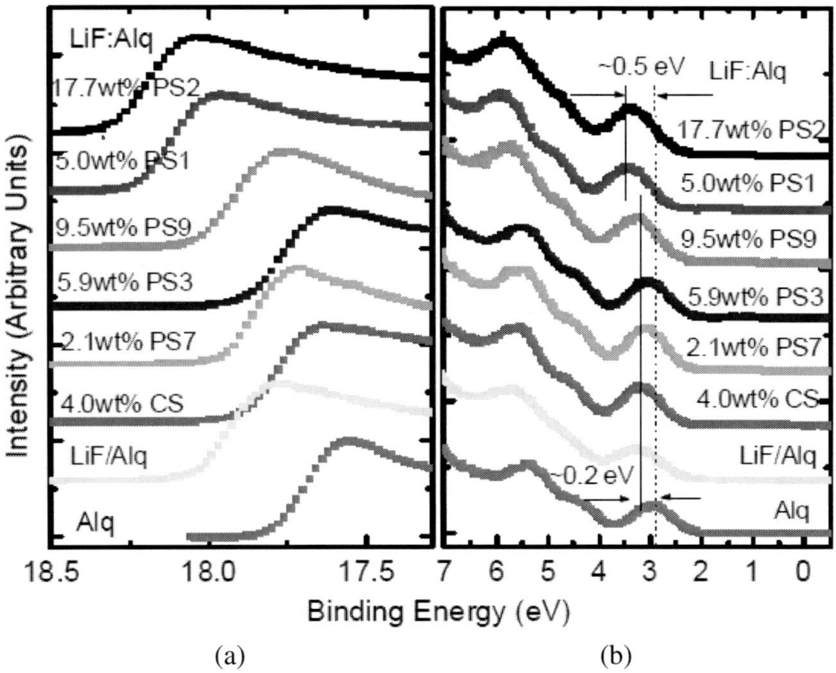

Figure 84. Evolution of the cut-off (a) and the HOMO (b) region in the UPS spectra for LiF-doped Alq film with different weight percentage. The spectra of pristine Alq and LiF-coated Alq films are also plotted at the bottom. The co-evaporated films prepared from crystal LiF source and powder source are noted by CS and PS, respectively. The evaporation sequences from the power LiF source are labeled as Ps1 to Ps9.

In Figure 84, [197] the UPS spectra of (a) the cut-off and (b) the HOMO region are presented for LiF-doped Alq films with various weight percentage (wt%). Given the molecular weight of LiF and Alq, 1 wt% LiF doping is equivalent to a LiF:Alq molecular ratio of ~ 0.18:1. The spectra for pristine Alq film and Alq film with a thin layer of 5 Å LiF evaporated from a crystalline source are also plotted at the bottom. All spectra are plotted as a function of binding energy relative to the Fermi level of the Au substrate. The pristine Alq film has a work function of 3.5 eV and a well-defined HOMO peak at about 3.0 eV below the Fermi level. The co-evaporated film with 3.5 wt% LiF from a crystalline LiF source (CS) resembles that of pristine Alq with negligible broadening, which suggests a well-preserved Alq surface, better than that of LiF-coated Alq where broadening of the peaks can be easily perceived. The HOMO of the 3.5 wt% LiF-doped Alq film shifts toward higher binding energy for about 0.2 eV, which is slightly less than that of LiF-coated Alq. The behavior of all LiF-doped Alq films from crystalline source is regardless of the doping ratio. This small shift of ~ 0.2 eV indicates limited n-doping, most likely induced by Li released during the evaporation process, and the LiF-coated Alq sets the upper limit of this doping effect when a crystalline LiF source is used. The situation is different for co-evaporated films from powder LiF sources (PS). In Figure 84, the spectra of the co-evaporated films from powder LiF source (PS) are labeled by the number of evaporation time sequence. Including degassing and rate calibration, each evaporation sequence takes about 45 minutes. It can be seen that the first two evaporations from powder source result in a comparably large shift toward higher binding energy for all energy levels in Alq. It is also important to note that although the weight percentages are quite different for these two samples, the mount of the shift for Alq HOMO is about the same, approximately 0.5 eV, corresponding to a stronger n-doping effect in comparison to that from a crystalline LiF source. Interestingly, the stronger doping effect diminishes after the first two evaporation sequences. By then, the total evaporation time for LiF powder source is estimated to be 1.5 hours. After that, the valence region resembles to that of LiF-doped Alq film from a crystalline LiF source. As one can see from curves of PS7, PS3, and PS9 in Figure 84, the shift of the HOMO is ~ 0.2 eV in a wide range of LiF doping ratio.

The lack of change in the UPS spectra of LiF-doped Alq film clearly suggests that there are no strong interactions between LiF and Alq. This is understandable, since LiF is one of the strongest ionic bonded materials. Although there is distortion of electron clouds centered at each hetero-nucleus Li and F, the molecular structure is too strong to be decomposed in large quantity by thermal evaporation. This means that LiF itself cannot supply Alq with many electrons. Nevertheless, though not much, the HOMO does shift toward higher binding energy for around 0.2 eV, which is an indication of n-type doping. [13] Moreover, this n-doping effect becomes stronger right after the crystalline source is crushed mechanically into powder. It is thus concluded that the LiF induced n-doping effect is mainly caused by the defects inside the source, which is increased as the crystal is crushed into powder. Li atoms can then be released from those defects relatively easily, and subsequently dope the Alq film. Such defect induced doping has also been found at Au/LiF/Alq interface, as we discussed in the previous section. [83] However, the mount of such defects is limited even in the powder source, and they are usually easier to evaporate. As a result, they can be quickly exhausted after a few evaporations. In fact, the spectra of later evaporations (PS3, PS7 and PS9), which are lack from a strong doping effect, further corroborate the above conclusion.

Figure 85 (a) and (b) show the evolution of the UPS spectra in the valence region of LiF-doped Alq films when metals are deposited on top. [197] Here, both LiF-doped Alq films are prepared from the crystalline LiF source. When Al is deposited onto the LiF-doped Alq film with LiF 3.5 wt%, the HOMO of Alq shifts toward higher binding energy dramatically, which is consistent with the behavior for Al on LiF-coated Alq film. [99] For thicker Al overlayers, the Alq features are diminished and broadened. At the final step, the HOMO shifts slightly back, but is still about 1.0 eV lower than that of pristine Alq film, resulting in a dramatic reduction of the electron injection barrier. Another piece of important information is that a gap state forms right after the Al deposition, whose position relative to the HOMO peak agrees very well with that of alkali-metal-doped Alq film (1.6 eV). [13] The appearance of the gap state and the diminishing of the Alq HOMO features clearly suggest that there are interactions between the deposited Al and the co-evaporated film. In contrast, when Au is used as the metal overlayer, the Alq features are well preserved with a gradual attenuation by the Au. The initial Au evaporation also introduces a small shift toward higher binding energy for the Alq HOMO. However, the shift is comparably small (~ 0.3 eV), and is quickly reversed by the subsequent depositions. Upon 5 Å of Au deposition, the HOMO is seen to shift back to the pristine Alq position.

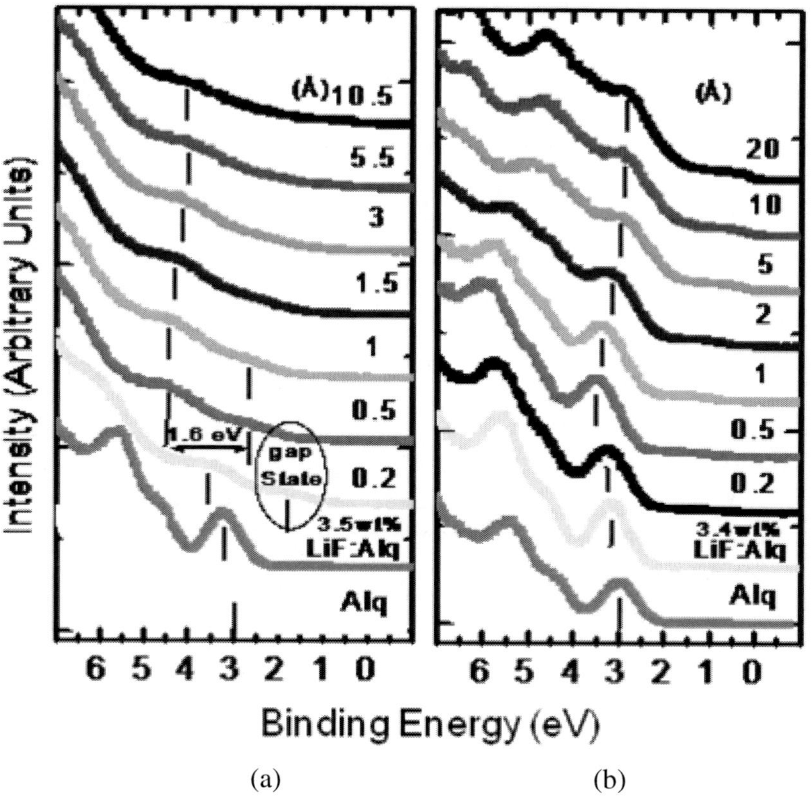

Figure 85. Evolution of the HOMO region in the UPS spectra for Al (a) and Au (b) on Li-doped Alq film as a function of metal coverage. When Al is deposited, a gap state appears whose position is about 1.6 eV above the HOMO.

The UPS measurements clearly suggest that the deposition of the metal overlayer can further influence the energy levels in Li-doped Alq in the way that is similar to LiF-coated Alq. [83, 99] Al is known to help the decomposition of LiF with the coexisting of Alq. [18] The released Li atoms then act as n-dopants, which donate electrons to Alq molecules, leading to a downwards shift for all the energy levels. Furthermore, as the electrons are transferred into Alq, they can fill in the former LUMO, leading to the formation of the gap state, as being reported for alkali-metal–doped Alq. [13] A recent J-V measurement of LiF-doped Alq with Al cathode indicates that the current density with 4 wt% doping is four orders higher than that in the undoped Alq film. [195] The photoemission results suggest that the increase of the current density can be attributed to the dramatic reduction of the electron injection barrier at the interface, due to the large downwards shift of about 1.0 eV for all energy levels. The other issue one may notice is that, after Al deposition, the valence features of LiF-doped Alq look quite different from those of the pristine Alq film (Figure 78 [99]). A very small amount of Al deposition (~ 0.2 Å) on pristine Alq leads to a significant modification of the valence features in the UPS spectrum. The Alq features are completely destroyed by the subsequent Al depositions. [99, 130] Shen *et al.* have also reported their UPS studies of Al/Alq interface, [198] where a small gap state is observed at the similar energy position as the one shown in Figure 85 (a). However, the intensity of the reported gap state at the Al/Alq interface is much less than the case of Al on LiF-doped Alq for the similar amount of Al deposition. More importantly, at the initial Al deposition (~ 1 Å), the LiF-doped Alq exhibits a much larger downwards shift (~ 1.5 eV), in comparison to that from the pristine Alq film. [198] These discrepancies indicate that the interface between Al and LiF-doped Alq film is more complicated, consisting of not only interactions that can be observed at Al/Alq interface, but a stronger n-type doping effect by the released Li atoms. In fact, the UPS spectra of Al on LiF-doped Alq resemble those of Al on LiF-coated Alq film. [99] The resemblance of the behaviors of Al on LiF-doped and LiF-covered Alq films indicates that the existence of LiF in either a mixed or overlayer form can inhibits the strong chemical reactions between Al and Alq, resulting in an n-doped interface region. The releasing of the Li atoms from LiF crystal is observed to depend greatly on the reactive nature of the metal overlayer. When a noble metal like Au is used instead of Al, strong interactions are not expected. A small amount of Li atoms may still be released from the existing defects by the hot Au atoms arriving at the surface, which explains the small shift at the initial Au deposition. However, the p-doping nature of Au can reverse the doping quickly, resulting in a fat-band situation. [139, 169]

Besides the formation of the new state in the former energy gap, it is also reported that in alkali-metal-doped Alq films, the doped electrons are localized at the pyridyl side of one of the three quinolate rings. [18] As a result, the N 1s core levels will be greatly affected. Figure 86 shows the N 1s core level spectra for (a) 3 Å Al and (b) 2 Å Au on LiF-doped Alq. [197] It is clear that Au deposition does not introduce any new component in the N 1s peak. However, when Al is deposited, a new component appears at the lower binding energy side of the original N 1s peak. The new component is characteristic of n-doping of Alq and the separation between the two peaks is about 1.8 eV, which is consistent with the earlier report. [18] Both peaks are broadened in comparison to the case of Au deposition. The experimental data are fitted by Gaussian peaks after careful deduction of the background. The area of the doping-induced component is estimated to be 0.32 relative to the total N 1s area, which suggests that about one additional electron has been transferred to each Alq molecule. The molecular ratio of Li to Alq is 0.61:1 for the co-evaporated film with 3.4 wt% of LiF. Since each alkali metal atom donates only one electron and there are three pyridyl groups in an Alq molecule, the maximum ratio of the doping-induced to the total N 1s peak area should be 0.63/3 = 0.21 even if all LiF molecules have been decomposed by Al deposition within the XPS probing depth of about 50 Å. The discrepancy of the maximum available Li and the actual amount of doping may have two possible reasons. First, Al may have participated in the doping process. It has been shown that Al deposition on pristine Alq produces a broad feature at the lower binding energy side of N 1s, which can be the contribution of both doping and destructive interactions. [99, 198] However, the intensity of the feature in the N 1s spectra at Al/Alq interface is usually very weak in comparison to the main component. Therefore, it should not have a dominate contribution to the low binding energy component of the N 1s spectra for Al on LiF-doped Alq. The second possibility is the charge accumulation at the interface region, as the large downwards shift of the energy levels by Al deposition may attract some electrons to the region.

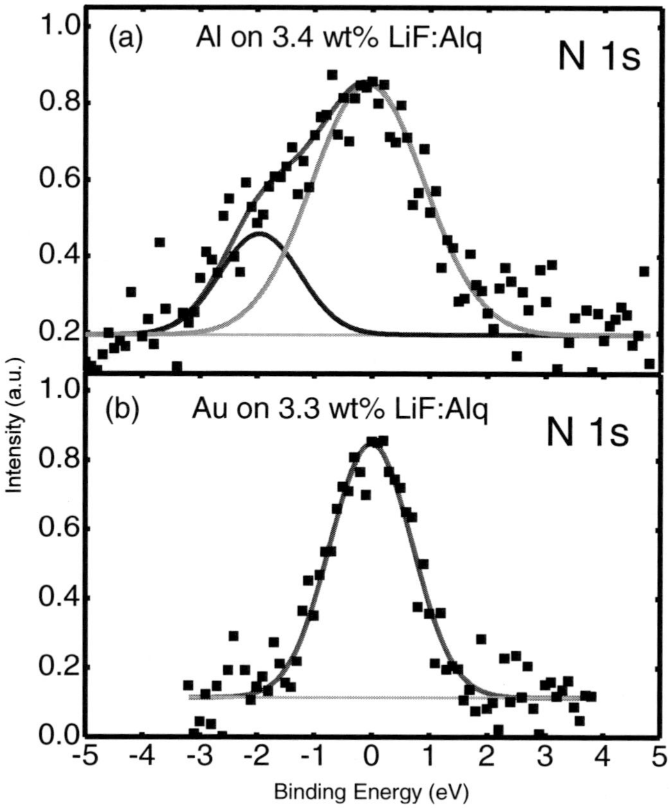

Figure 86. Peak decomposition of N 1s core level in XPS spectra from Al (a) and Au (b) on Li-doped Alq film. A new component is observed at the lower binding energy side for Al/LiF:Alq.

In the previous section, we have summarized the photoemission investigations on the electronic structure of LiF interlayer, either in its thin film form or as a dopant in Alq film, and the further interactions with different metal overlayers. In conclusion, LiF itself turns out to be an inefficient n-dopant in either thin film or dopant form. The weak n-doping effect is believed to be the result of the defects in LiF, mostly introduced during the preparation of the LiF source. However, Li can be released and subsequent dopes Alq film upon Al deposition. The strong n-doping effect substantially shifts the Alq energy levels downwards for about 1 eV, which significantly reduces the electron injection barrier. The thin LiF insertion layer also works as a buffer layer, which prevents Al diffusion and further destructive reactions with Alq molecules. In comparison, discernable interactions can be found when Al is deposited on to LiF-doped Alq film. Au does not cause extensive doping and results in an overall flat band in the interface region. Au deposition produced a small downwards shift of the Alq energy levels at the initial coverage. However, subsequent Au depositions reversed the shift to reach a flat-band situation. The reversal is attributed to the p-doping by Au that compensates for any Li release. The results here therefore explain the strong dependence on the metal overlayer material for the LiF/Alq system and the thickness requirement for polymer based devices prepared in ambient condition.

REFERENCES

[1] A. Yamamori, C. Adachi, T. Koyama and Y. Taniguchi, "Doped organic light emitting diodes having a 650-nm-thick hole transport layer," *Appl. Phys. Lett.* 72, 2147-2149 (1998).

[2] J. Kido, K. Hongawa, K. Okuyama and K. Nagai, "White Light-Emitting Organic Electroluminescent Devices Using the Poly(N-Vinylcarbazole) Emitter Layer Doped with 3 Fluorescent Dyes," *Appl. Phys. Lett.* 64, 815-817 (1994).

[3] H. Fujii, T. Sano, Y. Nishio, Y. Hamada and K. Shibata, "Improved stability of molecular organic EL devices," *Macromolecular Symposia* 125, 77-82 (1998).

[4] J. Kido and T. Matsumoto, "Bright organic electroluminescent devices having a metal-doped electron-injecting layer," *Appl. Phys. Lett.* 73, 2866-2868 (1998).

[5] E. J. Lous, P. W. M. Blom, L. W. Molenkamp and D. M. Deleeuw, "Schottky Contacts on a Highly Doped Organic Semiconductor," *Physical Review* B 51, 17251-17254 (1995).

[6] C. M. Heller and I. H. Campbell, "Chemical potential pinning due to equilibrium electron transfer at metal/C-60-doped polymer interfaces," *J. Appl. Phys.* 81, 3227-3231 (1997).

[7] Y. Yang and A. J. Heeger, "A New Architecture for Polymer Transistors," *Nature* 372, 344-346 (1994).

[8] J. R. Sheats, H. Antoniadis, M. Hueschen, W. Leonard, J. Miller, R. Moon, D. Roitman and A. Stocking, "Organic electroluminescent devices," *Science* 273, 884-888 (1996).

[9] J. Xue and S. R. Forrest, "Bipolar doping between a molecular organic donor-acceptor couple," *Physical Review* B 69, 245322 (2004).

[10] T. Nishi, K. Kanai, Y. Ouchi, M. R. Willis and K. Seki, "Oxygen effects on the interfacial electronic structure of titanyl phthalocyanine film: p-type doping, band bending and Fermi level alignment," *Chemical Physics* 325, 121-128 (2006).

[11] G. Parthasarathy, C. Shen, A. Kahn and S. R. Forrest, "Lithium doping of semiconducting organic charge transport materials," *J. Appl. Phys.* 89, 4986-4992 (2001).

[12] L. Yan, N. J. Watkins, S. Zorba, Y. L. Gao and C. W. Tang, "Direct observation of fermi-level pinning in Cs-doped CuPc film," *Applied Physics Letters* 79, 4148-4150 (2001).

[13] H. Ding and Y. Gao, "Electronic structure of Cs-doped tris(8-hydroxyquinoline) aluminum," *Appl. Phys. Lett.* 86, 213508 (2005).

[14] H. Ding and Y. Gao, "Alkali metal doping and energy level shift in organic semiconductors," *Applied Surface Science* 252, 3943-3947 (2006).

[15] H. Ding and Y. Gao, "Evolution of the unoccupied states in Cs-doped copper phthalocyanine," *Appl. Phys. Lett.* 92, 053309 (2008).

[16] H. Ding, K. W. Park, K. Green and Y. Gao, "Electronic structure modification of copper phthalocyanine (CuPc) induced by intensive Na doping," *Chemical Physics Letters* 454, 229-232 (2008).

[17] Y. L. Gao and L. Yan, "Cs doping and energy level shift in CuPc," *Chemical Physics Letters* 380, 451-455 (2003).

[18] M. G. Mason, C. W. Tang, L. S. Hung, P. Raychaudhuri, J. Madathil, D. J. Giesen, L. Yan, Q. T. Le, Y. Gao, S. T. Lee, L. S. Liao, L. F. Cheng, W. R. Salaneck, D. A. dos Santos and J. L. Bredas, "Interfacial chemistry of Alq(3) and LiF with reactive metals," *J. Appl. Phys.* 89, 2756-2765 (2001).

[19] H. J. Ding and Y. L. Gao, "Alkali metal doping and energy level shift in organic semiconductors," *Applied Surface Science* 252, 3943-3947 (2006).

[20] J. Wusten, S. Berger, K. Heimer, S. Lach and C. Ziegler, *Journal of Applied Physics* 98 (2005).

[21] J. Wusten, K. Heimer, S. Lach and C. Ziegler, *Journal of Applied Physics* 102 (2007).

[22] N. Koch, H. Oji, E. Ito, E. Zojer, H. Ishii, G. Leising and K. Seki, "Combined photoelectron and metastable atom electron spectroscopy study of n-doped oligophenylene thin films," *Applied Surface Science* 175, 764-768 (2001).

[23] N. Koch, G. Leising, L. M. Yu, A. Rajagopal, J. J. Pireaux and R. L. Johnson, "Bipolaron formation in para-sexiphenyl thin films upon Cs doping," *Journal of Vacuum Science and Technology a-Vacuum Surfaces and Films* 18, 295-298 (2000).

[24] Y. Zhiyong, M. Bensimon, V. Sarria, I. Stolitchnov, W. Jardim, D. Laub, E. Mielczarski, J. Mielczarski, L. Kiwi-Minsker and J. Kiwi, "ZnSO4-TiO2 doped catalyst with higher activity in photocatalytic processes," *Applied Catalysis B-Environmental* 76, 185-195 (2007).

[25] W. F. Zhang, P. Schmidt-Zhang and G. Kossmehl, "Analysis of photocurrent spectra at polybithienyl film coated on Pt," *Journal of Solid State Electrochemistry* 4, 225-230 (2000).

[26] C. K. Yang, C. M. Yang, H. H. Liao, S. F. Horng and H. F. Meng, "Current injection and transport in polyfluorene," *Applied Physics Letters* 91 (2007).

[27] L. Yan, N. J. Watkins, C. W. Tang and Y. L. Gao, "Fermi level pinning in Cs doped CuPc," *Synthetic Metals* 137, 1037-1038 (2003).

[28] J. G. Xue and S. R. Forrest, "Bipolar doping between a molecular organic donor-acceptor couple," *Physical Review* B 69 (2004).

[29] C. I. Wu, C. T. Lin, Y. H. Chen, M. H. Chen, Y. J. Lu and C. C. Wu, "Electronic structures and electron-injection mechanisms of cesium-carbonate-incorporated cathode structures for organic light-emitting devices," *Applied Physics Letters* 88 (2006).

[30] M. Willander, A. Assadi and C. Svensson, "Polymer-Based Devices Their Function and Characterization," *Synthetic Metals* 57, 4099-4104 (1993).

[31] Y. Wang, W. Y. Gao, S. Braun, W. R. Salaneck, F. Amy, C. Chan and A. Kahn, "Enhancement of iridium-based organic light-emitting diodes by spatial doping of the hole transport layer," *Applied Physics Letters* 87 (2005).

[32] S. D. Wang, K. Kanai, E. Kawabe, Y. Ouchi and K. Seki, "Current characteristics of pristine and tetrathianaphthacene-doped tris(8-hydroxyquinoline) aluminum (ALQ3) thin films," *Molecular Crystals and Liquid Crystals* 455, 339-346 (2006).

[33] S. D. Wang, K. Kanai, E. Kawabe, Y. Ouchi and K. Seki, "Enhanced electron injection into tris(8-hydroxyquinoline) aluminum (Alq(3)) thin films by tetrathianaphthacene (TTN) doping revealed by current-voltage characteristics," *Chemical Physics Letters* 423, 170-173 (2006).

[34] A. Vollmer, O. D. Jurchescu, I. Arfaoui, I. Salzmann, T. T. M. Palstra, P. Rudolf, J. Niemax, J. Pflaum, J. P. Rabe and N. Koch, "The effect of oxygen exposure on pentacene electronic structure," *European Physical Journal* E 17, 339-343 (2005).

[35] D. M. Taylor and N. Alves, "Separating interface state response from parasitic effects in conductance measurements on organic metal-insulator-semiconductor capacitors," *Journal of Applied Physics* 103 (2008).

[36] M. C. Tanese, D. Fine, A. Dodabalapur and L. Torsi, "Interface and gate bias dependence responses of sensing organic thin-film transistors," *Biosensors and Bioelectronics* 21, 782-788 (2005).

[37] S. Tanaka, K. Kanai, E. Kawabe, T. Iwahashi, T. Nishi, Y. Ouchi and K. Seki, "Doping effect of tetrathianaphthacene molecule in organic semiconductors on their interfacial electronic structures studied by UV photoemission spectroscopy," *Japanese Journal of Applied Physics Part 1-Regular Papers Short Notes and Review Papers* 44, 3760-3763 (2005).

[38] T. T. Tan, J. W. Gardner, J. Farrington and P. N. Bartlett, "Electronic-Properties of Metal Poly(Pyrrole) Junctions," *International Journal of Electronics* 77, 173-184 (1994).

[39] W. Takashima, T. Murasaki, S. Nagamatsu, T. Morita and K. Kaneto, "Unipolarization of ambipolar organic field effect transistors toward high-impedance complementary metal-oxide-semiconductor circuits," *Applied Physics Letters* 91 (2007).

[40] Y. L. Shen, A. R. Hosseini, M. H. Wong and G. G. Malliaras, "How to make ohmic contacts to organic semiconductors," *Chemphyschem* 5, 16-25 (2004).

[41] C. Shen and A. Kahn, "Electronic structure, diffusion, and p-doping at the Au/F16CuPc interface," *Journal of Applied Physics* 90, 4549-4554 (2001).

[42] G. D. Sharma, S. G. Sangodkar and M. S. Roy, "Effect of rare-earth doping on the electrical and photoelectrical properties of furazano[3,4-b]piperazine (FP) thin-film devices," *Synthetic Metals* 80, 249-256 (1996).

[43] G. D. Sharma, S. K. Gupta and M. S. Roy, "Investigation of the effect of rare earth doping on the electrical and photovoltaic properties of chromotrope 2R thin film devices," *Journal of Materials Science-Materials in Electronics* 9, 91-97 (1998).

[44] K. Seki, T. Tani and H. Ishii, "Electronic structures of organic-inorganic interfaces studied by UV photoemission," *Thin Solid Films* 273, 20-26 (1996).

[45] R. Schlaf, C. D. Merritt, L. A. Crisafulli and Z. H. Kafafi, "Organic semiconductor interfaces: Discrimination between charging and band bending related shifts in frontier orbital line-up measurements with photoemission spectroscopy," *Journal of Applied Physics* 86, 5678-5686 (1999).

[46] T. J. Savenije, E. Moons, G. K. Boschloo, A. Goossens and T. J. Schaafsma, "Photogeneration and transport of charge carriers in a porphyrin p/n heterojunction," *Physical Review* B 55, 9685-9692 (1997).

[47] N. Sato, H. Yoshida and K. Tsutsumi, "Unoccupied electronic structure in organic thin films studied by inverse photoemission spectroscopy," *Journal of Materials Chemistry* 10, 85-89 (2000).

[48] W. R. Salaneck, *"Conjugated polymer surfaces and interfaces,"* Philosophical Transactions of the Royal Society of London Series a-Mathematical Physical and Engineering Sciences 355, 789-799 (1997).

[49] Y. Sakurai, T. Yokoyama, Y. Hosoi, H. Ishii, Y. Ouchi, G. Salvan, A. Kobitski, T. U. Kampen, D. R. T. Zahn and K. Seki, "Study of the interaction of tris-(8-hydroxyquinoline) aluminum (Alq(3)) with potassium using vibrational spectroscopy: Examination of the possible isomerization upon K-doping," *Synthetic Metals* 154, 161-164 (2005).

[50] B. Remaki, G. Guillaud and D. Mayes, "Space charge analysis in doped zinc phthalocyanine thin films," *Optical Materials* 9, 240-244 (1998).

[51] R. Q. Png, P. J. Chia, S. Sivaramakrishnan, L. Y. Wong, M. Zhou, L. L. Chua and P. K. H. Ho, "Electromigration of the conducting polymer in organic semiconductor devices and its stabilization by cross-linking," *Applied Physics Letters* 91 (2007).

[52] T. W. Pi and T. C. Yu, "The organic led surface: A synchrotron radiation photoemission study," *Surface Review and Letters* 14, 377-385 (2007).

[53] H. Peisert, M. Knupfer, F. Zhang, A. Petr, L. Dunsch and J. Fink, "The interface between phthalocyanines and PEDOT: PSS: evidence for charge transfer and doping," *Surface Science* 566, 554-559 (2004).

[54] H. Peisert, M. Knupfer, F. Zhang, A. Petr, L. Dunsch and J. Fink, "Charge transfer and doping at organic/organic interfaces," *Applied Physics Letters* 83, 3930-3932 (2003).

[55] G. Paasch and S. Scheinert, "Space charge layers in organic field-effect transistors with Gaussian or exponential semiconductor density of states," *Journal of Applied Physics* 101 (2007).

[56] T. Osada, P. Barta, N. Johansson, T. Kugler, P. Broms and W. R. Salaneck, "XPS and UPS study of charge transport material electrode interface of light emitting diodes," *Synthetic Metals* 102, 1103-1104 (1999).

[57] B. S. Ong, Y. L. Wu and P. Liu, "Design of high-performance regioregular polythiophenes for organic thin-film transistors," *Proceedings of the Ieee* 93, 1412-1419 (2005).

[58] T. Naito, T. Inabe, H. Niimi and K. Asakura, "A new optical doping method toward molecular electronics," *Synthetic Metals* 152, 289-292 (2005).

[59] I. Muzikante, V. Parra, R. Dobulans, E. Fonavs, J. Latvels and M. Bouvet, "A novel gas sensor transducer based on phthalocyanine heterojunction devices," *Sensors* 7, 2984-2996 (2007).

[60] T. Miyadera, T. Minari, K. Tsukagoshi, H. Ito and Y. Aoyagi, "Frequency response analysis of pentacene thin-film transistors with low impedance contact by interface molecular doping," *Applied Physics Letters* 91 (2007).

[61] A. Liu, S. Zhao, S. B. Rim, J. Wu, M. Konemann, P. Erk and P. Peumans, "Control of electric field strength and orientation at the donor-acceptor interface in organic solar cells," *Advanced Materials* 20, 1065-+ (2008).

[62] D. S. Leem, S. Y. Kim, J. J. Kim, M. H. Chen and C. I. Wu, "Rubidium-Carbonate-Doped 4,7-Diphenyl-1,10-phenanthroline Electron Transporting Layer for High-Efficiency p-i-n Organic Light Emitting Diodes," *Electrochemical and Solid State Letters* 12, J8-J10 (2009).

[63] K. Hong, J. W. Lee, S. Y. Yang, K. Shin, H. Jeon, S. H. Kim, C. Yang and C. E. Park, "Lower hole-injection barrier between pentacene and a 1-hexadecanethiol-modified gold substrate with a lowered work function," *Organic Electronics* 9, 21-29 (2008).

[64] S. Kwon, S. C. Kim, Y. Kim, J. G. Lee, S. Kim and K. Jeong, "Photoemission spectroscopy study of Alq3 and metal mixed interfaces," *Applied Physics Letters* 79, 4595-4597 (2001).

[65] A. Kahn, W. Zhao, W. Y. Gao, H. Vazquez and F. Flores, "Doping-induced realignment of molecular levels at organic-organic heterojunctions," *Chemical Physics* 325, 129-137 (2006).

[66] D. B. Janes, M. Batistuta, S. Datta, M. R. Melloch, R. P. Andres, J. Liu, N. P. Chen, T. Lee, R. Reifenberger, E. H. Chen and J. M. Woodall, "Interface and contact structures for nanoelectronic devices using assemblies of metallic nanoclusters, conjugated organic molecules and chemically stable semiconductor layers," *Superlattices and Microstructures* 27, 555-563 (2000).

[67] H. Ishii, N. Hayashi, E. Ito, Y. Washizu, K. Sugi, Y. Kimura, M. Niwano, Y. Ouchi and K. Seki, "Kelvin probe study of band bending at organic semiconductor/metal interfaces: examination of Fermi level alignment," *Physica Status Solidi a-Applied Research* 201, 1075-1094 (2004).

[68] K. Ikegami, H. Ohnuki and M. Izumi, "Field-effect doping phenomenon in an organic semiconductor layer deposited by the Langmuir-Blodgett technique," *Current Applied Physics* 6, 808-812 (2006).

[69] J. Hwang and A. Kahn, "Electrical doping of poly(9,9-dioctylfluorenyl-2,7-diyl) with tetrafluorotetracyanoquinodimethane by solution method," *Journal of Applied Physics* 97 (2005).

[70] A. R. Hosseini, M. H. Wong, Y. L. Shen and G. G. Malliaras, "Charge injection in doped organic semiconductors," *Journal of Applied Physics* 97 (2005).

[71] G. Greczynski, W. R. Salaneck and M. Fahlman, "Hybrid interfaces in polymer-based electronics," *Synthetic Metals* 121, 1625-1628 (2001).

[72] G. Greczynski, M. Fahlman and W. R. Salaneck, "Hybrid interfaces of poly(9,9-dioctylfluorene) employing thin insulating layers of CsF: A photoelectron spectroscopy study," *Journal of Chemical Physics* 114, 8628-8636 (2001).

[73] F. Garnier, "Thin-film transistors based on organic conjugated semiconductors," *Chemical Physics* 227, 253-262 (1998).

[74] F. Garnier, "Specificity and limits of organic-based electronic devices," *Photonic and Optoelectronic Polymers* 672, 420-436 (1997).

[75] W. Y. Gao and A. Kahn, "Effect of electrical doping on molecular level alignment at organic-organic heterojunctions," *Applied Physics Letters* 82, 4815-4817 (2003).

[76] W. Y. Gao and A. Kahn, "Controlled p doping of the hole-transport molecular material N,N '-diphenyl-N,N '-bis(1-naphthyl)-1,1(')-biphenyl-4,4 '-diamine with tetrafluorotetracyanoquinodimethane," *Journal of Applied Physics* 94, 359-366 (2003).

[77] W. Y. Gao and A. Kahn, "Electronic structure and current injection in zinc phthalocyanine doped with tetrafluorotetracyanoquinodimethane: Interface versus bulk effects," *Organic Electronics* 3, 53-63 (2002).

[78] W. Y. Gao and A. Kahn, "Controlled p-doping of zinc phthalocyanine by coevaporation with tetrafluorotetracyanoquinodimethane: A direct and inverse photoemission study," *Applied Physics Letters* 79, 4040-4042 (2001).

[79] M. K. Fung, K. M. Lau, S. L. Lai, C. W. Law, M. Y. Chan, C. S. Lee and S. T. Lee, "Charge generation layer in stacked organic light-emitting devices," *Journal of Applied Physics* 104 (2008).

[80] M. K. Fung, S. L. Lai, S. N. Bao, C. S. Lee, S. T. Lee, W. W. Wu, M. Inbasekaran and J. J. O'Brien, "Interface between poly (9,9-dioctylfluorene) and alkali metals: cesium, potassium, sodium, and lithium," *Journal of Vacuum Science and Technology a-Vacuum Surfaces and Films* 20, 911-918 (2002).

[81] K. Fehse, S. Olthof, K. Walzer, K. Leo, R. L. Johnson, H. Glowatzki, B. Broker and N. Koch, "Energy level alignment of electrically doped hole transport layers with transparent and conductive indium tin oxide and polymer anodes," *Journal of Applied Physics* 102 (2007).

[82] H. J. Ding, K. W. Park, K. Green and Y. L. Gao, "Electronic structure modification of copper phthalocyanine (CuPc) induced by intensive Na doping," *Chemical Physics Letters* 454, 229-232 (2008).

[83] H. Ding and Y. Gao, "Au/LiF/tris(8-hydroxyquinoline) aluminum interfaces," *Applied Physics Letters* 91 (2007).

[84] N. R. Dennington, A. C. Wright and J. O. Williams, "Use of Phenylarsine in the Atmospheric-Pressure Metal Organic-Chemical Vapor-Deposition of GAAS on SI(100)," *Journal of Materials Chemistry* 1, 663-666 (1991).

[85] P. Cova, A. Singh, A. Medina and R. A. Masut, "Effect of doping on the forward current-transport mechanisms in a metal-insulator-semiconductor contact to InP : Zn grown by metal organic vapor phase epitaxy," *Solid-State Electronics* 42, 477-485 (1998).

[86] W. Chen, X. Y. Gao, D. C. Qi, S. Chen, Z. K. Chen and A. T. S. Wee, "Surface-transfer doping of organic semiconductors using functionalized self-assembled monolayers," *Advanced Functional Materials* 17, 1339-1344 (2007).

[87] F. C. Chen, J. L. Wu, S. S. Yang, K. H. Hsieh and W. C. Chen, "Cesium carbonate as a functional interlayer for polymer photovoltaic devices," *Journal of Applied Physics* 103 (2008).

[88] M. Y. Chan, S. L. Lai, K. M. Lau, C. S. Lee and S. T. Lee, "Application of metal-doped organic layer both as exciton blocker and optical spacer for organic photovoltaic devices," *Applied Physics Letters* 89 (2006).

[89] C. K. Chan, W. Zhao, S. Barlow, S. Marder and A. Kahn, "Decamethylcobaltocene as an efficient n-dopant in organic electronic materials and devices," *Organic Electronics* 9, 575-581 (2008).

[90] C. K. Chan, E. G. Kim, J. L. Bredas and A. Kahn, "Molecular n-type doping of 1,4,5,8-naphthalene tetracarboxylic dianhydride by pyronin B studied using direct and inverse photoelectron spectroscopies," *Advanced Functional Materials* 16, 831-837 (2006).

[91] D. P. Brunco, B. De Jaeger, G. Eneman, J. Mitard, G. Hellings, A. Satta, V. Terzieva, L. Souriau, F. E. Leys, G. Pourtois, M. Houssa, G. Winderickx, E. Vrancken, S. Sioncke, K. Opsomer, G. Nicholas, M. Caymax, A. Stesmans, J. Van Steenbergen, P. W. Mertens, M. Meuris and M. M. Heyns, "Germanium MOSFET devices: Advances in materials understanding, process development, and electrical performance," *Journal of the Electrochemical Society* 155, H552-H561 (2008).

[92] J. Bisquert, G. Garcia-Belmonte, A. Munar, M. Sessolo, A. Soriano and H. J. Bolink, "Band unpinning and photovoltaic model for P3HT:PCBM organic bulk heterojunctions under illumination," *Chemical Physics Letters* 465, 57-62 (2008).
[93] H. Benmansour, F. A. Castro, M. Nagel, J. Heier, R. Hany and F. Nuesch, "Ionic space charge driven organic photovoltaic devices," *Chimia* 61, 787-791 (2007).
[94] P. Bauerle, G. Gotz, M. Hiller, S. Scheib, T. Fischer, U. Segelbacher, M. Bennati, A. Grupp, M. Mehring, M. Stoldt, C. Seidel, F. Geiger, H. Schweizer, E. Umbach, M. Schmelzer, S. Roth, H. J. Egelhaaf, D. Oelkrug, P. Emele and H. Port, "Design, Synthesis and Assembly of New Thiophene-Based Molecular Functional Units with Controlled Properties," *Synthetic Metals* 61, 71-79 (1993).
[95] V. Y. Aristov, O. V. Molodtsova, V. M. Zhilin, D. V. Vyalikh and M. Knupfer, "Characterisation of metal-organic semiconductor interfaces: In and Sn on CuPc," *Journal De Physique Iv* 132, 101-104 (2006).
[96] M. Ahles, R. Schmechel and H. von Seggern, "Complementary inverter based on interface doped pentacene," *Applied Physics Letters* 87 (2005).
[97] M. Ahles, R. Schmechel and H. von Seggern, "n-type organic field-effect transistor based on interface-doped pentacene," *Applied Physics Letters* 85, 4499-4501 (2004).
[98] J. Ackermann, C. Videlot, A. El Kassmi, R. Guglielmetti and F. Fages, "Highly efficient hybrid solar cells based on an octithiophene-GaAs heterojunction," *Advanced Functional Materials* 15, 810-817 (2005).
[99] Q. T. Le, L. Yan, Y. G. Gao, M. G. Mason, D. J. Giesen and C. W. Tang, "Photoemission study of aluminum/tris-(8-hydroxyquinoline) aluminum and aluminum/LiF/tris-(8-hydroxyquinoline) aluminum interfaces," *J. Appl. Phys.* 87, 375-379 (2000).
[100] T. D. Anthopoulos, J. P. J. Markham, E. B. Namdas, J. R. Lawrence, I. D. W. Samuel, S. C. Lo and P. L. Burn, "Influence of molecular structure on the properties of dendrimer light-emitting diodes," *Organic Electronics* 4, 71-76 (2003).
[101] T. D. Anthopoulos, J. P. J. Markham, E. B. Namdas, I. D. W. Samuel, S. C. Lo and P. L. Burn, "Highly efficient single-layer dendrimer light-emitting diodes with balanced charge transport," *Appl. Phys. Lett.* 82, 4824-4826 (2003).
[102] S. M. Sze, *Physics of Semiconductor Devices*. (John Wiley, New York, 1981).
[103] H. Ishii, K. Sugiyama, E. Ito and K. Seki, "Energy level alignment and interfacial electronic structures at organic metal and organic organic interfaces," *Adv. Mater.* 11, 605-+ (1999).
[104] *Semiconducting Polymer*. (Wiley-VCH, Weinheim, 2007).
[105] D. P. Woodruff and T. A. Delchar, *Modern Techniques of Surface Science*. (Cambridge University Press, Cambridge, 1986).
[106] E. A. Silinsh and V. Capek, *Organic Molecular Crystal*. (AIP Press, New York, 1994).
[107] Y. Toyozawa, *Progr. Theor. Phys*. (Tokyo) 12, 241 (1954).
[108] A. Einstein, "On a Heuristic Viewpoint Concerning the Production and Transformation of Light," *Annalen der Physik* 17, 132-148 (1905).
[109] M. Cardona and L. Ley, *Photoemission in Solids*. (Springer-Verlag, Berlin, Germany, 1978).
[110] S. Hufner, *Photoelectron Spectroscopy: Principles and applications*. (Springer-Verlag, Berlin, Germany, 1996).

[111] M. Pope and C. E. Swenberg, *Electronic Processes in Organic Crystals and Polymers*, 2nd Edition ed. (Oxford Univ. Press, New York, 1999).
[112] R. L. Park and M. G. Lagally, *Solid State Physics: Surfaces*. (Academic Press, Orlando, 1985).
[113] P. W. Palmberg, "Combined Esca-Auger System Based on DOuble Pass Cylindrical Mirror Analyzer," *Journal of Electron Spectroscopy and Related Phenomena* 5, 691-703 (1974).
[114] S. D. Kevan, "Design of a High-Resolution Angle-Resolving Electron-Energy Analyzer," *Review of Scientific Instruments* 54, 1441-1445 (1983).
[115] T. K. Koopmans, *Physica* 1, 104 (1933).
[116] L. E. Davis, J. F. Moulder, C. D. Wagner, W. M. Riggs and G. E. Muilenberg, *Handbook of X-ray Photoemission Spectroscopy*. (Perkin-Elmer Corporation, Minnesota, 1978).
[117] M. P. Seah and W. A. Dench, *Surface and Interface Analysis* 1 (1979).
[118] R. F. Reilman, A. Msezane and S. T. Manson, "Relative Intensities in Photoelectron-Spectroscopy of Atoms and Molecules," *Journal of Electron Spectroscopy and Related Phenomena* 8, 389-394 (1976).
[119] J. H. Scofield, "Hartree-Slater Subshell Photoionization Cross-Sections at 1254 and 1487EV," *Journal of Electron Spectroscopy and Related Phenomena* 8, 129-137 (1976).
[120] N. V. Smith and D. P. Woodruff, "Inverse Photoemission from Metal-Surfaces," *Progress in Surface Science* 21, 295-370 (1986).
[121] G. Borstel and G. Thorner, "Inverse Photoemission from Solids - Theoretical Aspects and Applications," *Surface Science Reports* 8, 1-41 (1988).
[122] N. V. Smith, "Inverse Photoemission," *Reports on Progress in Physics* 51, 1227-1294 (1988).
[123] F. J. Himpsel, "Inverse Photoemission from Semiconductors," *Surface Science Reports* 12, 1-48 (1990).
[124] P. D. Johnson and S. L. Hulbert, "Inverse Photoemission," *Review of Scientific Instruments* 61, 2277-2288 (1990).
[125] L. S. Hung, L. S. Liao, C. S. Lee and S. T. Lee, "Sputter deposition of cathodes in organic light emitting diodes," *Journal of Applied Physics* 86, 4607-4612 (1999).
[126] R. Treusch, F. J. Himpsel, S. Kakar, L. J. Terminello, C. Heske, T. van Buuren, V. V. Dinh, H. W. Lee, K. Pakbaz, G. Fox and I. Jimenez, "X-ray photoemission and photoabsorption of organic electroluminescent materials," *Journal of Applied Physics* 86, 88-93 (1999).
[127] N. Isomura, T. Satoh, M. Suzuki, T. Ohwaki and Y. Taga, "Photoemission spectroscopy of the interface between indium-tin-oxide and copper phthalocyanine for transparent organic light-emitting devices," *Japanese Journal of Applied Physics* Part 2-Letters 40, L1038-L1039 (2001).
[128] R. I. R. Blyth, J. Thompson, V. Arima, Y. Zou, R. Fink, E. Umbach, G. Gigli and R. Cingolani, "Occupied and unoccupied states of the organic infrared emitters Yb- and Er-tris(8-hydroxyquinoline) studied by photoemission and X-ray absorption," *Synthetic Metals* 142, 293-298 (2004).
[129] L. J. Brillson, "*Handbook on Semiconductors* , New York, 1992," *Vol. 1,* edited by P. T. Landsberg (Elsevier Science Publishers, New York, 1992).

[130] L. Yan, M. G. Mason, C. W. Tang and Y. L. Gao, "Photoemission study of energy alignment at the metal/Alq(3) interfaces," *Applied Surface Science* 175, 412-418 (2001).
[131] P. E. Burrows, Z. Shen, V. Bulovic, D. M. McCarty, S. R. Forrest, J. A. Cronin and M. E. Thompson, "Relationship between electroluminescence and current transport in organic heterojunction light-emitting devices," *J. Appl. Phys.* 79, 7991-8006 (1996).
[132] S. Krause, M. B. Casu, A. Scholl and E. Umbach, "Determination of transport levels of organic semiconductors by UPS and IPS," *New Journal of Physics* 10, - (2008).
[133] *CRC handbook on Chemistry and Physics*. (The Chemical Rubber Co., Cleveland, OH, 2008).
[134] N. Johansson, T. Osada, S. Stafstrom, W. R. Salaneck, V. Parente, D. A. dos Santos, X. Crispin and J. L. Bredas, "Electronic structure of tris(8-hydroxyquinoline) aluminum thin films in the pristine and reduced states," *Journal of Chemical Physics* 111, 2157-2163 (1999).
[135] M. S. Liao and S. Scheiner, "Electronic structure and bonding in metal phthalocyanines, Metal=Fe, Co, Ni, Cu, Zn, Mg," *Journal of Chemical Physics* 114, 9780-9791 (2001).
[136] Y. Harima, H. Okazaki, Y. Kunugi, K. Yamashita, H. Ishii and K. Seki, "Formation of Schottky barriers at interfaces between metals and molecular semiconductors of p- and n-type conductances," *Appl. Phys. Lett.* 69, 1059-1061 (1996).
[137] I. G. Hill, A. Kahn, J. Cornil, D. A. dos Santos and J. L. Bredas, "Occupied and unoccupied electronic levels in organic pi-conjugated molecules: comparison between experiment and theory," *Chemical Physics Letters* 317, 444-450 (2000).
[138] I. G. Hill, A. Kahn, Z. G. Soos and R. A. Pascal, "Charge-separation energy in films of pi-conjugated organic molecules," *Chemical Physics Letters* 327, 181-188 (2000).
[139] H. Ding and Y. Gao, "Photoemission study of oxygen and Au modification of doped copper phthalocyanine," *J. Appl. Phys.* 102, 043703 (2007).
[140] T. Schwieger, H. Peisert, M. S. Golden, M. Knupfer and J. Fink, "Electronic structure of the organic semiconductor copper phthalocyanine and K-CuPc studied using photoemission spectroscopy," *Physical Review* B 66, 155207 (2002).
[141] A. Rosa and E. J. Baerends, "Origin and Relevance of the Staggering in One-Dimensional Molecular-Metals - a Density Functional-Study of Metallophthalocyanine Model Dimers," *Inorganic Chemistry* 31, 4717-4726 (1992).
[142] O. V. Molodtsova, V. M. Zhilin, D. V. Vyalikh, V. Y. Aristov and M. Knupfer, "Electronic properties of potassium-doped CuPc," *J. Appl. Phys.* 98, 093702 (2005).
[143] H. Peisert, M. Knupfer, T. Schwieger, G. G. Fuentes, D. Olligs, J. Fink and T. Schmidt, "Fluorination of copper phthalocyanines: Electronic structure and interface properties," *J. Appl. Phys.* 93, 9683-9692 (2003).
[144] J. X. Tang, C. S. Lee and S. T. Lee, "Chemical bonding and electronic structures at magnesium/copper phthalocyanine interfaces," *Applied Surface Science* 252, 3948-3952 (2006).
[145] F. Evangelista, A. Ruocco, D. Pasca, C. Baldacchini, M. G. Betti, V. Corradini and C. Mariani, "Au(110) induced reconstruction by pi conjugated molecules adsorption investigated by photoemission spectroscopy and low energy electron diffraction," *Surface Science* 566, 79-83 (2004).

[146] F. Evangelista, A. Ruocco, V. Corradini, M. P. Donzello, C. Mariani and M. G. Betti, "CuPc molecules adsorbed on Au(110)-(1 x 2): Growth morphology and evolution of valence band states," *Surface Science* 531, 123-130 (2003).

[147] M. G. Betti, F. Crispoldi, A. Ruocco and C. Mariani, "Insulating state of electron-doped Cu-phthalocyanine layers," *Physical Review* B 76, 125407 (2007).

[148] H. Overhof, J. Knecht, R. Fischer and F. Hensel, "Photoelectron-Spectroscopy Studies and Calculation of Energy-Bands of Csau and Rbau," *J. Phys. F Met. Phys.* 8, 1607-1616 (1978).

[149] S. Berger, K. Heimer, H. G. Mack and C. Ziegler, "IR and SFM study of PTCDA thin films on different substrates," *Applied Surface Science* 252, 81-84 (2005).

[150] S. R. Forrest, P. E. Burrows, E. I. Haskal and F. F. So, "Ultrahigh-Vacuum Quasiepitaxial Growth of Model Van-Der-Waals Thin-Films .2. Experiment," *Physical Review* B 49, 11309-11321 (1994).

[151] M. Schneider, E. Umbach and M. Sokolowski, "Growth-dependent optical properties of 3,4,9,10-perylenetetracarboxylicacid-dianhydride (PTCDA) films on Ag(111)," *Chemical Physics* 325, 185-192 (2006).

[152] J. Wusten, S. Berger, K. Heimer, S. Lach and C. Ziegler, "Interaction of alkali metals with perylene-3,4,9,10-tetracarboxylic-dianhydride thin films," *J. Appl. Phys.* 98, 023708 (2005).

[153] S. R. Forrest, "Ultrathin organic films grown by organic molecular beam deposition and related techniques," *Chemical Reviews* 97, 1793-1896 (1997).

[154] M. Hoffmann, K. Schmidt, T. Fritz, T. Hasche, V. M. Agranovich and K. Leo, "The lowest energy Frenkel and charge-transfer excitons in quasi-one-dimensional structures: application to MePTCDI and PTCDA crystals," *Chemical Physics* 258, 73-96 (2000).

[155] J. Wuesten, K. Heimer, S. Lach and C. Ziegler, "Alkali metals in perylene-3,4,9,10-tetracarboxylicdianhydride thin films," *J. Appl. Phys.* 102, 023708 (2007).

[156] K. Heimer, J. Wuesten, S. Lach and C. Ziegler, "Interaction of alkali metals with perylene-3,4,9,10-tetracarboxylic-dianhydride thin films studied by IR spectroscopy," *Journal of Chemical Physics* 126 (2007).

[157] J. Wuesten, C. Ziegler and T. Ertl, "Electron transport in pristine and alkali metal doped perylene-3,4,9,10-tetracarboxylicdianhydride (PTCDA) thin films," *Physical Review* B (Condensed Matter and Materials Physics) 74 (2006).

[158] G. Grem, G. Leditzky, B. Ullrich and G. Leising, "Realization of a Blue-Light-Emitting Device Using Poly(Para-PHenylene)," *Adv. Mater.* 4, 36-37 (1992).

[159] W. Graupner, G. Grem, F. Meghdadi, C. Paar, G. Leising, U. Scherf, K. Mullen, W. Fischer and F. Stelzer, *"Electroluminescence with Conjugated Polymers and Oligomers,"* Molecular Crystals and Liquid Crystals Science and Technology Section a-Molecular Crystals and Liquid Crystals 256, 549-554 (1994).

[160] F. Meghdadi, G. Leising, W. Fischer and F. Stelzer, "Multicolour electroluminescence diodes using oligophenylene and oligophenylenevinylene multilayers," *Synthetic Metals* 76, 113-115 (1996).

[161] N. Koch, A. Rajagopal, E. Zojer, J. Ghijsen, X. Crispin, G. Pourtois, J. L. Bredas, R. L. Johnson, J. J. Pireaux and G. Leising, "The influence of the counterion on the electronic structure in doped phenylene-based materials," *Surface Science* 454, 1000-1004 (2000).

[162] N. Koch, A. Rajagopal, J. Ghijsen, R. L. Johnson, G. Leising and J. J. Pireaux, "Bipolaron: The stable charged species in n-doped p-sexiphenyl," *Journal of Physical Chemistry* B 104, 1434-1438 (2000).

[163] E. Ito, N. Koch, H. Oji, H. Ishii, G. Leising and K. Seki, "Photoemission study of metal deposited p-sexiphenyl film," Electrical, Optical, and Magnetic Properties of Organic Solid-State Materials V. Symposium (Materials Research Society Proceedings Vol. 598), BB3.65.61-BB63. 65.66BB63.65.66 (2000).

[164] N. Koch, L. M. Yu, V. Parente, R. Lazzaroni, R. L. Johnson, G. Leising, J. J. Pireaux and J. L. Bredas, "Evidence for physisorption of aluminum on the surface of electroluminescent sexiphenyl," *Adv. Mater.* 10, 1038-+ (1998).

[165] M. G. Ramsey, M. Schatzmayr, S. Stafstrom and F. P. Netzer, "Interacting Bipolarons in N-doped Sexiphenyl Films," *Europhys. Lett.* 28, 85-90 (1994).

[166] R. Resel, N. Koch, F. Meghdadi, G. Leising, W. Unzog and K. Reichmann, "Growth and preferred crystallographic orientation of hexaphenyl thin films," *Thin Solid Films* 305, 232-242 (1997).

[167] H. Ding, *Ph. D. Thesis*, University of Rochester, 2010.

[168] H. Ding and Y. Gao, "Evolution of the electronic structure of alkali metal-doped copper-phthalocyanine (CuPc) on different metal substrates," *Organic Electronics* 11, 1786-1791 (2010).

[169] H. Ding and Y. Gao, "Reversal of doping-induced energy level shift: Au on Cs-doped tris(8-hydroxyquinoline) aluminum," *Appl. Phys. Lett.* 87, 051918 (2005).

[170] S. T. Lee, X. Y. Hou, M. G. Mason and C. W. Tang, "Energy level alignment at Alq/metal interfaces," *Appl. Phys. Lett.* 72, 1593-1595 (1998).

[171] A. Turak, D. Grozea, X. D. Feng, Z. H. Lu, H. Aziz and A. M. Hor, "Metal/AlQ(3) interface structures," *Appl. Phys. Lett.* 81, 766-768 (2002).

[172] A. C. Durr, F. Schreiber, M. Kelsch, H. D. Carstanjen and H. Dosch, "Morphology and thermal stability of metal contacts on crystalline organic thin films," *Adv. Mater.* 14, 961-+ (2002).

[173] H. Ding and Y. Gao, "Reversal of Doping Induced Energy Level Shift in Organic Semiconductors," *International Journal of Nanoscience* 6, 125 (2007).

[174] T. Nishi, K. Kanai, Y. Ouchi, M. R. Willis and K. Seki, "Evidence for the atmospheric p-type doping of titanyl phthalocyanine thin film by oxygen observed as the change of interfacial electronic structure," *Chemical Physics Letters* 414, 479-482 (2005).

[175] L. S. Hung, C. W. Tang and M. G. Mason, "Enhanced electron injection in organic electroluminescence devices using an Al/LiF electrode," *Appl. Phys. Lett.* 70, 152-154 (1997).

[176] G. E. Jabbour, Y. Kawabe, S. E. Shaheen, J. F. Wang, M. M. Morrell, B. Kippelen and N. Peyghambarian, "Highly efficient and bright organic electroluminescent devices with an aluminum cathode," *Appl. Phys. Lett.* 71, 1762-1764 (1997).

[177] S. E. Shaheen, G. E. Jabbour, M. M. Morrell, Y. Kawabe, B. Kippelen, N. Peyghambarian, M. F. Nabor, R. Schlaf, E. A. Mash and N. R. Armstrong, "Bright blue organic light-emitting diode with improved color purity using a LiF/Al cathode," *J. Appl. Phys.* 84, 2324-2327 (1998).

[178] C. H. Lee, "Enhanced efficiency and durability of organic electroluminescent devices by inserting a thin insulating layer at the Alq(3)/cathode interface," *Synthetic Metals* 91, 125-127 (1997).

[179] H. Tang, F. Li and J. Shinar, "Bright high efficiency blue organic light-emitting diodes with Al2O3/Al cathodes," *Appl. Phys. Lett.* 71, 2560-2562 (1997).
[180] X. H. Yang, Y. Q. Mo, W. Yang, G. Yu and Y. Cao, "Efficient polymer light emitting diodes with metal fluoride/Al cathodes," *Appl. Phys. Lett.* 79, 563-565 (2001).
[181] R. Schlaf, B. A. Parkinson, P. A. Lee, K. W. Nebesny and N. R. Armstrong, "Determination of frontier orbital alignment and band bending at an organic semiconductor heterointerface by combined X-ray and ultraviolet photoemission measurements," *Appl. Phys. Lett.* 73, 1026-1028 (1998).
[182] J. M. Zhao, Y. Q. Zhan, S. T. Zhang, X. J. Wang, Y. C. Zhou, Y. Wu, Z. J. Wang, X. M. Ding and X. Y. Hou, "Mechanisms of injection enhancement in organic light-emitting diodes through insulating buffer," *Appl. Phys. Lett.* 84, 5377-5379 (2004).
[183] J. M. Zhao, S. T. Zhang, X. J. Wang, Y. Q. Zhan, X. Z. Wang, G. Y. Zhong, Z. J. Wang, X. M. Ding, W. Huang and X. Y. Hou, "Dual role of LiF as a hole-injection buffer in organic light-emitting diodes," *Appl. Phys. Lett.* 84, 2913-2915 (2004).
[184] Y. Gao, K. T. Park and B. R. Hsieh, "Interface Formation of CA with Poly(P-Phenylene Vinylene)," *J. Appl. Phys.* 73, 7894-7899 (1993).
[185] K. L. Wang, B. Lai, M. Lu, X. Zhou, L. S. Liao, X. M. Ding, X. Y. Hou and S. T. Lee, "Electronic structure and energy level alignment of Alq(3)/Al2O3/Al and Alq(3)/Al interfaces studied by ultraviolet photoemission spectroscopy," *Thin Solid Films* 363, 178-181 (2000).
[186] C. I. Wu, G. R. Lee and T. W. Pi, "Energy structures and chemical reactions at the Al/LiF/Alq(3) interfaces studied by synchrotron-radiation photoemission spectroscopy," *Appl. Phys. Lett.* 87 (2005).
[187] A. Rajagopal and A. Kahn, "Photoemission spectroscopy investigation of magnesium-Alq(3) interfaces," *J. Appl. Phys.* 84, 355-358 (1998).
[188] H. Ishii and K. Seki, "Energy level alignment at organic/metal interfaces studied by UV photoemission: Breakdown of traditional assumption of a common vacuum level at the interface," *Ieee Transactions on Electron Devices* 44, 1295-1301 (1997).
[189] T. M. Brown, R. H. Friend, I. S. Millard, D. J. Lacey, J. H. Burroughes and F. Cacialli, "LiF/Al cathodes and the effect of LiF thickness on the device characteristics and built-in potential of polymer light-emitting diodes," *Appl. Phys. Lett.* 77, 3096-3098 (2000).
[190] A. Curioni, W. Andreoni, R. Treusch, F. J. Himpsel, E. Haskal, P. Seidler, C. Heske, S. Kakar, T. van Buuren and L. J. Terminello, "Atom-resolved electronic spectra for Alq(3) from theory and experiment," *Appl. Phys. Lett.* 72, 1575-1577 (1998).
[191] L. Yan, N. J. Watkins, S. Zorba, Y. L. Gao and C. W. Tang, "Thermodynamic equilibrium and metal-organic interface dipole," *Appl. Phys. Lett.* 81, 2752-2754 (2002).
[192] N. J. Watkins, L. Yan and Y. L. Gao, "Electronic structure symmetry of interfaces between pentacene and metals," *Appl. Phys. Lett.* 80, 4384-4386 (2002).
[193] N. Koch, A. C. Durr, J. Ghijsen, R. L. Johnson, J. J. Pireaux, J. Schwartz, F. Schreiber, H. Dosch and A. Kahn, "Optically induced electron transfer from conjugated organic molecules to charged metal clusters," *Thin Solid Films* 441, 145-149 (2003).
[194] V. E. Choong, S. Shi, J. Curless and F. So, "Bipolar transport organic light emitting diodes with enhanced reliability by LiF doping," *Appl. Phys. Lett.* 76, 958-960 (2000).
[195] K. R. Choudhury, J. H. Yoon and F. So, "LiF as an n-dopant in tris(8-hydroxyquinoline) aluminum thin films," *Adv. Mater.* 20, 1456-+ (2008).

[196] Y. Yuan, D. Grozea, S. Han and Z. H. Lu, "Interaction between organic semiconductors and LiF dopant," *Appl. Phys. Lett.* 85, 4959-4961 (2004).
[197] H. Ding, K. Park, Y. Gao, D. Y. Kim and F. So, "Electronic structure and interactions of LiF doped tris (8-hydroxyquinoline) aluminum (Alq)," *Chemical Physics Letters* 473, 92-95 (2009).
[198] C. Shen, A. Kahn and J. Schwartz, "Chemical and electrical properties of interfaces between magnesium and aluminum and tris-(8-hydroxy quinoline) aluminum," *J. Appl. Phys.* 89, 449-459 (2001).

In: Doping
Editor: Lixin Yu

ISBN: 978-1-62618-097-0

Chapter 6

CERAMIC MATERIALS FOR ENERGY AND ENVIRONMENTAL APPLICATIONS: FUNCTIONALIZING OF PROPERTIES BY TAILORED COMPOSITIONS

Mariya Ivanova,[1,*] Sandrine Ricote,[2] Stefan Baumann,[1] Wilhelm A. Meulenberg,[1] Frank Tietz,[1] Jose M. Serra[3,†] and Hannes Richter[4,‡]

[1]Forschungszentrum Jülich GmbH, Institute of Energy and Climate Research IEK-1, Jülich, Germany
[2]Department of Energy Conversion and Storage, Risø campus, Technical University of Denmark, Frederiksborgvej, Roskilde, Denmark
[3]Instituto de Tecnología Química. Universidad Politécnica de Valencia, Consejo Superior de Investigaciones Científicas), Valencia, Spain
[4]Fraunhofer Institute for Ceramic Technologies and Systems, Hermsdorf branch, Hermsdorf, Germany

ABSTRACT

Stable social development requires novel approaches for energy production, distribution and storage combined with reasonable restrictions of the environmental impact. The fuel cell-based technologies, as well as the separation of gases from mixtures, particularly implemented into innovative power plant concepts, are therefore in the scientific and political focus.

* Corresponding author: Dr. Mariya E. Ivanova. Forschungszentrum Jülich GmbH, Institute of Energy and Climate Research IEK-1, Leo-Brandt-Str. 52425 Jülich, Germany. Ph.: 0049(0)2461 615194; Fax: 0049(0)2461 612455; E-mail: m.ivanova@fz-juelich.de.

† Jose M. Serra: Instituto de Tecnología Química (Universidad Politécnica de Valencia – Consejo Superior de Investigaciones Científicas), av. Los Naranjos s/n, E-46022 Valencia, Spain.

‡ Hannes Richter: Fraunhofer Institute for Ceramic Technologies and Systems – Hermsdorf branch, Michael-Faraday-Str. 1, 07629 Hermsdorf, Germany.

At the very core of these technologies are selected materials from a number of classes ranging from purely organic- and inorganic-based matrices to the precious metals and alloys that all ought to meet well defined operation-relevant conditions and efficiencies.

Depending on the targeted application, each material foreseen as a base for component development has to be evaluated individually and its properties have to be modified in terms of corresponding application requirements. In order to elaborate tailored ceramic materials for gas separation and ion/electron transport at the relevant operating conditions and stability ranges, improved electrical or ionic conductivities and permeation rates are required. That can be achieved by doping and substitution which are actors on a nano-scale that usually lead to macroscopic impacts.

This chapter is dedicated to the fascinating world of tailoring ceramic materials for energy and environmental applications. Selected approaches to tune ceramics will be discussed to illustrate the versatile effects that compositional variation can have on the macroscopic properties, e.g. the conductivity of protons, oxygen ions and/or electronic carriers, stability, etc. The present chapter will therefore consider the structural features of selected material classes, as well as the principles of transport in bulk and microporous solids. It will furthermore illustrate and discuss the effects of selected additives and substituents on sinterability, electrical/ electrochemical properties and stability of selected ceramic materials for energy and environmental applications. The material variety will cover ceramic materials with different crystal structures like fluorites, perovskites, pyrochlores, fergusonites, as well as selected zeolite structures.

Keywords: Doping, substitution, ceramic materials, fluorites, perovskites, pyrochlores, fergusonites, zeolites, dense ceramic membranes, microporous membranes, ionic conductors, mixed ionic-electronic conductors, molecular sieve, gas separation from mixtures, solid oxide fuel cells

1. INTRODUCTION

Sustainable social development requires novel technologies in electricity production, distribution and storage combined with less environmental impact. Wider use of renewable energy with linked and adopted energy storage, more efficient energy conversion with fuel cell technologies as well as the separation of gases from mixtures, particularly implemented into innovative power plant concepts, are therefore in the scientific and political focus. The reduction of CO_2 emissions from power plants fired by coal or gas can be achieved with gas separation techniques based on the membrane technology with significantly lower efficiency losses compared to the conventional separation techniques [1].

In addition to the gas separation, ceramic membranes are also used as a dense solid electrolyte in high-temperature fuel cells, either as an oxygen ionic conductor in solid oxide fuel cells (SOFCs) or as a proton conductor in proton-conducting SOFCs (PC-SOFCs).

Despite the relatively mature state of SOFC technology, PC-SOFCs represent an alternative technology for clean electricity production that offers several benefits regarding to fuel utilization, overall efficiency, and system simplicity. Both types of fuel cells can be operated with reformed fossil fuels and with hydrogen from renewables.

Depending on the targeted application, each material foreseen as a base for component development has to be evaluated individually and its properties have to be modified in terms of the corresponding application requirements. Usually, elevated temperatures are required to use ceramic materials as support, electrolyte or electrode material in solid oxide cells (SOCs) applied as electrolyser cell, as fuel cell or as membrane system.

However, the strict requirements limit the range of candidate materials that can operate efficiently under the relevant conditions maintaining their functionality and durability in the long term without significant performance degradation.

Briefly, these requirements for a ceramic electrolyte material in a PC-SOC (Figure 1A [2]) are high protonic conductivity, very low electronic contribution, good compatibility to the anode and cathode materials, proper thermal expansion coefficient and mechanical strength.

In order to be implemented in the H_2 separation technology, a reasonable mixed protonic-electronic conductivity is required (Figure 1B [2]) together with a number of similar demands on durability, compatibility, and mechanical stability (with respect to re-oxidation).

High-temperature proton conductors can also be applied as sensors for quantification of H_2 and humidity. Gas sensing with sufficient accuracy can be performed not only in hot gas streams but also in molten salts or metals, covering a wide detection range of H_2 concentration. The EMF voltage and the degree of hydration together with the kinetic effects of the electrocatalysts on surface reactions allow the development of sensors for detection of H_2, CO and water for very particular applications [3].

In order to elaborate tailored ceramic materials for specific gas separation and ion/electron transport at the relevant operating conditions and required stability ranges, improved permeation rates and electrical or ionic conductivities are needed which can be achieved by doping and substitution. Often a compromise between the relevant operating conditions and material's functional properties has to be found, which can only be realized by changing the composition. In this context, minor additives or substitution of larger amounts of one or more host elements can act as a powerful tool for attaining the necessary material properties. Doping and substitution are actors on the atomic scale that usually lead to macroscopic impacts. *Doping* does not modify the crystal lattice structure of the host material and the atoms, the so-called "dopants", do not interact with each other. In the case of *substitution*, a large number of atoms are exchanged leading to crystallographic changes (lattice parameters, distortions, order/disorder on sub-lattices) and playing a significant role in the defect-chemical relations as a function of numerous internal and external parameters.

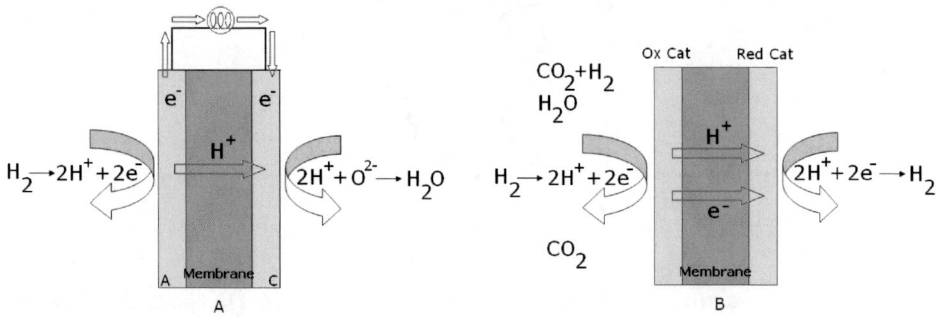

Reproduced from [2] by permission of Woodhead Publishing Ltd.

Figure 1. Schematic illustration of the operating principle of A) PC-SOFC and B) H_2-separating membrane.

However, new functional properties are achieved as a macroscopic outcome either by creating disorder, introducing additional energy levels to the originally available or stabilizing specific local structures, and electronic or energetic states.

The focus of the present chapter is tailoring ceramic materials for energy and environmental applications. Selected approaches to tune ceramics will be discussed to illustrate the versatile effects that compositional variation can have on the macroscopic properties, e.g. the conductivity of protons, oxygen ions and/or electrons, the catalytic activity, the mechanical stability etc. The present chapter will therefore focus on and discuss the structural features of different material classes, as well as the transport principles governing the mass transfer in dense and microporous solids. It will furthermore give examples of tailored ceramic systems in order to illustrate the influences of selected additives and substituents on properties like sinterability, mechanical and chemical stability, as well as electrical and electrochemical behavior. The variety of materials covers oxide ceramics of different structural types (e.g. fluorites, perovskites, pyrochlores, etc.), as well as representatives of microporous structures, e.g. selected zeolites, all having potential use in different energy and environmental applications. With respect to the transport properties of the materials, the chapter gives an insight into crystalline matrices with dominant anionic (O^{2-}), protonic or mixed ionic-/protonic-electronic conductivity.

2. Structural Features of Selected Material Classes

2.1. Oxides with Fluorite Structure (MO_2)

In the oxides with fluorite structure, the cations are 8-fold coordinated in face-centered cubic arrangement, while the anions are 4-fold coordinated and occupy tetrahedral sites. Ionic conductivity is usually promoted by incorporation of di- or trivalent cations and the associated formation of oxygen vacancies as shown in Figure 2 [4]. Fluorite-structured oxides usually show low levels of activation energies for migration of oxygen ions, therefore high ionic conductivity is observed.

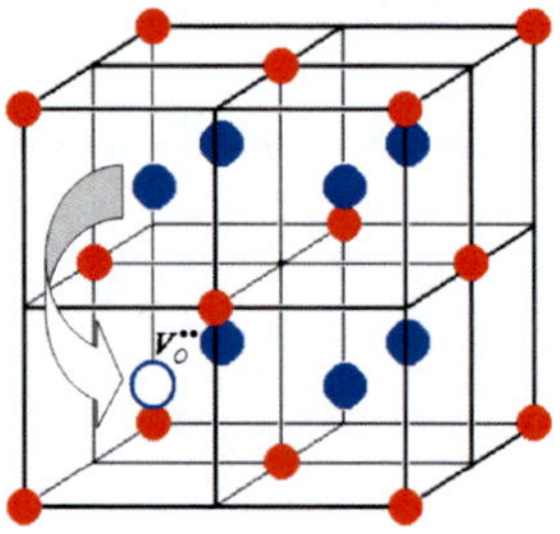

Adapted from [4].

Figure 2. Schematic illustration of the fluorite (MO_2) lattice with visualized ion positions (blue: O^{2-} red: M^{4+}) and hopping mechanism of an oxygen ion to the neighboring oxygen ion vacancy in the anionic sublattice.

Compounds with fluorite structure like ZrO_2, CeO_2, δ-Bi_2O_3, HfO_2, ThO_2 [5-8] are known for their good oxygen ion conductivity. Typically, the highest ionic conductivity for substituted materials is achieved using cations with ionic radius matching well the radius of the tetravalent host cations. Furthermore, in some cases, as for example ZrO_2, the substituent is necessary to stabilize the cubic fluorite structure at room temperature.

From a practical point of view, the most studied oxides with fluorite structure are ZrO_2, CeO_2 and δ-Bi_2O_3 in the last decade.

Their high conductivity is ascribed to the structure tolerance to some degree of disorder and the high ionic mobility due to the open character of the structure. Further properties depending on substitution will be discussed in the next sections.

2.2. Oxides with Perovskite Structure (ABO_3)

Perovskites are oxides with the general formula ABO_3. The crystal structure is named after the Russian mineralogist Lew Alexejewitsch Perowski (1792–1856) and is shown in Figure 3 with the origin on the A-site cation.

The ideal perovskite structure can be regarded as a cubic close-packed structure of the oxygen ions and the A-site cations layers stacked along the cubic [111] direction. The resulting sites within the oxygen octahedra are occupied by B-site cations [9].

Apart from the mineral *perovskite* ($SrTiO_3$) with ideal perovskite structure, a wide range of oxides belong to this structure type. The A-cation can be monovalent, divalent, or trivalent, while the B-cation is tri- to hexavalent compensating the charges of the A-cation.

Moreover, the cations can be widely substituted by other aliovalent cations with stability ranges at values of Goldschmidt tolerance factor, t, between 0.8 and 1.0 (Equation 1):

$$t = \frac{R_A + R_B}{\sqrt{2} \cdot (R_B + R_O)} \quad (1)$$

where R_A, R_B, and R_O are the radii of the A-cation, B-cation, and oxygen anion, respectively.

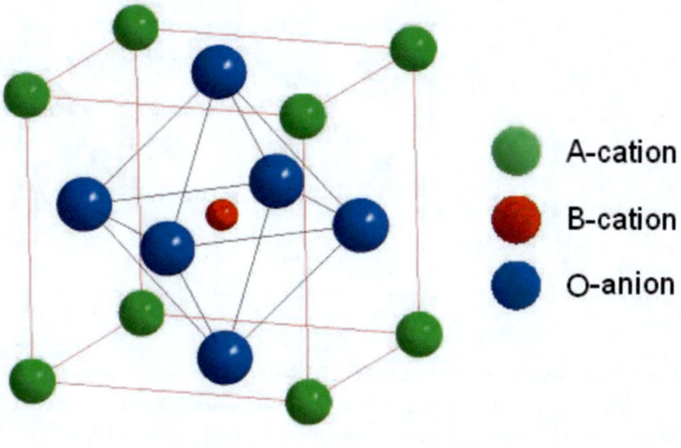

Figure 3. Schematic illustration of the perovskite lattice with visualized ion positions.

With decreasing *t*-value the structure becomes more and more disordered changing from cubic to tetragonal or orthorhombic symmetry. *t*-values above 1 result in the formation of the hexagonal (rhombohedral) structure.

Due to the crystallographic flexibility and the fact that the electrical properties can be tuned by selective substitution, perovskites are widely studied for application in electroceramics, catalysis, energy conversion processes, thermoelectrics, ion conductors, superconductors, etc. [10].

2.3. Oxides with Pyrochlore Structure ($A_2B_2O_7$)

The general formula of the pyrochlore family is $A_2B_2O_7$, with A^{3+} and B^{4+} or A^{2+} and B^{5+}. The A-cations are located within distorted cubes of oxygen ions (6 A-O and 2 A-O' with a shorter A-O' bond), while the B-cations reside in trigonal antiprism of equidistant oxygen ions (6 B-O). From a crystallographic point of view, the formula can be written $A_2B_2O_6O'$ as there are four crystallographic non-equivalent kinds of atoms [11]. As mentioned in [11], one often finds in the literature that the polyhedra for the A- and B-cations are cubic and octahedral, respectively, which is not entirely correct. Many pyrochlore compositions crystallize in the cubic structure, with the Fd3m space group, and a lattice parameter (*a*) of about 10 Å (with 8 molecules per unit cell).

There are two ways of determining the positions of the atoms in the pyrochlore lattice [11-13]. If the origin is set on the B-cation (position 16c), the A-cations are at 16d, O at 48f and O' at 8b. Instead, if the origin is set on the A-cation (16c), the positions of B-cation, O and O' are respectively 16d, 48f and 8a, respectively. The pyrochlore can be visualized as a disordered fluorite. The fluorite (AO_2) is formed of a face-centered (FCC) cubic lattice of A-cations, with the 8 oxygen anions occupying the tetrahedral sites as shown in Figure 2 and Figure 4A. For the oxygen-deficient fluorite, A- and B-cations are randomly positioned on the cation sublattice and 7 oxygen ions are distributed over the 8 equivalent tetrahedral sites (Figure 4B) leaving one vacant site. However, in the case of the pyrochlore (Figure 4C), the A- and B-cations have distinct positions on the cation sublattice, leading to three kinds of tetrahedral positions for the oxygen anions (4 A-cations as neighbors (8b), 4 B-cations (8a) or 2 A- and 2 B-cations (48f)).

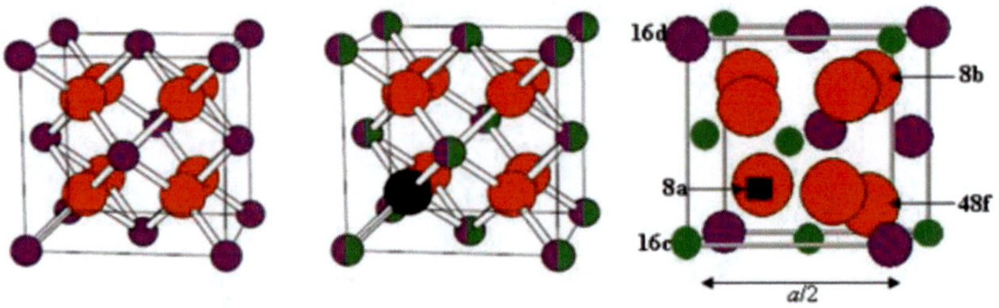

(a) (b) (c)

Figure 4. Schematic visualization of: A) fluorite AO_2; B) disordered deficient fluorite $A_2B_2O_7$ and C) pyrochlore $A_2B_2O_6O'$. The purple, green, red and black colors correspond to the A-cations, the B-cations, the oxygen anions and the oxygen vacancies, respectively. For the sake of illustration the random occupation of A- and B-cations has been marked with the purple-green balls [14]. Reproduced from [14] by permission of Elsevier Ltd.

Six oxygen atoms occupy the 48f sites, the seventh on the 8b site and the 8a site is unoccupied. The oxygen atoms on the position 48f are shifted towards B-cation by a parameter called "x", which is equal to 0.375 in the case of a perfect pyrochlore.

Structural disorder occurs easily in pyrochlores by partial reversion of the pyrochlore structure towards the fluorite structure [15]. The stability field of the pyrochlore structure depends on the ratio of the ionic radius of the A-site ($r_{A^{3+}}$) to that of the B-site ($r_{B^{4+}}$) and it can vary in the range of $1.46 \le r_{A^{3+}} / r_{B^{4+}} \le 1.80$ [15, 16] for the pyrochlore structure of zirconates and titanates.

2.4. Oxides with Lower Structural Symmetry (ABO_4, A=Ln, B=Nb, Ta)

Rare earth ortho-niobates ($LnNbO_4$) and tantalates ($LnTaO_4$) occur in two structure types. At room temperature the niobates and most of the tantalates have fergusonite structure (referred to as monoclinic distortion of the scheelite structure) which transforms to the scheelite structure at high temperatures (tetragonal symmetry) [17].

As previously reported [18-20], the scheelite-to-fergusonite transition can be described as a shear transformation in which the scheelite structure is partially conserved while certain sheets are slightly shifted. The scheelite structure can be described as two intercalated diamond lattices, one for the A-site cation and the other for the B-site cation [21]. The A-site cation is coordinated by eight oxygen ions and the B-site cation is tetrahedrally coordinated by the oxygen ions.

In the fergusonite polymorph, A- and B-sites are coordinated similarly to the scheelite form, with two additional near-oxygen sites at each B-site cation. The temperature of phase transition for $LaNbO_4$ is in the range of 495–520 °C [17-19, 22, 23] as indicated in Figure 5.

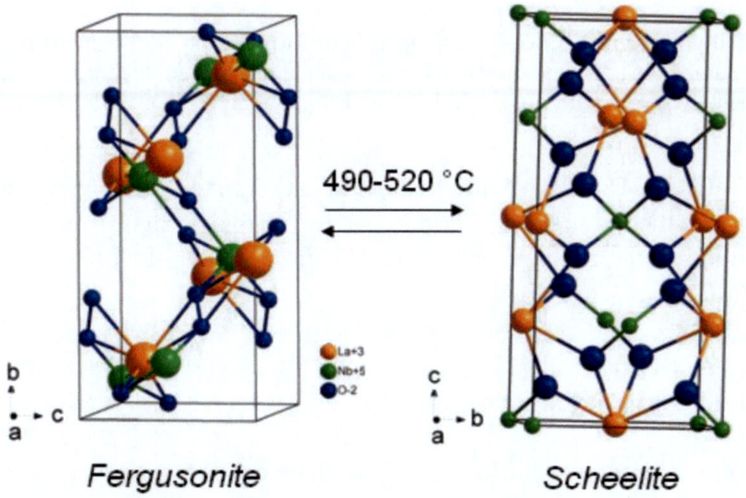

Figure 5. Schematic illustration of the Fergusonite and Scheelite structure with phase transition temperature for $LaNbO_4$ [24].

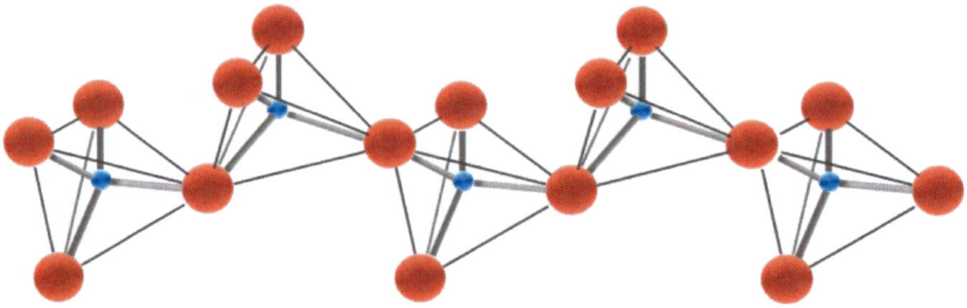

Figure 6. Chain of tetrahedrally configurated silicon and oxygen.

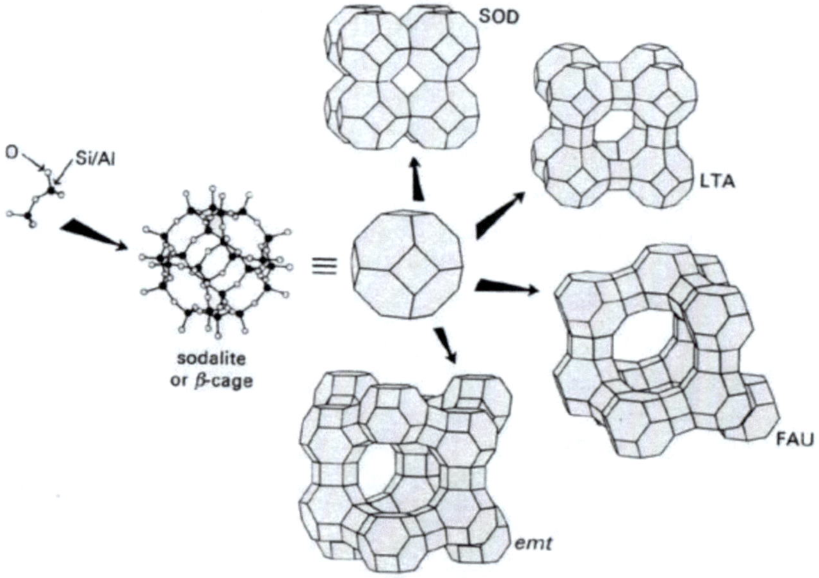

Figure 7. Tetrahedrally configured silicon and oxygen (left) forming sodalite or the so-called β-cage. Depending on the number of units and their combination, a number of zeolite structures can thus be formed: SOD: sodalite; LTA: zeolite type A; FAU: faujasite; emt: a hexagonal variant of FAU present in zeolite EMC-2 [26].

$LaTaO_4$, $CeTaO_4$ and $PrTaO_4$ are also monoclinic at room temperature but they transform to orthorhombic symmetry and the phase transition displays significant hysteresis. The transition temperatures on heating/cooling are 175/150 °C for $LaTaO_4$, 818/729 °C for $CeTaO_4$, and 1300/1205 °C for $PrTaO_4$ [25].

2.5. Microporous Networks: Zeolites ($(M^{2+}M^{+}_2)O \cdot Al_2O_3 \cdot N\ SiO_2 \cdot M\ H_2O$)

Zeolites are crystalline tectosilicates and their chemical composition may be given with the general formula $(M^{2+}M^{+}_2)O \cdot Al_2O_3 \cdot n\ SiO_2 \cdot m\ H_2O$. Silicon is tetrahedrally coordinated with oxygens. The $[SiO_4]^{4-}$-tetrahedra are connected by vertex-shared oxygens (Figure 6) thus creating a three-dimensional fully cross-linked solid framework.

In Figure 7 the β-cage as repeating element of the sodalite 3D framework is shown as an example.

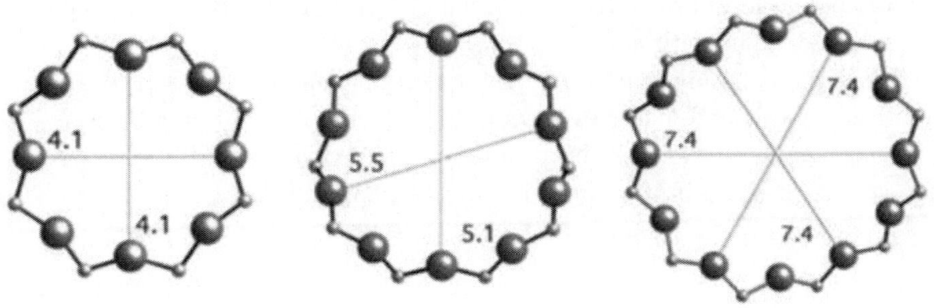

Figure 8. Window openings composed of A) 8 $[SiO_4]^{4-}$-tetrahedra in LTA, B) 10 $[SiO_4]^{4-}$-tetrahedra in MFI and C) 12 $[SiO_4]^{4-}$-tetrahedra in FAU. Diameters are given in Å [27].

Inside the crystalline structure open cages and channels are formed with entrances called *windows*. Examples of windows with different diameters are shown in Figure 8. The size of the windows differs for the numerous zeolites and is determined by the number of $[SiO_4]^{4-}$-tetrahedra. Depending on the crystal structure windows in the range of 0.3 and 0.9 nm can be created, which fit the size of small gas molecules. This structural particularity makes zeolites attractive for numerous of industrial applications, among them gas separation by adsorption and membrane separation.

Up to 50% aluminum can be incorporated into the framework instead of silicon (e.g. zeolite LTA). The missing positive charge of the trivalent aluminum is compensated by additional alkali or alkaline earth metal ions. These extra-framework cations can be easily exchanged which is practically used in washing agents for water softening. The extra-framework cations influence the open space in the cages and the size of the windows as well. Polar sites in zeolites with high alumina content make the crystals hydrophilic and are the reason for pore filling with water at ambient conditions. The water can be completely desorbed by heating (at temperature of 200 °C or higher, depending on the zeolite type) without losing the crystalline structure of the zeolites. In low Al-containing zeolites with neutral frameworks, the cavities can be completely empty. The crystal structure of some zeolites is stable even at temperatures up to 800 °C [28], an attractive feature for certain applications, e.g. catalytic cracking, at elevated temperatures.

3. Principles of Transport through the Bulk Solids

3.1. Oxygen Ionic and Mixed Oxygen Ionic-Electronic Conduction

Inter-ionic forces retaining ions at their regular positions in the crystal lattice allow certain fluctuations around their equilibrium state. The average energy of such heat fluctuations is about 0.03 eV at room temperature and about 0.1 eV at 600 °C. In order to break the bonds between the ions, energy of at least 0.5 eV is required. When this barrier is overcame, ions jump from their regular position in the lattice to an interstitial site or an adjacent vacant position, thus being the reason for ionic conduction. The energetic states of an interstitial ion and an ion positioned in a regular crystal lattice site are not equal: less energy is required to move the interstitial ion.

In addition, ionic mobility is also promoted when open layered crystalline structures, easily polarized M-O bonds and lower coordination numbers of mobile ions are present. These structural requirements are best fulfilled for oxides with fluorite, perovskite, and pyrochlore structure. Equilibrium concentration of the point defects increases exponentially with temperature. The ionic conductivity, σ_{ion}, can be written according to Equation 2:

$$\sigma_{ion} = n_c \cdot C \cdot q \cdot \mu = A \cdot \exp\left(-\frac{E_{act}}{kT}\right) \tag{2}$$

where n_c is the concentration of the vacant positions, C is the concentration of the adjacent equivalent positions, q and μ are the charge and the mobility of the charge carriers, A is a pre-exponential term, E_{act} is the activation energy for ionic conductivity, and finally, k and T are the Boltzmann constant and the absolute temperature, respectively. The pre-exponential coefficient is a complex term and takes several parameters into account as given with Equation 3:

$$A = f \cdot \left(\frac{z}{6}\right) \cdot \left(\frac{Cq^2}{k}\right) \cdot n_c \cdot (1 - n_c) \cdot l^2 \cdot v_0 \cdot \exp\left(\frac{\Delta S_m}{k}\right) \tag{3}$$

In this equation, l is the distance to z nearest equivalent positions with the probability $(1-n_c)$ of being vacant; v_0 is the jump frequency for ions (10^{12}-10^{13} Hz) and f is a geometrical factor depending on the jump trajectory. C and n_c have the same meaning as explained above and ΔS_m is the entropy of migration, m standing for *migration*. The mobility of ionic defects, μ_{ion}, can be described by the Nernst-Einstein relation (Equation 4), where F, R, T and D_i are the Faraday constant, the gas constant ($J \cdot mol^{-1} \cdot K^{-1}$), the absolute temperature and the diffusivity ($cm^2 \cdot s^{-1}$), respectively:

$$\mu_{ion} = \frac{F \cdot D_i}{R \cdot T} \tag{4}$$

Mixed oxygen ionic-electronic conductors (MIEC) combine ionic and electronic conductivity. Therefore, an external electrical circuit can be avoided in MIEC-based systems and operating in an oxygen chemical potential gradient (Figure 9). For their practical applicability, MIECs have attracted great interest [29].

The oxygen transport through dense solid membranes can be described using the Wagner equation [30]:

$$j_{O2} = -\frac{RT}{16F^2} \cdot \frac{1}{L} \int_{\ln P'_{O2}}^{\ln P''_{O2}} \frac{\sigma_{el}\sigma_{ion}}{\sigma_{el} + \sigma_{ion}} d \ln P_{O2} \tag{5}$$

where L is the membrane thickness, σ_{el} and σ_{ion} are the electronic and ionic conductivity, respectively, p'_{O2} and p''_{O2} are the oxygen partial pressures at the two sides of the membrane.

Equation 5 implies that the oxygen flux becomes infinite when reducing the membrane thickness towards zero and it only considers solid state diffusion. In any case, the surface exchange of oxygen also has to be taken into account. Depending on the membrane thickness and the specific operating conditions, surface kinetics can partially control the transport process. Therefore, Bouwmeester et al. [30] proposed a reduction factor to be taken into account (Equation 6):

$$\frac{1}{1+\left(\frac{2L_c}{L}\right)} \quad with \quad L_c = \frac{D_s}{k_s} \tag{6}$$

where L_c is the so-called characteristic thickness, D_s the self-diffusion coefficient, and k_s the surface exchange coefficient. Then Equation 6 can be re-written as:

$$j_{O2} = -\frac{RT}{16F^2} \cdot \frac{1}{L+2L_c} \int_{\ln P'_{O2}}^{\ln P''_{O2}} \frac{\sigma_{el}\sigma_{ion}}{\sigma_{el}+\sigma_{ion}} d\ln P_{O2} \tag{7}$$

This reduction factor leads to a limiting value of the oxygen flux depending on the material properties, as shown in Figure 10.

In order to manufacture membranes for practical use, which are favourably designed as a thin functional layer, a reasonable mechanical stability is also required. This can be achieved by using of a substrate with sufficient porosity and gas permeability. As shown recently by Baumann *et al.* [31] the support plays a significant role in the oxygen permeation because of the limited gas diffusion through the pores.

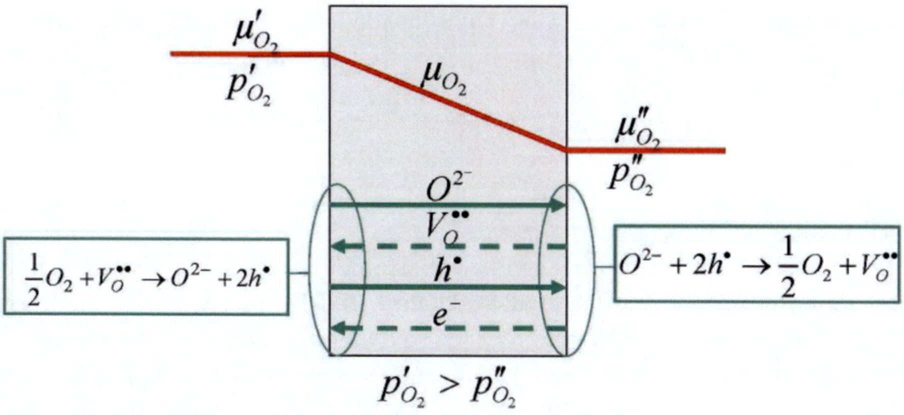

Figure 9. Oxygen transport in mixed ionic-electronic conductors (MIECs).

When high oxygen permeation rates are realized, the concentration polarization in the adjacent gas phase can become an important parameter [32, 33].

3.2. Protonic and Mixed Protonic-Electronic Conduction

High temperature proton-conducting (HTPC) oxides take up protons from water vapor or hydrogen-containing atmospheres. According to Kreuer [34], in large band gap oxide materials (e.g. perovskites) protons are formed by dissociative absorption of water which requires the presence of oxygen ion vacancies ($V_O^{\bullet\bullet}$).

Water from the atmosphere dissociates into hydroxide ions and protons; the hydroxide ion fills the oxygen vacancy, and the proton forms a covalent bond with the lattice oxygen [34-42].

The successive formation and annihilation of protonic defects, according to Kreuer, may be written with the following equations in Kröger-Vink notation:

$$H_2O_{(g)} + V_O^{\bullet\bullet} + O_O^X \leftrightarrow 2OH_O^{\bullet}, \; K = (K_1K_2)^{-1} \quad (8)$$

$$2OH_O^{\bullet} + 1/2O_2 \leftrightarrow H_2O_{(g)} + 2h^{\bullet} + 2O_O^X, \; (K_1) \quad (9)$$

$$2h^{\bullet} + O_O^X \leftrightarrow V_O^{\bullet\bullet} + 1/2O_2, \; (K_2) \quad (10)$$

where $V_O^{\bullet\bullet}$ stands for oxygen vacancy, O_O^X is oxygen in a regular crystal lattice site, $OH_O^{\bullet}$ denotes the protonic defect, $h^{\bullet}$ is an electronic hole, and K_1 and K_2 are the equilibrium constants for hole and oxygen vacancy formation. Subscript *(g)* stands for gaseous phase. The formation of protonic defects given in Equation 8 is a linear combination of Equation 9 and Equation 10.

In the presence of hydrogen in the environment (Equation 11) protons are formed accordingly:

$$H_{2(g)} + 2O_O^X \leftrightarrow 2OH_O^{\bullet} + 2e' \quad (11)$$

where e' is the Kröger-Vink notation for electrons.

The formation of protonic defects can be considered as an amphoteric reaction, because the water molecule splits into a hydroxide ion and a proton. The oxide material simultaneously acts as an acid (absorption of hydroxide ion by oxygen ion vacancy) and as a base (protonation of lattice oxygen ions). Kreuer suggests that the adsorption of hydroxide ion by an oxygen vacancy is the dominating step in binary rare earth oxides possibly due to the small local variation of the lattice basicity in rare earth oxides. The protonation of the lattice oxygen ions seems to dominate in perovskite-type oxides due to the low formation enthalpies of oxygen ions as a result of low bond strengths and strong relaxation effects.

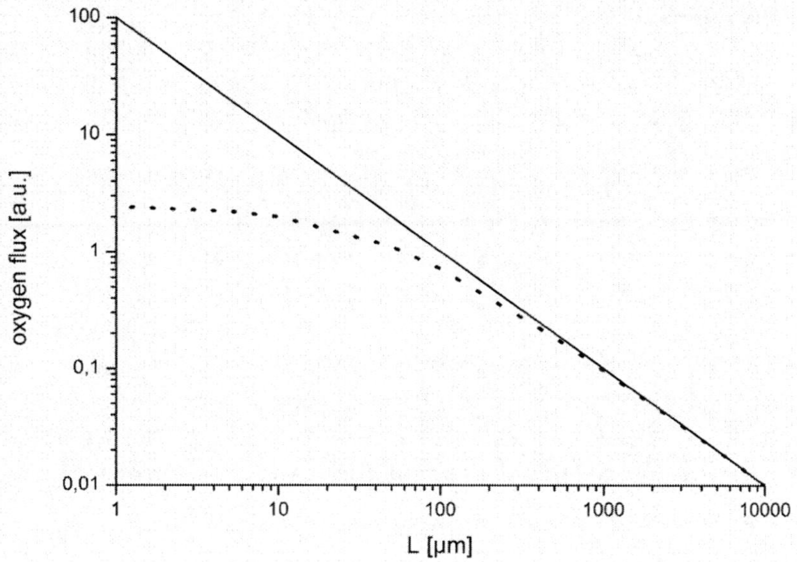

Figure 10. Oxygen permeation rate as a function of the membrane thickness according to Wagner theory (solid line) and considering an arbitrary L_c of 20 μm (dotted line).

Since protons are located in the electronic cloud of the host oxygen ions and move freely, they are often considered as interstitial defects notified by ($H_i^{\bullet}$) [43, 44].

If not intrinsically present in the material, the oxygen vacancies are created by doping/substituting some of the host cations by lowervalent cations.

An example is given in Equation 12 when the substitution of B^{4+} cations by M^{3+} cations causes formation of oxygen vacancies:

$$2B_B^X + O_O^X + M_2O_3 \leftrightarrow 2M_B' + V_O^{\bullet\bullet} + 2BO_2 \qquad (12)$$

where B_B^X is B^{4+} cation positioned in an ordinary crystal lattice site, M_B' stands for M^{3+} cation doped/substituted on B site thus generating an effective negative charge.

When the host oxide lattice is modified by acceptor doping or substitution (depending on the level) as shown in Equation 12, protonic defects are then competing with other positively charged defects as electron holes ($h^{\bullet}$) and oxygen vacancies ($V_O^{\bullet\bullet}$) to maintain electroneutrality (Equation 13):

$$[OH_O^{\bullet}] + [h^{\bullet}] + 2[V_O^{\bullet\bullet}] = [M_B'] + [e'] \qquad (13)$$

The formation of protonic defects is accompanied by a significant weight increase. The proton concentration can be calculated using a thermobalance to determine the water uptake [34, 45-47]. The dependence of protonic defect concentration on temperature and water partial pressure gives information about the protonic defect stability. The equilibrium of the hydration reaction (Equation 8) shifts to the left with increasing temperature since the reaction is exothermic (negative hydration enthalpy). Thus the hydration enthalpy is a measure for the proton stability: the more negative the hydration enthalpy, the higher proton content and dehydration temperature. Based on thermodynamic data available for several perovskites [48, 49] the electronegativity of the cations that interact with the lattice oxygen is correlated to the enthalpy of the hydration reaction.

It was found that the hydration reaction becomes more exothermic with decreasing electronegativity of the cations (with decreasing Brønsted basicity). Moreover, the most negative enthalpies are reported for similar electronegativities of A- and B-site cations.

Furthermore, the crystal symmetry and chemical nature strongly affect the stability and mobility of protonic defects. Structures of higher symmetry are known to exhibit the highest values of conductivity, which usually decreases with decreasing the symmetry [50].

Furthermore, closely packed structures are considered to be beneficial for the protonic transport due to the shorter O-O bonds that are the paths for protonic transport. However, some exceptions exist, for example the case of Al_2O_3 which is a proton blocker. Molecular dynamics studies on perovskite compounds show that a symmetry decrease leads inevitably to lower proton stability. Basic oxides stabilize the protonic defects but their basicity definitely is not beneficial in terms of chemical stability under reduction environments [49].

Two basic mechanisms describe the proton diffusion [51]. *Vehicle facilitated mechanism:* Proton migration is assisted by the translational dynamics of bigger species, and thus protons move together with "a vehicle" (molecular diffusion). The electroneutrality is maintained by the counter-diffusion of un-protonated vehicles. *Grotthuss mechanism:* The protons are the only mobiles species, while the oxygen ions are located in the vicinity of their crystallographic position. The principal features of this transport mechanism are rotational diffusion of protonic defects and proton transfer to an adjacent oxygen ion as shown in Figure 11 and they both contribute to the mobility enthalpy [50, 52-54].

For structural reasons, the Grotthuss mechanism is most likely to take place in oxides than migration of hydroxyl ions. For perovskites as the most studied proton-conducting materials, it is considered that the proton transfer is the rate limiting step.

However, there is still discrepancy on the kinetics of the reaction mechanism. Some experiments [55, 56] and quantum molecular dynamics simulations [57, 58] revealed that the proton transfer reaction is the rate-limiting step, while infrared measurements favour fast proton transfer rather than fast reorientation process [54].

With increasing temperature a gradual transition from Grotthuss to vehicle facilitated mechanism can be expected due to elongation and breaking of the hydrogen bonds which may suppress proton transfer and release translational degree of freedom.

The protonic conductivity given in Equation 14 is a product of the charge ($zF = F$ for a proton), the concentration ($C_H, mol \cdot cm^{-3}$) and the mobility of the protonic defects ($\mu_H, cm^2 \cdot s^{-1} \cdot V^{-1}$) as follow:

$$\sigma_{H^+} = F \cdot C_{H^+} \cdot \mu_{H^+} \tag{14}$$

The mobility of protonic defects and the Arrhenius expression for protonic conductivity can be expressed with Equation 4 and Equation 2, respectively, in which subscript *i* for *ionic* is replaced by subscript H^+ for *protonic*.

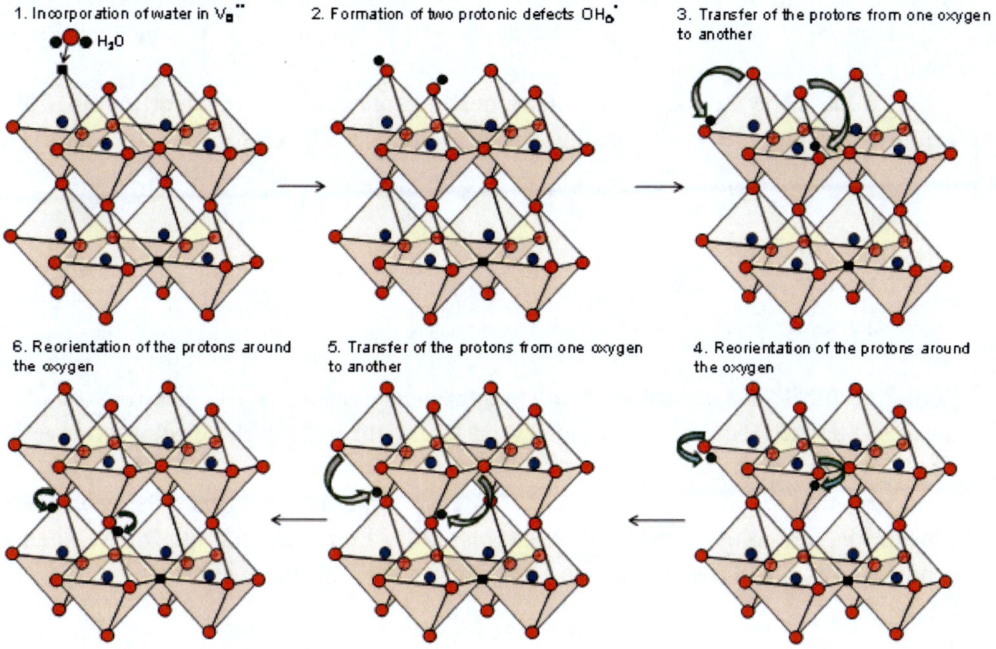

Figure 11. Mechanism of the incorporation of water and proton transport via the Grotthuss mechanism. The red circles, dark green circles, blue circles and black squares correspond to oxygen ions, protons, B-site cations and oxygen vacancies, respectively.

In proton-conducting perovskites an activation enthalpy of long-range proton diffusion not more than 0.15 eV can be expected based on thermodynamic consideration regarding to the dynamic hydrogen bonds. However, protonic defects in cubic perovskite-type oxides have mobility activation enthalpies in the range of 0.4-0.6 eV [50, 34]. This raises a question about the kind of interactions contributing to the increased value of activation enthalpy of proton transfer. The reason can be the strong repulsive force between the protons and highly charged B-site cations which results in non-linear hydrogen bond formation (elongated bond) and in creating a proton transfer barrier. Furthermore, any distortion of the crystal lattice can have a pronounced impact on the energetic properties of the host oxygen ions in terms of different binding energies to protons. Along with reduction of crystallographic symmetry, local structural and chemical perturbations induced by dopants, mixed occupancy of lattice sites, trapping effects, lattice relaxations around oxygen ion vacancies etc., can also affect significantly the proton mobility and conductivity. In mixed proton-electron conductors (MPECs) both protons and electrons contribute as charge carriers to the total conductivity. The total conductivity, σ_{tot}, of MPECs therefore consists of a protonic (σ_{H+}) and electronic (σ_e) conductivity term, according to Equation 15:

$$\sigma_{tot} = \sigma_{H+} + \sigma_{e-} \qquad (15)$$

The total as well as the ionic conductivity of the materials can be directly measured, while the electronic conductivity can be calculated by subtracting the ionic from the total conductivity term.

In order to describe the partial conductivities of a material, the transport number is used and it can be determined for every charge carrier of interest. The proton and electron transport numbers, t_{H+} and t_{e-} respectively, can be defined according to Equation 16:

$$t_{H+} = \frac{\sigma_{H+}}{\sigma_{tot}} \quad \text{and} \quad t_{e-} = \frac{\sigma_{e-}}{\sigma_{tot}} \tag{16}$$

In order to fulfill the electroneutrality requirement (Equation 13) electrons or electron holes must additionally be created in the crystal lattice. In the MPEC, electrons move in the same direction as the protons.

The permeation of hydrogen through a dense ceramic membrane is proportionally dependent on the ambipolar conductivity, expressed by Equation 15, with σ_{H+} replacing the ionic conductivity, σ_{ion}. Thus the proton flux density through the membrane can be written accordingly:

$$j_{H^+} = -\frac{RT}{2F^2} \cdot \frac{1}{L} \int_{\ln p'_{H2}}^{\ln p''_{H2}} \sigma_{H^+} t_e d \ln p_{H_2} \tag{17}$$

Mixed protonic and electronic conductivity in ceramic oxides is required for the application as hydrogen membranes but also as electrode in PC-SOFC. The promotion of the electronic conductivity in classical oxide-based proton conductors by selective doping is very challenging due to potential interference of the new electronic carriers with ionic defects. For practical application as separation membranes, ambipolar conductivity is required in the temperature range from 600 to 1000 °C.

4. Microporous Solids: Gas Flow Diffusion and Molecular Size Exclusion

Zeolites are characterized by pores with the size of small gas molecules. Once a gas flow through the pores starts, a strong interaction between the gas molecules and the pore walls of the zeolite lattice occurs, which can be described by the process known as configurational diffusion.

Depending on the temperature and the gas molecule involved, the process is distinguished between *solid vibration model (SV)* and *gas translation model (GT)* [59].

At low temperatures and strong gas molecule-zeolite interaction, the permeating molecules are losing their gaseous character: they vibrate with the zeolite lattice and jump from one to the next adsorption site when enough energy is available. In this case the mass transfer, J_{surf}, in this case can be described by the surface diffusion model [60] (Equation 18):

$$J_{surf} = -D_{surf} \nabla q \tag{18}$$

The driving force in this process is the gradient of surface loading across the pore length or the thickness of the membrane, ∇q.

The surface diffusion coefficient, D_{surf}, is dominated by the jump frequency $v_{surf,0}$ and jump distance λ of the molecule, as well as the activation energy E_{act} and the temperature T, according to Equation 19:

$$D_{surf} = \frac{1}{4} \cdot v_{surf,0} \cdot \lambda_{surf}^2 \cdot \exp\left(-\frac{E_{act}}{RT}\right) = D_0 \cdot \exp\left(-\frac{E_{act}}{RT}\right) \tag{19}$$

The activation energy correlates linearly with the differential heat of adsorption ΔH. When ΔH of an adsorbed component q is known, then the surface diffusion coefficient can be calculated using Equation 20:

$$D_{surf}(q) = D_{surf,0} \cdot \exp\left(-\frac{a\Delta H_{ad}(q)}{RT}\right) \tag{20}$$

At higher temperatures and weak molecule-zeolite interactions the molecules inside the pores retain their gaseous character and consecutively the mass transfer can be described by the gas translation (GT) model.

The GT diffusion coefficient D_{GT} defined with Equation 21 is characterized by the mean free path, λ, of the molecule, the reliability factor y, molar weight M, temperature T and activation energy $E_{A,GT}$ [61]:

$$D_{GT} = \frac{\lambda}{y} \cdot \sqrt{\frac{8RT}{\pi M}} \cdot \exp\left[-\frac{E_{act,GT}}{RT}\right] \tag{21}$$

The overall mass transfer through a zeolite material, J_{tot}, is the sum of the mass transfer caused by the gas translation, J_{GT}, and the solid vibration, J_{SV}: $J_{tot} = J_{GT} + J_{SV}$.

The selective separation of gases through the zeolite pores is based on size exclusion (molecular sieving) and different adsorption of the components (adsorption selective separation).

In the case of adsorption selective separation, all components of the targeted mixture are entering the pores of the zeolite. As previously mentioned, the adsorption of the single gases inside the pores depends on the temperature, pressure and component concentration.

Therefore, the selectivity varies with the separation conditions. In the case of molecular sieving, only the gas smaller in size is able to enter the pores. Hence, the selectivity is independent on the separation conditions.

Molecular sieving membranes are preferred candidates for selective membranes at high temperature and high pressure.

5. STRUCTURAL, CHEMICAL AND FUNCTIONAL EFFECTS OF ADDITIVES AND SUBSTITUENTS. SELECTED EXAMPLES

5.1. Effect of Additives/Substituents on Sintering Barium Cerate and Barium Zirconate

Although oxides may have the required electrical properties, their application can be limited by the high sintering temperature for manufacturing dense samples, especially when they have to be co-fired with materials which restrict the sintering temperature. The sintering temperature can be lowered, in some cases significantly, by adding small amounts of sintering aids. Different mechanisms have been proposed, depending on the sintering aid, which will not be discussed here. However, for example i) rearrangement of the grains and boundary layer diffusion [62-64] or ii) formation of a new phase with low melting point [64] (liquid phase sintering) should be mentioned here.

In order to illustrate the impact of additives on sintering, several examples are considered below.

$BaZrO_3$-based ceramics have been widely studied in terms of protonic conductivity. It was reported that these materials require very high sintering temperatures to obtain dense samples. Usually temperatures of about 1700 °C or even higher are required for powders prepared via the solid state reaction [65-67].

These extremely high manufacturing temperatures can be decreased by tailoring the morphology of the synthesized powder, for example via the glycine-nitrate combustion synthesis [68], or using low temperature synthesis techniques [69], as well as using of milling [70] or sintering aids. The later first was used by Babilo and Haile [71] who prepared a 93% dense $BaZr_{0.85}Y_{0.15}O_{3-\delta}$ (BZY15) ceramics at 1300 °C by adding 4 mol.% of ZnO to the synthesized BZY15 powder. They reported that ZnO accumulated at the grain boundaries. It was furthermore shown that other additives, e.g. Sc, Ti, V, Cr, Mn, Fe, Co, do not improve the sinterability of the material, whereas Ni and Cu were as beneficial as Zn. Tao and Irvine [72], who investigated the effect of ZnO on $BaZr_{0.80}Y_{0.20}O_{3-\delta}$ (BZY20), claimed that zinc enters the lattice and occupies the B-site of the perovskite. Figure 12 represents the shrinkage of the BZY20 samples with and without sintering aid.

The sintering temperature could also be lowered by adding the sintering aid in the stoichiometry of the compound during synthesis, i.e. by doping the B-site of the perovskite with Ni or Co [73].

$BaZr_{0.9}Y_{0.1-x}Ni_xO_{3-\delta}$ and $BaZr_{0.9}Y_{0.1-x}Co_xO_{3-\delta}$ (x=0.01 and 0.02) were prepared by solid state reaction at 1400 °C and sintered at 1450 °C, leading to 94% dense samples (compared to 80% without Ni or Co doping).

Figure 13 shows a photograph of pellets before (a) and after (b) sintering: there is a much higher shrinkage for the compounds with Ni and Co, as similarly was illustrated in Figure 12 for BZY20.

Since the use of sintering aids in barium zirconate led to a significant reduction of the sintering temperature, similar attempts on solid solutions of barium zirconate and barium cerate were investigated. $BaCe_{0.5}Zr_{0.4}(Y,Yb)_{0.1-x}Co_xO_{3-\delta}$ was sintered at temperatures between 1300 and 1425 °C and were shown to be stable for $x < 0.05$, with no obvious segregation of Co and without secondary phases that could react with air or wet atmosphere [74].

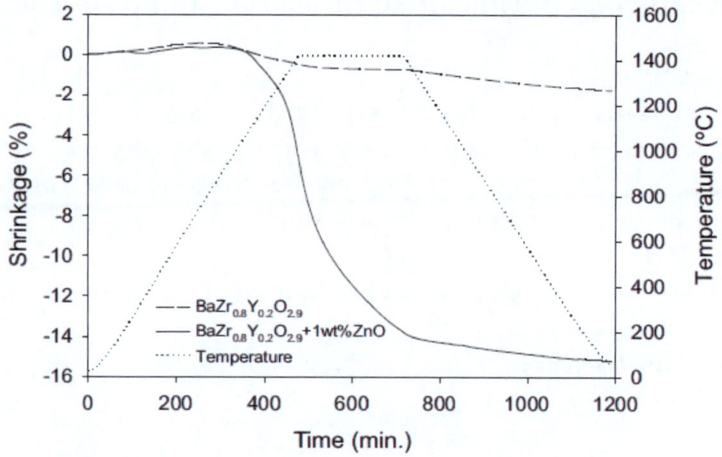

Figure 12. Shrinkage of $BaZr_{0.80}Y_{0.20}O_{3-\delta}$ and $BaZr_{0.80}Y_{0.20}O_{3-\delta}$ + 1wt% ZnO. Reproduced from [72] by permission of Elsevier Ltd.

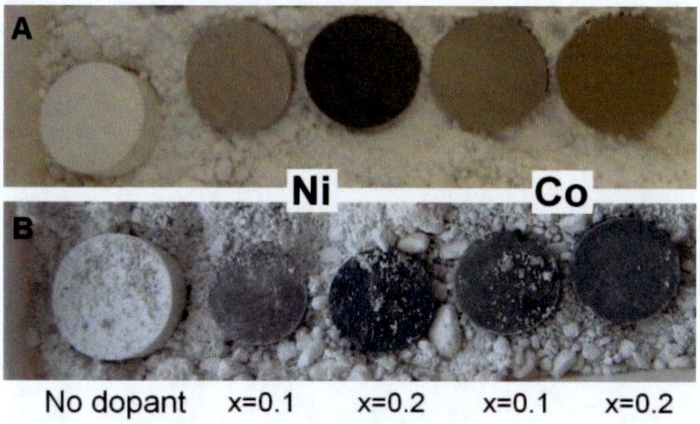

Figure 13. Photograph of BZY10, $BaZr_{0.9}Y_{0.1-x}Ni_xO_{3-\delta}$ and $BaZr_{0.9}Y_{0.1-x}Co_xO_{3-\delta}$ (x=0.01 and 0.02) before (a) and after sintering (b) at 1450 °C.

$BaCe_{0.5}Zr_{0.3}Y_{0.15}Zn_{0.04}O_{3-\delta}$ was prepared at 1325 °C [75]. $BaCe_{0.2}Zr_{0.7}Y_{0.1-x}Co_xO_{3-\delta}$ and $BaCe_{0.2}Zr_{0.7}Y_{0.1-x}Ni_xO_{3-\delta}$ (x= 0.01 and 0.02) were prepared by sintering at 1450 °C [73, 76]. Guo *et al.* [77] investigated the effect of zinc on the sintering of $BaZr_{0.4}Ce_{0.4}Y_{0.2}O_{3-\delta}$ and showed that it was more effective when added as oxide to the synthesized BCZY powder than when added as nitrate during the sol-gel process. Apart from using a number of additives or modifying the powder properties by powder synthesis methods, another way to influence the sinterability and to lower the sintering temperature is the development of a proper technology during component manufacturing, for example the so-called *in situ* solid state transformation [78-81]. Here dense layers of substituted $BaZrO_3$ or $BaCeO_3$ were manufactured at 1300-1400 °C after screen printing a $BaCO_3$ layer onto layers of 8 mol.% Y_2O_3-stabilized ZrO_2 (YSZ) or 20 mol.% Gd_2O_3-substituted CeO_2 (CGO). As a result, a dense gas-tight $BaZr_{1-x}Y_xO_{3-\delta}$ or $BaCe_{1-x}Gd_xO_{3-\delta}$ layer was formed.

5.2. Effect of Additives/Substituents on Electrical and Electrochemical Properties

The electrical properties of materials are influenced by a number of factors: the type and concentration of the dopant/substituent (defect concentration), the mobility of charge carriers, temperature, atmosphere etc. In this paragraph we focus on the conductivity-dopant correlation for various ionic and protonic conductors.

5.2.1. Oxygen Ionic and Mixed Oxygen Ionic-Electronic Conductors

Oxides with Fluorite Structure

ZrO_2

Yttria-stabilized zirconia (YSZ) is a widely used material for manufacturing of SOFC electrolytes due to several criteria: i) its pure ionic conductivity at elevated temperatures (800-1000 °C); ii) its very good mechanical and red-ox stability in the range of p_{O2} from 1 to 10^{-18} bar, as well as iii) its moderate price [6, 82, 83].

Introducing substituents in ZrO_2 usually has two major functions [84]: stabilization of the high temperature cubic polymorph down to room temperature and generation of oxygen vacancies responsible for the ionic conductivity of YSZ at high temperatures.

Formation of oxygen vacancies can be written according to the quasi-chemical reaction in Equation 22:

$$Y_2O_3 \xrightarrow{ZrO_2} = 2Y'_{Zr} + 3O_O^X + V_O^{\bullet\bullet} \tag{22}$$

The result of rare earth (RE) substitution on the structural stability is summarized in Table 1 [85].

Table 1. Influence of different RE elements on the structure of ZrO_2. *C, M, T* and *Rh* refer to cubic, monoclinic, tetragonal and rhombohedral phases, respectively, formed after high temperature sintering

<table>
<tr><td rowspan="2">x</td><td colspan="7">Crystal structure of $Zr_{1-x}RE_xO_{2-x/2}$</td></tr>
<tr><td>Sc</td><td>Yb</td><td>Er</td><td>Y</td><td>Dy</td><td>Gd</td><td>Eu</td></tr>
<tr><td>0.05</td><td rowspan="3">C. + T.</td><td rowspan="3">C. + T.</td><td rowspan="3">C. + M.</td><td rowspan="2">C. + T.</td><td rowspan="2">M.</td><td rowspan="3">C. + T.</td><td rowspan="6">T.</td></tr>
<tr><td>0.06</td></tr>
<tr><td>0.07</td><td rowspan="9">C.</td><td>C. + M.</td></tr>
<tr><td>0.08</td><td rowspan="3">C.</td><td rowspan="8">C.</td><td rowspan="8">C.</td><td rowspan="8">C.</td><td rowspan="8">C.</td></tr>
<tr><td>0.09</td></tr>
<tr><td>0.10</td></tr>
<tr><td>0.11</td><td rowspan="5">Rh.</td><td rowspan="5">C.</td></tr>
<tr><td>0.12</td></tr>
<tr><td>0.13</td></tr>
<tr><td>0.14</td></tr>
<tr><td>0.15</td></tr>
</table>

Ionic conductivity of ZrO_2 substituted with rare earths (Y_2O_3, Yb_2O_3, Gd_2O_3) ranges from about $1\cdot10^{-1}$ $S\cdot cm^{-1}$to $1\cdot10^{-2}$ $S\cdot cm^{-1}$ [86]. Ionic conductivity (0.1 $S\cdot cm^{-1}$) was measured for 10 mol % Y-stabilized ZrO_2 at 1000 °C [83]. At 800 °C this material exhibits ionic conductivity of $1.5\cdot10^{-2}$ $S\cdot cm^{-1}$, so that thin membranes can assure reasonable current densities at these temperature [87, 88]. Apart from Y_2O_3, ZrO_2 can be structurally stabilized by means of e.g. NaO, CaO, MgO, Al_2O_3, and RE-sesquioxides with bigger cation than Y [86, 89, 90] but this affects negatively the electrical properties.

CeO_2

Doped/substituted ceria has been proposed and investigated for several applications, e.g. as electrolyte [91, 92] and anode [93, 94] in SOFCs, mixed-conductive membranes for oxygen separation [29, 95], catalyst for partial oxidation of hydrocarbons [96, 97], oxygen sensors [98], oxygen storage materials and catalysts for other industrial processes [99, 100]. Ceria-based compounds are especially attractive for these applications since they combine high oxygen-ion mobility, catalytic properties, chemical compatibility with water and carbon dioxide at high temperatures and sufficient resistance to reduction under relatively low oxygen partial pressures below 500 °C. The ion conduction of ceria-based materials is one of the highest reported for oxide ceramics. Therefore these materials received much attention as candidates for SOFC electrolyte operating in the intermediate temperature range of 450-550 °C.

The prototype compound $Ce_{0.8}Gd_{0.2}O_{1.9}$ (CGO) exhibits ionic conductivity of 0.1 $S\cdot cm^{-1}$ at 750 °C, while similar conductivity is achieved for YSZ at 1000 °C. The application as solid electrolyte for SOFC based on $Ce_{1-x}Gd_xO_{2-0.5x}$ ($0<x<0.3$) was reported in [101]. It was found that the Gd content strongly affects the size of crystal grains and the grain boundaries blocking the charge carriers which is not only due to the grain size but also due to the microstructure and chemical characteristics of the boundaries. Apart from the high ionic conductivity of this oxide, the relatively low reactivity with other cell components led to the use of these materials as protection layers [102] (or cation diffusion barrier) and as a component in composite electrodes, e.g. LSFC-CGO cathodes.

At low oxygen partial pressures, Ce^{4+} cations are easily reduced to the Ce^{3+} for charge compensating the oxygen vacancies formed while the fluorite structure is retained. The material exhibits *n*-type electronic conductivity due to the polaron migration by thermally activated hopping, and the induced mixed conductivity limit its application as SOFC anode and oxygen membranes subjected to very high p_{O2} gradients to temperatures not higher than 550 °C. The main reason for this temperature limitation is a volume expansion accompanied by the partial reduction of the cerium ions. The *chemical expansion* leads to internal stresses in the structure and the components which easily result in strong warpage or cracking.

Substitution with La increases the lattice constant of CeO_2 promoting the mobility of oxygen ions, although the material becomes more reducible [103]. Dysprosium has been proposed as yttrium co-substituent in ceria systems [104] and erbium has been used as a substituent in the ceria-zirconia systems for catalytic applications [105]. The addition of elements with multiple valence states resulted in some cases in development of mixed ionic-electronic conductors, i.e. ceria substituted with Tb [106] and Pr [107]. In fact, at high oxygen partial pressures both ionic and *p*-type electronic conductivities are present in Pr/Tb-containing materials which make them very suitable for O_2-separation membranes and SOFC cathodes. Specifically, praseodymium has been employed as a substituent for gravimetric gas sensors because it increases the level of non-stoichiometry of ceria in oxidizing conditions [108]. However, it has a substitution limit of nearly 30% arising from the high and non-linear thermal expansion and the large chemical expansion upon reduction [109]. Furthermore, when Y_2O_3 or CaO are added to CeO_2, solid solutions are formed having high stability in a broad temperature range [110-114].

Ceria co-substitution with Tb and Co allowed polaron hopping between Tb ions of different oxidation states and enabled, in turn, achieving an oxygen flux at 1000 °C of ca. 0.02 and 0.08 ml·min^{-1}·cm^{2} for Ar-sweeping, through a thick membrane made of $Ce_{0.9}Tb_{0.1}O_{2-\delta}$ + Co and $Ce_{0.8}Tb_{0.2}O_{2-\delta}$ + Co, respectively [106]. These permeation values were increased to 0.5 ml·min^{-1}·cm^{-2} at 1000 °C ($Ce_{0.9}Tb_{0.1}O_{2-\delta}$ + Co sample) by using methane as a sweep gas. The additional cobalt content enhanced the sintering and densification of the specimens and led to significant increase of the total conductivity mainly due to (i) the improved material reducibility at high temperatures and the associated raise in the electronic conductivity; and (ii) the increased concentration of oxygen vacancies.

Despite all beneficial features of CeO_2-based materials, main disadvantage of these materials is the significant change of stoichiometry leading to chemical/compositional and structural instabilities, especially at high temperatures and in reducing atmospheres.

Bi_2O_3

The high-temperature modification of Bi_2O_3 (δ-Bi_2O_3) exhibits fluorite structure. However, δ-Bi_2O_3 is stable only in a narrow temperature range of 730 °C to its melting point, 825 °C. Below 730 °C, the α-Bi_2O_3 phase having monoclinic structure is stable and electronic conductivity prevails.

Bi_2O_3 forms a large variety of solid solutions with other metal oxides having very high oxygen ion conductivity. The most promising systems are Bi_2O_3-M_2O_3, with M = Y and RE elements. Most of the Bi_2O_3-M_2O_3 solid solutions crystallize either as *fcc* δ-Bi_2O_3, or as rhombohedral phase.

Stabilization of Bi_2O_3 can be realized either by addition of alkaline-earth oxides (SrO, CaO) or RE oxides (La_2O_3, Y_2O_3, Er_2O_3) [115]. The stabilization of δ-Bi_2O_3 at room temperature by tailored substitution with Er and Y is possible. However, thermal cycling leads to transformation to the tetragonal phase at temperatures below 600 °C [116].

According to [117], the conductivity of Y_2O_3-stabilized Bi_2O_3 is $1.0 \cdot 10^{-1}$ S·cm^{-1} and $1.0 \cdot 10^{-2}$ S·cm^{-1}at 700 and 500 °C, respectively.

The combination of Dy and W enabled to obtain the cubic phase in samples with substituent levels of 11-17 mol.%, while typically the substitution level is above 25mol.%. The conductivity of (Dy,W)-substituted Bi_2O_3 was reported to be higher (0.4 S·cm^{-1} at 700°C) that that of single Dy- (0.15 S·cm^{-1} at 700°C) or W- (0.08 S·cm^{-1} at 700°C) substituted sample [118].

Similar results were reported for the combination of W-Er or W-Y [119], although the cubic δ-Bi_2O_3 phase was not stable after long-term tests.

Despite its high ionic conductivity, especially at low/intermediate temperatures (400 – 700 °C), the ionic conductivity dominates only in a very restricted range of p_{O2}. At low p_{O2} ($p_{O2} \approx 10^{-13}$ bar) and temperatures of 600 °C Bi_2O_3 reduces to metallic Bi. The limited stability of these materials in reducing environment restricts their practical application. However, the most suited fields of application are SOFC cathodes or oxygen-transport membranes for the oxyfuel process [120].

Oxides with Perovskite Structure

Compounds with perovskite structure and dominant ionic conductivity find applications as solid electrolytes for SOCs at elevated temperatures. Materials operating efficiently at lower temperatures, e.g. below 700-800 °C, are especially attractive [121-124].

$La_{1-X}Sr_xGa_{1-Y}Mg_yO_{3-\Delta}$ (LSGM)

$LaGaO_3$ has been an object of numerous studies since about 15 years. Ishihara et al. [125, 126], K. Huang *et al.* [127], P. Huang *et al.* [128], Feng *et al.* [129] and Goodenough *et al.* [130] identified compositions in the system La_2O_3-SrO-Ga_2O_3-MgO as excellent oxygen ion conductors with promising application in SOFCs. Materials based on $LaGaO_3$ were considered as alternative to the conventional high temperature materials, e.g. YSZ used for manufacturing of SOFC electrolytes, mainly due to their higher ionic conductivity at lower operating temperatures [131-137]. Partial substitution in $LaGaO_3$ with Sr^{2+} and Mg^{2+} on La^{3+} and Ga^{3+} positions in the perovskite crystal lattice, respectively, was found to improve significantly the oxygen ion conductivity as a result of increased concentration of oxygen vacancies [125]. Since SrO and MgO are thermodynamically stable oxides and the cations do not change their valence, oxygen vacancy generation is the charge compensating mechanism leading to predominant ionic conductivity [125, 126, 136, 138-140] in a broad p_{O2} range (10^{-20} – 1 bar) and at intermediate temperatures.

The following defect-chemical reaction (Equation 23) describes the substitution of La by Sr:

$$SrO + La_{La}^{X} + \frac{1}{2}O_{O}^{X} \rightarrow Sr_{La}^{\bullet} + \frac{1}{2}V_{O}^{\bullet\bullet} + \frac{1}{2}La_2O_3 \tag{23}$$

Electrical properties of substituted $LaGaO_3$ are investigated by number of groups and plenty of papers are available on that topic [6, 123-128, 130, 132, 134-137, 139, 141-146].

Partial substitution on both A- and B-positions in $LaGaO_3$, with rare earths and alkaline earths, respectively, and characterization of their properties was carried out in [125-128, 143, 144, 147-152].

Substitution at A-site in $LaGaO_3$ with Ba, Ca, Sr gave the highest ionic conductivities in the case of Sr, according to [125]. Furthermore, when Al, In or Mg were added on the B-site in $La_{0.9}Sr_{0.1}GaO_{3-\delta}$, best conductivities were measured for substitution with Mg. The conductivity of $La_{0.9}Sr_{0.1}Ga_{0.8}Mg_{0.2}O_{3-\delta}$ at 800 °C is $1.5 \cdot 10^{-1}$ S·cm^{-1} [125], while the conductivity of YSZ at the same temperature is $3.4 \cdot 10^{-2}$ S·cm^{-1} [86]. Huang et al. [127, 153, 154] published thorough investigations on LSGM. The authors considered the phase equilibrium in the system and characterized the electrical properties of LSGM along with a study of the performance of a fuel cell with LSGM electrolyte.

Kharton et al. [149] investigated ionic and electronic transport in perovskites of the type La(Ga,M)$O_{3-\delta}$ with M = Mg, Cr, Fe, Co, Ni or Nb. Studies of Trofimenko and Ullmann [150] were focused on the correlation between the composition, structure and conductivity in Co-substituted LSGM.

Along with the extensive studies on the ionic conductivity as a function of substituent and concentration, trapping effects were thoroughly enlightened usually induced by the complex co-existence of more than one substituent in the crystal lattice. Khan et al. [148] performed energetic calculations of *trapping clusters* formed between the substituting ion and the oxygen vacancy for $LaGaO_3$ in both cubic and orthorhombic symmetry after the partial replacement of La with Sr and Ga with Mg, Cu or Hg. The lowest values for the cluster binding energy were found for Sr-$V_O^{\bullet\bullet}$ clusters in the cubic modification. The calculations also showed that the activation energy changes are negligible for different concentrations of Sr. Therefore, oxygen vacancies formed as a result of substitution with Sr would improve the ionic conductivity of the material. On the contrary, Mg substituted for Ga traps more effectively the oxygen vacancies, as the binding cluster energy is much higher. These results find confirmation in experimental studies on $La_{0.8}Sr_{0.2}Ga_{1-y}Mg_yO_{3-\delta}$ where activation energy increases and conductivity substantially decreases with increasing Mg content [128]. Based on a theoretical model, the authors were able to predict the behavior of trapping clusters calculated for different substituents on A- and B-sites in $La_{1-x}A^{2+}{}_xGa_{1-y}B^{2+}{}_yO_{3-\delta}$, where A^{2+} = Sr, Mn, Hg, Fe, Ni, Pb, Cd and B^{2+} = Cu, Hg, Ni, Mg, Co, Mn, Sr, Pb or Cd.

We refrain from discussing the phase purity and the electrical properties of the possible secondary phases that may be formed along with the LSGM. However, an important issue is the Ga stoichiometry which should be briefly mentioned, because it is a quite sensitive parameter. Ga evaporates at high processing and manufacturing temperatures [145] and even small changes in its content affect the phase formation, sintering behavior, and morphology along with the electrical properties. It was found that Ga evaporation as $Ga_2O_{(g)}$ is mainly related to the presence of $LaSrGaO_4$ which has a melting point of about 1400 °C. Usually, with the increase of the Ga content, $LaSrGa_3O_7$ forms having much lower ionic conductivity ($2.7 \cdot 10^{-4}$ $S \cdot cm^{-1}$ at 800 °C). It was found that the deficiency of gallium helps to form additional oxygen vacancies along with those formed due to the aliovalent substitution. On the contrary, excess of Ga results in decreasing vacancies concentration and ionic conductivity. Table 2 presents conductivities and activation energies of solid ionic conductors from the literature [155].

Table 2. Conductivity data for various solid electrolyte materials

Material	Synthesis route	T (°C)	σ ($S \cdot cm^{-1}$)	E_{act}, eV (kcal/mol)	Author [Ref.]
10YSZ		800 1000	$1.5 \cdot 10^{-2}$ $1.0 \cdot 10^{-1}$	1.05 (24.2)	[7,157]
8YSZ		800	$2.0 \cdot 10^{-2}$		[131]
c-YSZ				1.22 (28.1)	[158]
Bi_2O_3:Nd_2O_3		700	$0.85 \cdot 10^{-1}$		[114]
CeO_2:Sm_2O_3		800	$1.4 \cdot 10^{-1}$		
CeO_2:Gd_2O_3		800	$1.1 \cdot 10^{-1}$		
$Pr_{0.93}Sr_{0.07}Ga_{0.85}Mg_{0.15}O_{3-\delta}$		800	$0.9 \cdot 10^{-1}$		
$BaCeO_{2.9}$:20%Gd		800	$4.1 \cdot 10^{-2}$		
$Gd_2(Zr_xTi_{1-x})_2O_7$ (x>0.4)		800	$3.2 \cdot 10^{-4}$		
Bi_2O_3:Y_2O_3		500 700	$1.0 \cdot 10^{-2}$ $1.0 \cdot 10^{-1}$		[117]

Material	Synthesis route	T (°C)	σ (S·cm-1)	Eact, eV (kcal/mol)	Author [Ref.]
$La_{0.9}Sr_{0.1}Ga_{0.8}Mg_{0.2}O_{3-\delta}$	Solid state	600	$1.98\cdot10^{-2}$	1.13 (26.06)	[128]
		800	$1.21\cdot10^{-1}$		
		600	$2.21\cdot10^{-2}$		
$La_{0.8}Sr_{0.2}Ga_{0.8}Mg_{0.2}O_{3-\delta}$		800	$1.37\cdot10^{-1}$	1.15 (26.5)	
$La_{0.9}Sr_{0.1}Ga_{0.8}Mg_{0.2}O_{3-\delta}$	Solid state	700	$8.0\cdot10^{-2}$	0.86 (19.83)	[152]
		800	$1.0\cdot10^{-1}$	0.84 (19.37)	
$La_{0.8}Sr_{0.2}Ga_{0.85}Mg_{0.15}O_{3-\delta}$	Glycine-nitrate Tsinter=1400 °C	600	$1.74\cdot10^{-3}$		[156]
		700	$5.55\cdot10^{-3}$		
		800	$1.79\cdot10^{-2}$		
		850	$3.36\cdot10^{-2}$		
	Glycine-nitrate Tsinter=1500 °C	600	$8.09\cdot10^{-3}$		
		700	$2.63\cdot10^{-3}$		
		800	$6.06\cdot10^{-2}$		
		850	$7.8\cdot10^{-2}$		
$La_{0.9}Sr_{0.1}Ga_{0.8}Mg_{0.2}O_{3-\delta}$		600	$1.8\cdot10^{-2}$	1.11±0.003 (25.6±0.07)	[127]
		700	$6.0\cdot10^{-2}$		
		800	$1.3\cdot10^{-1}$		
$La_{0.8}Sr_{0.2}Ga_{0.83}Mg_{0.17}O_{3-\delta}$		600	$2.6\cdot10^{-2}$	1.076±0.008 (24.8±0.2)	
		700	$8.0\cdot10^{2}$		
		800	$1.7\cdot10^{-1}$		
$La_{0.8}Sr_{0.2}Ga_{0.8}Mg_{0.2}O_{3-\delta}$		600	$2.1\cdot10^{-2}$	1.074±0.007 (24.7±0.2)	
		700	$6.5\cdot10^{-2}$		
		800	$1.4\cdot10^{-1}$		
$La_{0.8}Sr_{0.2}Ga_{0.88}Mg_{0.1}O_{3-\delta}$	Solid state	600	$1.82\cdot10^{-2}$	(700-900)	[145]
		800	$9.31\cdot10^{-2}$	0.64 (14.7)	
$La_{0.8}Sr_{0.2}Ga_{0.89}Mg_{0.1}O_{3-\delta}$		600	$1.15\cdot10^{-2}$		
		800	$5.87\cdot10^{-2}$	0.73(16.8)	
$La_{0.8}Sr_{0.2}Ga_{0.92}Mg_{0.1}O_{3-\delta}$		600	$4.57\cdot10^{-3}$		
		800	$2.95\cdot10^{-2}$	0.82(18.9)	
$La_{0.8}Sr_{0.2}Ga_{0.94}Mg_{0.1}O_{3-\delta}$		600	$1.45\cdot10^{-3}$		
		800	$9.33\cdot10^{-3}$	0.87(20.1)	
$LaSrGa_3O_7$		800	$2.7\cdot10^{-4}$	1.09(25.1)	
$La_{0.9}Sr_{0.1}Ga_{0.8}Mg_{0.2}O_{3-\delta}$	Glycine-nitrate	600	$2.5\cdot10^{-2}$		[146]
		800	$1.5\cdot10^{-1}$		
		1000	$3.9\cdot10^{-1}$		
$La_{0.8}Sr_{0.2}Ga_{0.8}Mg_{0.2}O_{3-\delta}$		600	$2.5\cdot10^{-2}$		
		800	$1.7\cdot10^{-1}$		
		1000	$4.5\cdot10^{-1}$		
YSZ		600	$4.0\cdot10^{-3}$		
		800	$4.0\cdot10^{-2}$		
		1000	$1.5\cdot10^{-1}$		
$La_{0.9}Sr_{0.1}Ga_{0.8}Mg_{0.2}O_{3-\delta}$	Glycine-nitrate	600	$1.5\cdot10^{-2}$	LT/1.08(24.9) HT/0.74(17.06)	[147]
		800	$1.0\cdot10^{-1}$		
		1000	$2.6\cdot10^{-1}$		
$La_{0.8}Sr_{0.2}Ga_{0.8}Mg_{0.2}O_{3-\delta}$		600	$2.0\cdot10^{-2}$	LT/1.15 (26.5) HT/0.78 (17.9)	
		800	$1.2\cdot10^{-1}$		
		1000	$4.0\cdot10^{-1}$		
$La_{0.9}Sr_{0.1}Ga_{0.8}Mg_{0.2}O_{3-\delta}$	Solid state	700	$7.94\cdot10^{-2}$	0.79±0.04 (18.2±0.9)	[125]
		800	$1.51\cdot10^{-1}$		
		900	$2.34\cdot10^{-1}$		

$La_{1-x}Sr_xMnO_{3-\delta}$ (LSM), $La_{1-x}Sr_xCrO_{3-\delta}$ (LSCr), $Ba_{1-x}Sr_xCo_{1-y}Fe_yO_{3-\delta}$ (BSCF)

For the development of *mixed oxygen ionic-electronic conductors* there are in principle two possibilities: to introduce electronic conductivity in an oxygen ion conductor or to introduce oxygen ion conductivity in an electronic conductor.

One common approach is to prepare the so-called dual phase components (composites) by mixing a pure ionic conducting phase, e.g. stabilized zirconia or ceria with a pure electron conducting phase, e.g. noble metals or ceramic electronic conductors.

Apart from the difficulties to prepare such stable dual phase components ensuring the necessary percolation, this approach in principle limits the maximum conductivity or oxygen flux. First, the effective cross section of the ion conducting phase is significantly reduced due to the demand of 30-40 vol.% of electronic conducting phase. Moreover, triple phase boundaries (ion conductor, electron conductor and gas) are required for the reduction/ dissociation, as well as for oxidation/recombination of the oxygen ions. Nevertheless, SOFC composite cathodes like LSM/YSZ show superior electrochemical properties, especially small overvoltages, than pure LSM cathodes. Electrons and oxygen ions are transported separately in the different phases, e.g. O^{2-} move through the ceramic phase and the electrons through the metallic phase of a cermet. This again reduces the active volume, because the conduction pathways have a finite length. In LSM/YSZ cathodes the electrochemical active area is restricted to 10-20 μm related to the electrolyte interface. In the case of pure LSM cathodes, however, the electrochemical processes are restricted to the triple phase boundaries along the cathode/ electrolyte interface.

Another approach is the *heterogeneous doping*[1] of substituted ceria with cobalt, which was first reported by Fagg *et al.* [159]. Actually, cobalt was added as a sintering aid, but it turned out that the oxygen permeation rate is increased by an order of magnitude. Although the detailed mechanisms are not completely clarified, the increase in electronic conductivity was also proven for a wide range of gadolinium and praseodymium-substituted ceria [160]. However, there is still no breakthrough in enhancing the electronic conductivity to reach appreciable permeation rates.

Perovskites give the opportunity to tailor ionic and electronic conductivity intrinsically by doping or substitution. Considering a perovskite with the general formula $A^{3+}B^{3+}O_3$ with A = RE and B = Mn, Co, Fe, Cr, the A-site cation is partially substituted by divalent large earth-alkaline cations, e.g. Ca or Sr.

Two mechanisms exist to compensate this aliovalency. One possibility is a valence change of the B-site cations to a tetravalent state, e.g. in $La_{1-x}Sr_xMnO_{3-\delta}$ (LSM) or $La_{1-x}Sr_xCrO_{3-\delta}$ (LSCr). This leads to considerable semi-conducting behavior via polaron hopping [161, 162], i.e. electrons can move from a B^{3+} to a neighbouring B^{4+} ion. If the B-site cation cannot be oxidized, oxygen vacancies are formed, which can be expressed in Kröger-Vink notation with Equation 24:

$$SrO \xrightarrow{La_2O_3} Sr_{Ln}^{'} + O_O^x + \frac{1}{2}V_O^{\bullet\bullet} \tag{24}$$

[1] According to Maier [Maier, J. Prog. Solid St. Chem. 1995, 23, pp 171-263] homogeneous doping is the dissolution of suitable materials with aliovalent ions in the matrix/crystal lattice of the host material in order to influence the concentration of charge carriers. Heterogeneous doping is in an analogous manner to influence the conductivity by "physical" addition of coexisting second phases.

These vacancies are important for oxygen ion conductivity following Equation 25:

$$\sigma_{ion} = \frac{4 \cdot F^2 \cdot \left[V_O^{\bullet\bullet}\right] \cdot D_V}{R \cdot T \cdot V_m} \tag{25}$$

with σ_{ion} the ionic conductivity, D_V the diffusion coefficient of oxygen ions and V_m the perovskite molar volume.

The combination of both mechanisms, i.e. vacancy formation and polaron hopping, leads to mixed oxygen ionic-electronic conduction. This particular development started in the 1980's when Teraoka *et al.* first discovered the mixed conductivity of $La_{1-x}Sr_xCo_{1-y}Fe_yO_{3-\delta}$ (LSCF) [163]. Since only a limited fraction of Fe and Co cations can be oxidized to the tetravalent state for charge compensation of the Sr substitution, the mixed conductivity of LSCF was found to increase predominantly with increasing Sr and Co content.

Consequently, the highest permeability and conductivity should be obtained for $SrCoO_{3-\delta}$. However, this material does not have a perovskite structure, therefore exhibits poor oxygen permeability. Nevertheless, it was found that the perovskite structure is stabilized by partial substitution of Co with e.g. Fe, $SrCo_{0.8}Fe_{0.2}O_{3-\delta}$ [164].

A number of elements were investigated to stabilize the perovskite structure along with the improvement of oxygen permeation rates. Niobium was reported to be promising and stimulated certain interest [165]. Moreover, stabilization of the perovskite structure is also possible by partial substitution of Sr with Ba, leading to $Ba_{0.5}Sr_{0.5}Co_{0.8}Fe_{0.2}O_{3-\delta}$ (BSCF), which was first reported by Shao *et al.* [166] and has attracted great interest due to its very high oxygen permeability [31].

The average valence of Fe and Co at high temperature is 2.8 and 2.2, respectively [167]. This situation entails the high oxygen permeability due to a very high oxygen vacancy concentration of up to $\delta = 0.8$ [168] and the simultaneous existence of mixed oxidation states of the transition metals enabling the polaron hopping mechanism.

5.2.2. Protonic and Mixed Protonic-Electronic Conductors

The field of ceramic materials with protonic conductivity at elevated temperatures marked significant development since the first report on high-temperature protonic conduction in oxides published by Stotz and Wagner in 1966 [169].

In the early 1980s the protonic conductivity in oxides with perovskite structure was discovered by Iwahara *et al.* [170]. Since then the defect chemistry and the protonic conduction of a variety of oxide materials have received great attention [34, 39, 43,48-51, 171].

Different groups of materials with protonic conductivity were considered in the previous reviews [34, 172, 173] and comprise perovskites (acceptor doped $(Sr,Ba)ZrO_3$, $(Sr,Ba)CeO_3$; double perovskites as $Ba_3Ca_{1.18}Nb_{1.82}O_{8.73}$; elpasolite ($Me_4(Me_2'M_2^{5+})O_{11}[V_O]$) and cryolite ($Me_6M_2^{5+}O_{11}[V_O]$), where Me is an alkaline-earth ion, M^{5+} = Nb or Ta, and [VO] an intrinsic oxygen vacancy; brownmillerite compounds (e.g. Ti-substituted $Ba_2In_2O_5$); acceptor-doped rare-earth ortho-niobates and ortho-tantalates ($RE_{1-x}A_xMO_4$, with RE = La, Nd, Gd, Te, Er, Y, A = Ca, Sr, Ba and M = Nb or Ta; phosphates and solid acids.

Rare Earth Oxides

Norby and Kofstad found proton conduction in Y_2O_3 although the conductivity values were low (in the range of $10^{-4} – 10^{-5}$ S·cm^{-1}) [174, 175]. Furthermore, the proton conduction in numerous undoped and doped rare earth oxides with fluorite and perovskite structure were investigated and it was found that they also exhibit proton conduction at elevated temperatures [176-179]. An extensive study of equilibrium between water vapor, protons and oxygen vacancies in Nd_2O_3 and Sm_2O_3 doped with 1 and 2 mol.% Ca was conducted in [180]. These materials exhibit mixed protonic, O^{2-}-ionic and *p*-type electronic conductivity depending on the environmental conditions. They are nearly pure proton conductors at moderate temperatures and in wet atmospheres.

Moreover, the proton solubility at high temperatures increases with increasing stability and the packing density of the oxides. At the same time, the mobility of protonic defects decreases with increasing packing density which obviously influences negatively the proton conduction.

However, a different trend was tracked e.g. for $BaCeO_3$ which maintain its high proton solubility with high proton mobility.

Oxides with Perovskite Structure

Undoped and B-site doped/substituted oxides with perovskite structure as well as complex perovskites (double perovskites) of the types $A_2B'B''O_6$ and $A_3B'B_2''O_9$ are the most extensively studied materials that reveal protonic conduction at elevated temperatures. As discussed in the paragraph 3.2., the B-site doping/substitution with a lowervalent cation creates the oxygen vacancies which are hydrated in the presence of water. An overview of materials with perovskite structure and their properties in terms of fuel cell application is published by Norby [43]. Perovskites are also regarded as materials for hydrogen separating membranes [1, 44]. In a number of studies [34, 45, 47, 50, 51, 181-185] perovskites were investigated to better understand the proton conduction, e.g. the mechanisms of proton formation and diffusivity.

The best protonic conductivities $\geq 10^{-2}$ S·cm^{-1} were found for perovskites with the general formula $A^{2+}B^{4+}O_3$ (*type II-IV*) with A = Ba and B = Ce, Zr, Tb, Th. They have large negative hydration enthalpies and low activation energies for proton conduction [43]. The protonic conductivity, for example of $BaCe_{0.9}Nd_{0.1}O_{3-\delta}$ (BCNd), was $2.2 \cdot 10^{-2}$ S·cm^{-1} at 900 °C in pure H_2. This material contains considerable amounts of oxygen vacancies. However, BCNd exhibits also oxygen ion conductivity at temperatures higher than 800 °C. Similarly, $BaCe_{1-x}Y_xO_3$ (BCY) is a proton conductor at low and intermediate temperatures and an oxygen ion conductor at high temperature [186]. When the A-site is occupied by Sr instead of Ba, the conductivity drops down to 10^{-2} - 10^{-3} S·cm^{-1} ($SrZrO_3$, $SrCeO_3$ [170]). For instance, the measured protonic conductivity of $SrCe_{0.95}Yb_{0.05}O_{3-\delta}$ (SCYb) was $7 \cdot 10^{-3}$ S·cm^{-1} at 900 °C in pure H_2 and this material is mainly a protonic conductor. Calcium-based perovskites are also proton conductors, especially $CaZrO_3$.

Within the group of perovskites $A^{1+}B^{5+}O_3$ (*type I-V*) the proton conductivity is low, such as in $KTaO_3$ [187]. Therefore this sub-class of perovskites is not attractive in terms of the discussed applications. Furthermore, Ca-, Sr-, Ba-doped $LaErO_3$, $LaScO_3$, and $LaYO_3$ [188] were found to show protonic conductivity as well.

Typically, doping of Ba- or Sr-cerates and zirconates with multivalent cations has yielded the best ambipolar conductivities. For instance, the use of Eu, Sm, Tm or Yb seems to promote the electronic conductivity of Y-doped/substituted (Sr,Ba)CeO_3 oxides. A recent work reported [189] a supported 20 μm-thick membrane of $SrZr_{0.2}Ce_{0.8-x}Eu_xO_{3-\delta}$ for which a hydrogen flow of up to 0.5 $ml \cdot min^{-1} \cdot cm^{-2}$ was measured at 900 °C. Furthermore, studies [190, 191] on $BaZrO_3$ substituted with Pr and Mn revealed the high potential of these materials as hydrogen membranes and ceramic proton conducting fuel cell components. In [191] a number of compounds in the system $BaZr_{1-x-y}Y_xM_yO_{3-\delta}$ were developed and $BaZr_{0.8}Y_{0.15}Mn_{0.05}O_{3-\delta}$ was pointed out as a promising candidate for industrial hydrogen-oriented applications. A H_2 flux of around 7 $ml \cdot min^{-1} \cdot cm^{-2}$ was predicted for a 20 μm-thick membrane at 900 °C. Further data on the hydrogen fluxes for numerous cerates and zirconates can be found in the reviews by Meulenberg et al. [1, 2].

With respect to the cerates and zirconates it should be mentioned that despite their highest protonic conductivity, there are some disadvantages that are application-limiting for these materials. For example, the stability in reducing atmospheres has to be mentioned, especially in the case of the cerates. Due to their basicity they easily form carbonates when exposed to CO_2-containing atmospheres [192, 193]. Despite the better chemical stability of zirconates, their higher grain boundary resistance influences negatively the electrical performance [72, 78-81, 194, 195]. Furthermore, technological difficulties are experienced for the manufacturing of dense specimens of these state-of-the art proton conductors as already discussed in section 5.1.

Oxides with Pyrochlore Structure

Pyrochlore materials have been first studied for their oxygen ion conductivity [196-198]. Later on, the study of Shimura et al. [199] revealed the presence of protonic conductivity in $Ln_2Zr_{2-x}Y_xO_{7-\delta}$ (Ln = La, Nd, Sm, Gd, Er), whereas $Y_2Ti_{1.8}M_{0.2}O_{7-\delta}$ with M= In, Mg did not show measurable protonic transport [199]. $La_2Zr_2O_7$ was reported to show protonic conductivity, but not in the isostructural Ti-containing oxide [200]. Omata et al. studied the protonic conductivity in Ca-substituted $La_2Zr_2O_7$ [201, 202]. Eurenius et al. also showed protonic conductivity in $Sm_{1.92}Ca_{0.08}Ti_2O_{7-\delta}$, $Sm_2Ti_{1.92}Y_{0.08}O_{7-\delta}$, $Sm_{1.92}Ca_{0.08}Sn_2O_{7-\delta}$ and $Sm_2Ti_{1.92}Sn_{0.08}O_{7-\delta}$ [203-205]. The conductivity of $La_2Ti_2O_7$ was recently measured at relatively high temperatures (800-1000 °C) and with no emphasis on possible effects of protons and protonic transport [206]. Haugsrud et al. [207] studied 2% Ca-doped $La_2Ti_2O_7$ and protonic conductivity was determined to be ~$5 \cdot 10^{-5}$ $S \cdot cm^{-1}$ at 800 °C. The hydrogen flux across a Ca-doped $La_2Ti_2O_7$ film with a thickness of 10 μm was estimated on the basis of the partial conductivities. Such a membrane would yield a hydrogen flux in the order of 0.1 $ml_n \cdot min^{-1} \cdot cm^{-2}$ at 750 °C, a value which is lower by a factor of ~5 compared to that of 5% Yb-substituted $SrCeO_3$ under the same conditions. The conductivity of the three compounds $La_{2-x}Ca_xCe_{1.5}Zr_{0.5}O_{7-\delta}$ (x=0, 0.02, 0.1) was measured in wet oxidizing and reducing atmosphere (2.5 vol.% H_2O) by electrochemical impedance spectroscopy (EIS). The p_{O2} dependence revealed that these compounds exhibit an electronic conductivity at low oxygen partial pressure (below 10^{-20} bar) and are ionic conductors at $p_{O2} > 10^{-20}$ bar. Figure 14 is an example of p_{O2} dependence at 800 °C for $La_{1.98}Ca_{0.02}Ce_{1.5}Zr_{0.5}O_{7-\delta}$ [14]. Similar curves were obtained for other two compositions. The total conductivity can be then written as Equation 26:

$$\sigma_{tot} = \sigma_{ion} + \sigma_n \cdot p_{O2}^{-1/4} = \sigma_{H^+} + \sigma_{O^{2-}} + \sigma_n \cdot p_{O2}^{-1/4} \quad (26)$$

where σ_{ion} is the ionic conductivity, σ_{H+} the protonic conductivity, $\sigma_{O^{2-}}$ the oxygen ion conductivity and σ_n the electronic conductivity.

Due to the absence of electronic conduction at high oxygen partial pressure, the total conductivity measured in humidified oxygen corresponds to the ionic conductivity.

To determine the temperature range in which protons are the predominant charge carriers, the isotope effect on conductivity was studied [208] and EIS measurements were performed in deuterated O_2 (p_{D2O}=0.022 bar). It was found that the conductivity in wet O_2 was higher than that in deuterated O_2 below 500 °C.

We then can conclude that oxygen ion conductivity dominates at high temperatures while protons are the main charge carrier at temperatures below 500 °C.

The ionic conductivity in wet O_2 is plotted as a function of the inverse temperature for the three compositions in Figure 15, showing a decrease of the ionic conductivity with higher Ca^{2+} content in the whole temperature range. Unexpectedly, both the oxygen ionic and protonic conductivities are lowered with increasing Ca content.

The activation energy for protonic conduction (E_{act}), determined from the Arrhenius plots for temperatures below 500 °C (Table 3), is equal to the enthalpy of mobility of protons (ΔH_m) in the case of a constant proton concentration.

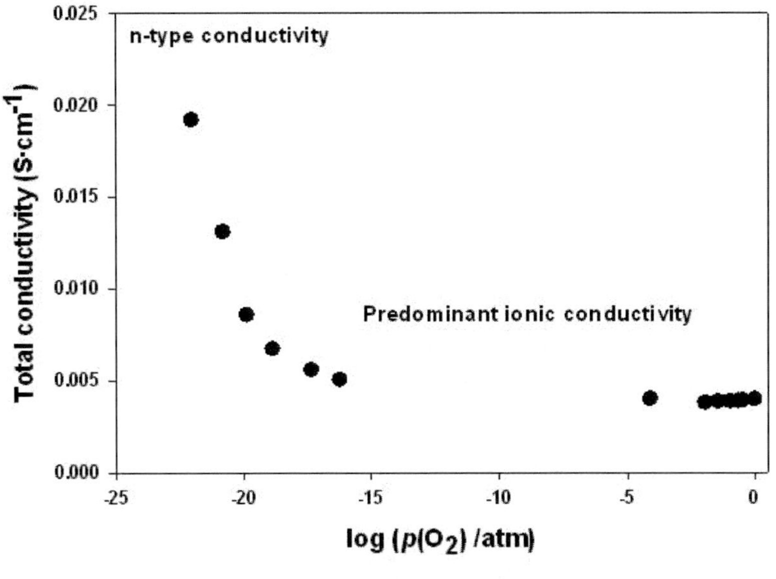

Figure 14. p_{O2} dependence of total conductivity of $La_{1.98}Ca_{0.02}Ce_{1.5}Zr_{0.5}O_{7-\delta}$ at 800 °C in wet atmosphere (p_{H2O}=0.025 bar) [14]. Reproduced from [14] by permission of Elsevier Ltd.

Table 3. Activation energy for protonic conduction in wet air (p_{H2O}=0.025 bar), T ≤ 500 °C

	$La_2Ce_{1.5}Zr_{0.5}O_{7-\delta}$	$La_{1.98}Ca_{0.02}Ce_{1.5}Zr_{0.5}O_{7-\delta}$	$La_{1.90}Ca_{0.10}Ce_{1.5}Zr_{0.5}O_{7-\delta}$
E_a (eV)	0.89	0.97	0.98

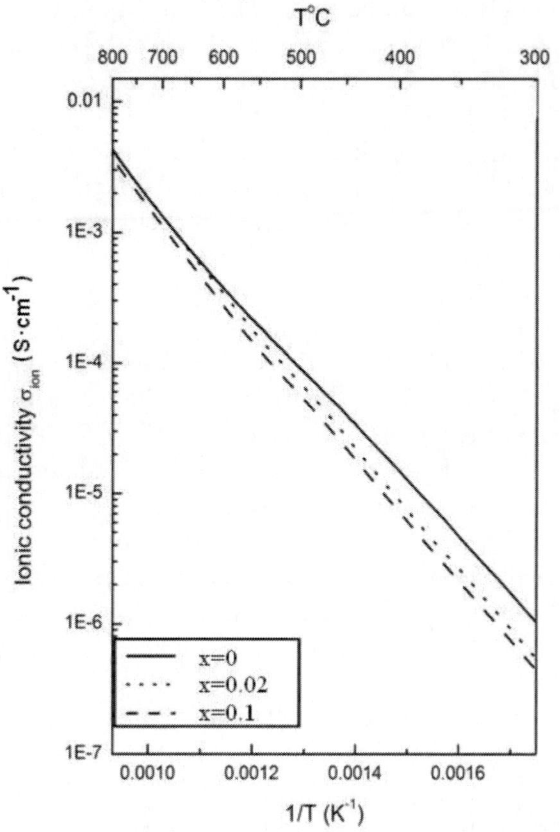

Figure 15. Ionic conductivity of $La_{2-x}Ca_xCe_{1.5}Zr_{0.5}O_{7-\delta}$ (x=0, 0.02, 0.1) in wet atmosphere (p_{H2O}=0.025 bar) [14]. Reproduced from [14] by permission of Elsevier Ltd.

If trapping effects are present, an additional energy-related term for proton trapping by the dopant (ΔH_t) has to be added (Equation 27) [209]:

$$E_{act} = \Delta H_m + \frac{1}{2}\Delta H_t \tag{27}$$

As the composition $La_2Ce_{1.5}Zr_{0.5}O_{7-\delta}$ exhibits a lower activation energy for protonic conduction than the substituted ones $La_{2-x}Ca_xCe_{1.5}Zr_{0.5}O_{7-\delta}$ (x=0.02, 0.1), it was speculated that the defects Ca'_{La} trap protons, resulting in a lower protonic conductivity.

Oxides with Defective Fluorite Structure: $Ln_{6-X}WO_{12-\Delta}$

Shimura et al. [210] studied $La_{5.8}WO_{11.7}$ in electrochemical hydrogen pumping experiments. For this material protonic conductivity of $5\cdot10^{-3}$ $S\cdot cm^{-1}$ was measured in wet H_2 at 900 °C and a protonic transport number between 0.7 and 0.9 was determined in H_2 atmosphere.

Haugsrud [211] studied defects and transport properties in undoped and Ca-doped Ln_6WO_{12} with Ln = La, Nd, Gd, and Er. As a general feature, these materials exhibit protonic conductivity in wet atmospheres and below ~900 °C. In oxidizing and reducing conditions and elevated temperatures Ln_6WO_{12} becomes predominantly a semiconductor. Maximum total conductivity in the range $3\text{-}5\cdot10^{-3}$ $S\cdot cm^{-1}$was observed at about 850 °C for undoped Ln_6WO_{12} (Figure 16).

1% Ca- doped La_6WO_{12} exhibits higher protonic conductivity ($6\cdot10^{-4}$ $S\cdot cm^{-1}$) than 1% Ca-doped Nd_6WO_{12} (~$4\cdot10^{-4}$ $S\cdot cm^{-1}$) and Er_6WO_{12} (~ $1\cdot10^{-4}$ $S\cdot cm^{-1}$) at 800 °C in wet H_2 [211].

An extensive research on the impact of different acceptor doping schemes in $Ln_{6-x}WO_{12-\delta}$ on the conductivity and H_2-flux was carried out [212-216].

In several publications, the feasibility of tungstates as membranes for H_2 separation was studied by flux measurements or theoretically estimated [210, 213, 217, 218].

A comparison of the ambipolar conductivity [217] of undoped $La_{6-x}WO_{12-\delta}$ and 5%-Yb-substituted $SrCeO_3$ (state-of-the art material) came to the conclusion that the undoped La_6WO_{12} is a good candidate for membrane applications.

Permeability data were calculated by Norby et al. [218] for La_6WO_{12}, $SrCeO_3$ and Er_6WO_{12}. The proton flux was calculated considering a 10 μm thick membrane with feed-side pressure of 10 atm H_2 and assuming that the flux is described by the Wagner equation, i.e. through the bulk diffusion mechanism. The predicted hydrogen permeation for La_6WO_{12} at 800 °C has a value of about 2.0 $ml_n\cdot min^{-1}\cdot cm^{-2}$ which is higher than $SrCeO_3$ and Er_6WO_{12} yielding hydrogen fluxes at the same temperature of about 1.0 $ml_n\cdot min^{-1}\cdot cm^{-2}$ and less than 0.1 $ml_n\cdot min^{-1}\cdot cm^{-2}$, respectively.

The hydrogen flux was experimentally determined by Escolástico et al. [213] through a 510 μm thick membrane of $Nd_{6-x}WO_{12-\delta}$. At the feed and sweep side, a mixture of humidified 20 % H_2/He and argon was used, respectively, both sides at atmospheric pressure. At temperatures below 900 ºC, the electron conductivity seems to be the limiting factor, while at high temperatures (above 900 ºC) the *p*-type conductivity becomes predominant and the protonic conductivity is the limiting step, likely due to the low concentration of protons in the oxide lattice.

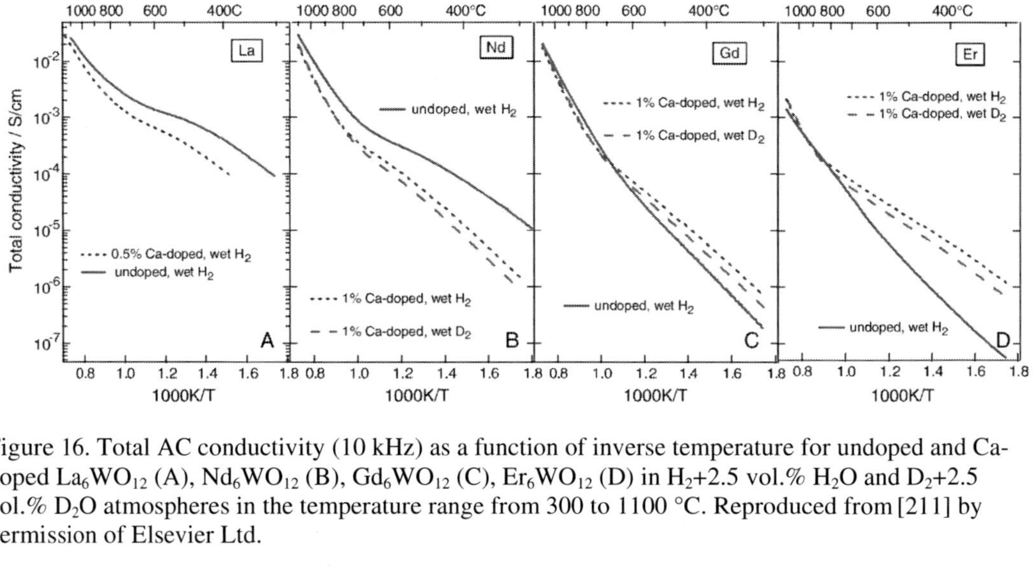

Figure 16. Total AC conductivity (10 kHz) as a function of inverse temperature for undoped and Ca-doped La_6WO_{12} (A), Nd_6WO_{12} (B), Gd_6WO_{12} (C), Er_6WO_{12} (D) in H_2+2.5 vol.% H_2O and D_2+2.5 vol.% D_2O atmospheres in the temperature range from 300 to 1100 °C. Reproduced from [211] by permission of Elsevier Ltd.

As reported furthermore by Escolástico et al. [214], the H_2 flux for $(La_{5/6}Nd_{1/6})_{5.5}WO_{12-\delta}$ measured at 1000 °C was in the range from $5\cdot10^{-3}$ up to $5\cdot10^{-2}$ ml·min^{-1}·cm^{-2} depending on the concentrations of H_2 and H_2O. However, these values are similar to those reported for $(Nd_{5/6}La_{1/6})_{5.5}WO_{12-\delta}$ membranes [215] (Figure 17) below 900 °C, although a substantial improvement was expected by considering the higher protonic conductivity of the La-rich compound.

As a consequence, the lack of flux improvement is due to the insufficient electronic conductivity. The significantly higher flux obtained for La-containing samples at temperatures above 900 °C (Figure 17, right) is also related to the positive contribution of oxygen ion transport in the hydrogen separation process via water splitting process.

As discussed with Escolástico and Serra (CSIC, Valencia, Spain) in a private communication, the H_2 flux of $La_{5.5}WO_{12-\delta}$ and $Nd_{5.5}WO_{12-\delta}$ can be significantly improved by doping with transition metal ions. Values of above 0.2-0.3 ml·min^{-1}·cm^{-2} at 700 °C were indicated for 0.4 μm-thick monolithic membranes.

Niobates/Tantalates

Protonic conductivity was found in several acceptor-doped rare earth ortho-niobates and ortho-tantalates ($RENbO_4$ and $RETaO_4$) at elevated temperatures and in a humid atmosphere [219]. These oxides exhibit mixed protonic, oxygen ionic and electronic conduction depending on the ambient conditions.

Both the low-temperature monoclinic and the high-temperature tetragonal polymorphs show protonic conduction with higher values for the tetragonal polymorph. Maximum protonic conductivity of about $9\cdot10^{-4}$ S·cm^{-1} was found for the A-site Ca-doped $LaNbO_4$ at 950 °C in wet H_2 [219].

This value is the highest measured so far for rare earth niobates. The investigations of Haugsrud and Norby [220] showed that the Ca substitution on the A-site plays a beneficial role and the conductivity is higher than that of pure $LaNbO_4$ studied in [221], and to that of a single B-site doped materials, e.g. $LaNb_{0.99}Ti_{0.01}O_{4-\delta}$ [222] and $LaNb_{1-x}M_xO_{4-\delta}$ with M = Ga, Ge, Si, B, Ti, Zr, P, Al [223].

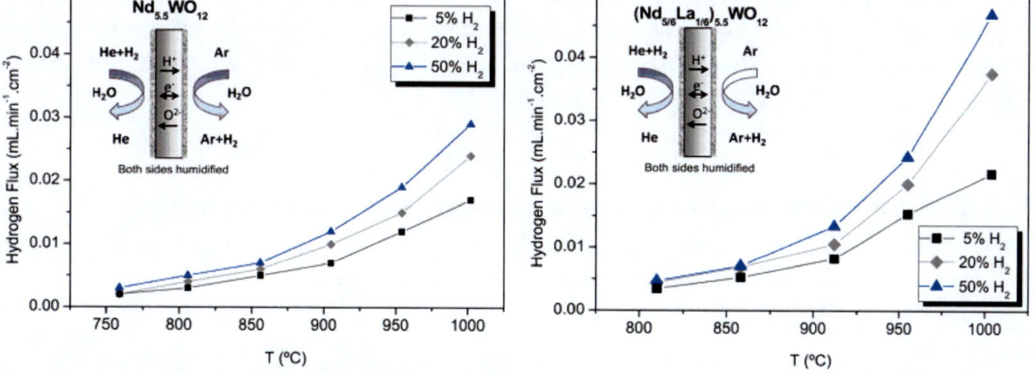

Reproduced from [215] by permission of Elsevier Ltd.

Figure 17. Hydrogen flux as a function of temperature for $Nd_{5.5}WO_{12-\delta}$ and $(Nd_{5/6}La_{1/6})_{5.5}WO_{12-\delta}$ at three different p_{H2} in the feed stream. Both membrane sides were humidified at room temperature.

Norby et al. [224] patented acceptor-doped rare earth ortho-niobate and ortho-tantalates with a general formula $Ln_{1-x}A_xM_{1-y}B_yO_4$ where Ln = Y, La-Lu, A = Mg, Ca or Sr, M = Nb or Ta and B = Mg, Sc, Al, Ti, Zr, Fe, Mn, Eu having a valence smaller than that of M; the substitution can be varied in the range $0 < x,y < 0.1$ with at least one of x or y > 0. The use of the materials as an electrolyte in a laboratory-scale fuel cell and water vapor sensor was also demonstrated in the patent. Due to its very close ionic radius to that of Nb, Ti was examined as either as a B-substituent [222, 223] or as additional substituent on the B-site in addition to Ca on the A-site [222]. For both cases, significant *n*-type contribution to the conductivity was found in reducing atmospheres, especially in the case of Ti and Ca-containing compound. Therefore, it was suggested for potential application as a H_2 separation membrane.

In a recent study by Ivanova et al. [225] on B-site doped ($LaNb_{0.99}B_{0.01}O_{4-\delta}$ with B = Ga, Ge or In) and co-doped $LaNbO_4$ ($La_{0.99}A_{0.01}Nb_{0.99}B_{0.01}O_{4-\delta}$ with A = Ca or Ba; B = Ga, Ge or In) using a humidified mixture of 9% H_2-N_2, it was shown that either the single B-site doping or the co-doping of $LaNbO_4$ were not as beneficial as the single A-site doping by Ca [219, 220]. The conductivity values for single B-site doped $LaNbO_4$ were much lower than those for the co-doped samples. However, the combination of Ca and Ga gave total conductivity of about $4\cdot10^{-4}$ $S\cdot cm^{-1}$ at 900 °C, which was the highest total conductivity within the investigated series.

Furthermore, Fjeld et al. [226] published an extensive impedance study of a single A-site doped $La_{0.995}Sr_{0.005}NbO_{4-\delta}$ supported by TEM-EDS analyses at the grain boundaries and triple grain junctions in order to elucidate the reason for the low grain boundary conductivities of this proton conductor. Based on the performed microstructural characterization and comparison of the electrical properties of samples sintered at different temperatures, it was shown that the grain boundary conductivity exhibits a higher functional dependence on p_{H2O} than the grain interior at 400 °C.

This is an indication for the role of space charge layers in the grain boundaries of acceptor-doped $LaNbO_4$. A comparison between the conductivity values of doped and undoped lanthanum niobates measured in H_2 atmosphere is presented in Table 4.

A variety of studies on rare earth ortho-tantalates has shown that these materials also exhibit protonic conductivity which generally is an order of magnitude lower than that of niobates [218]. With decreasing ionic radius in the row La→Nd→Gd→Er the activation energy for protonic conduction increases and this tendency holds both for ortho-niobates and ortho-tantalates. Table 5 summarizes the total conductivity and protonic conductivities of acceptor-doped $LnTaO_4$ at 1000 °C according to [218].

5.3. Effects of Additives on the Chemical and Mechanical Stability

5.3.1. Oxygen Ionic and Mixed Oxygen Ionic-Electronic Conductors

ZrO_2

Doping/substitution also affects the mechanical properties of the electrolyte material. The mechanical robustness is a determining factor for the manufacturing of thin solid electrolyte membranes (thickness less than 7-10 μm; for example, a dense 8YSZ electrolyte with thickness of 1 μm was developed at Research Center Jülich for SOFCs [88]).

Table 4. Conductivity values (in S·cm-1) for various lanthanum niobates

Compound	Atmosphere	T (°C)	$\sigma_{tot,}$10 kHz ($S\cdot cm^{-1}$)	σ_{bulk} ($S\cdot cm^{-1}$)	$\sigma_{prot.,}$emf ($S\cdot cm^{-1}$)	Ref.
$LaNbO_4$	H_2+2.5%H_2O	800 900 1000	~8-9·10^{-6} <2·10^{-5} ~4-5·10^{-5}	<2·10^{-5} ~4·10^{-5} ~6·10^{-5}		[221]
$La_{0.99}Ca_{0.01}NbO_{4-\delta}$	H_2+2% H_2O	800 900 1000	~7.5·10^{-4} ~9·10^{-4} ~1·10^{-3}		~7.5·10^{-4} ~8.5·10^{-4} ~9·10^{-4}	[220]
$La_{0.98}Ca_{0.02}NbO_{4-\delta}$	H_2+2.5%H_2O	800 900 1000	~3·10^{-4} ~4·10^{-4} ~5·10^{-4}	7·10^{-4}	8·10^{-4}	[221]
$La_{0.98}Sr_{0.02}NbO_{4-\delta}$	H_2+2.5%H_2O	800 900 1000	<4·10^{-4} ~4.5·10^{-4} <6·10^{-4}	8·10^{-4}		[221]
$La_{0.98}Ba_{0.02}NbO_{4-\delta}$	H_2+2.5%H_2O	800 900 1000	3·10^{-4} <4·10^{-4} <5·10^{-4}	5·10^{-4}		[221]
$LaNb_{0.99}Ti_{0.01}O_{4-\delta}$	H_2+2.5%H_2O	800 1000	7-8·10^{-5} ~1·10^{-4}			[222]
$LaNb_{0.99}Ga_{0.01}O_{4-\delta}$ $LaNb_{0.99}Ge_{0.01}O_{4-\delta}$ $LaNb_{0.99}In_{0.01}O_{4-\delta}$	wet 9%H_2+ N_2	800 900 800 900 800 900	7.7·10^{-5} 1.1·10^{-4} ~5·10^{-5} 7.6·10^{-5} 7.3·10^{-5} 1.4·10^{-4}			[225]
$La_{0.99}Ca_{0.01}Nb_{0.99}Ga_{0.01}O_{4-\delta}$ $La_{0.99}Ca_{0.01}Nb_{0.99}Ge_{0.01}O_{4-\delta}$ $La_{0.99}Ca_{0.01}Nb_{0.99}In_{0.01}O_{4-\delta}$	wet 9%H_2+ N_2	800 900 800 900 800 900	~3·10^{-4} ~4·10^{-4} 1.2·10^{-4} 1.7·10^{-4} 1.7·10^{-4} 2.5·10^{-4}			[225]
$La_{0.99}Ba_{0.01}Nb_{0.99}Ga_{0.01}O_{4-\delta}$ $La_{0.99}Ba_{0.01}Nb_{0.99}Ga_{0.01}O_{4-\delta}$ $La_{0.99}Ba_{0.01}Nb_{0.99}Ga_{0.01}O_{4-\delta}$	wet 9%H_2+ N	800 900 800 900 800 900	1.3·10^{-4} 1.8·10^{-4} 4.9·10^{-5} 7.6·10^{-5} 2.1·10^{-4} 2.9·10^{-4}			[225]
$La_{0.99}Ca_{0.01}Nb_{0.99}Ti_{0.01}O_{4-\delta}$	H_2+2.5%H_2O	800 1000	2·10^{-3} 4·10^{-3}			[222]

Yamamoto [227] studied the tetragonal polymorph of ZrO_2 (*t*-ZrO_2) aiming at improving the mechanical properties of ZrO_2-based electrolytes.

The fracture toughness of the *t*-ZrO_2 is much higher (1200 MPa) compared with the value of 300 MPa measured for the cubic modification (*c*-ZrO_2). *t*-ZrO_2 is stabilized with minor Y_2O_3 additions [228].

Table 5. Total and protonic conductivities (in S·cm^{-1}) of acceptor-doped $LnTaO_4$ at 1000 °C

Compound	Total conductivity (S·cm^{-1})	Protonic conductivity (S·cm^{-1})
$La_{0.99}Ca_{0.01}TaO_{4-\delta}$	$4 \cdot 10^{-4}$	$\sim 2 \cdot 10^{-4}$
$Nd_{0.99}Ca_{0.01}TaO_{4-\delta}$	$2 \cdot 10^{-4}$	$< 2 \cdot 10^{-4}$
$Gd_{0.99}Ca_{0.01}TaO_{4-\delta}$	$8 \cdot 10^{-5}$	$< 5 \cdot 10^{-5}$
$Er_{0.99}Ca_{0.01}TaO_{4-\delta}$	$4 \cdot 10^{-5}$	$4 \cdot 10^{-5}$

In [229], a composite solid electrolyte material was developed by mixing 3YSZ(t-ZrO_2) with 8YSZ(c-ZrO_2) in a weight ratio 1:1. The study was carried out to determine the effect of the tetragonal phase 3YSZ on mechanical and electrical properties of the material.

The composite had higher bending strength compared to the two starting phases. Furthermore, the longer the sintering time, the better the mechanical properties of the material. Certain ceria- and zirconia-based composites, e.g. $Ce_{0.9}Gd_{0.1}O_{1.95}$:8YSZ and $Sc_{0.1}Ce_{0.01}Zr_{0.89}O_{1.95}$:8YSZ exhibit appreciable mechanical stability combined with high ionic conductivity [229].

As it was mentioned in the previous paragraphs, despite its high ionic conductivity in the intermediate temperature range, Bi_2O_3 retains its ionic character only in a very restricted range of p_{O2}. This material suffers also from insufficient chemical and mechanical stability independent on stoichiometric variations: at low p_{O2} ($p_{O2} \approx 10^{-13}$ bar) and temperatures of 600 °C Bi_2O_3 reduces to metallic Bi.

Mixed Conductors

There is a principal trade-off between performance and stability in mixed oxygen ionic-electronic conductors. Highly permeable materials always suffer from low thermal, chemical, and/or mechanical stability. In BSCF showing the highest permeability so far, all of these stability issues are considerable disadvantages. Therefore, the different issues concerning this material are discussed below.

As already explained in section 5.2.1, the permeability is closely linked to the oxygen vacancy concentration and thus to the reducibility of the transition metal ions. Consequently, the highest permeability is achieved using perovskites with high Co-content.

On the contrary, high vacancy concentrations can lead to vacancy ordering, which is accompanied with a strongly reduced diffusivity of the oxygen ions. $SrCo_{0.8}Fe_{0.2}O_{3-\delta}$ for instance is prone to the formation of a brownmillerite structure, in which 1/6 of the oxygen sites of the perovskite are vacant, i.e. the oxygen deficiency δ is 0.5 already in pure N_2 atmosphere at temperatures of 600 – 900 °C, whereas in BSCF it is not.

The oxygen non-stoichiometry δ of BSCF was even found to be around 0.6 – 0.8 in this temperature range. Reasons might be the higher basicity and/or size of the Ba^{2+} ion compared to that of Sr^{2+} ion, potentially destabilizing highervalent B-site cations, i.e. tetravalent or trivalent transitions metals Co and Fe.

However, the mechanism is not completely understood so far [230]. At high temperatures BSCF exhibits a cubic perovskite structure enabling the high permeability. Below 840 °C a hexagonal polymorph occurs [231], first discovered by Švarcová et al. [232].

Recently, it was found that a miscibility gap exists [233] rather than a phase transition. This instability leads to degradation of the oxygen permeation rate in synthetic air as shown in Figure 18 (open squares). Stabilization of the cubic perovskite phase is therefore investigated by co-doping the B-site with higher valent cations such as Zr^{4+} [234] or Nb^{5+} [235]. A further degradation mechanism arises due to the presence of alkaline earth elements at the A-site. This leads to a pronounced reactivity with gaseous species such as CO_2, SO_x, H_2O, etc. As Ba and Sr tend to form very stable carbonates and sulphates this leads to severe degradation of the oxygen permeation rate using ambient air as shown in Figure 18 (solid squares). The mechanism of carbonate formation is ongoing [236] and substitution strategies have to be derived.nAnother issue is the phase stability in reducing atmospheres such as H_2 or CH_4. The transition metal ions are reduced to the metallic state and thus the perovskite structure is destroyed. Therefore, material development is ongoing to substitute Co by less reducible elements such as Al or Zn [237]. For industrial applications, the mechanical stability also has to be considered. High performance perovskites such as BSCF show several anomalies in mechanical behavior in the intermediate temperature regime, i.e. mechanical properties such as elastic modulus, fracture toughness, and strength of BSCF showed a minimum at 200 – 500 °C [238]. Furthermore, at temperatures above 800 – 850 °C, very high creep rates are observed, which are obviously dependent on the crystallographic phase [239].

5.3.2. Protonic and Mixed Protonic-Electronic Conductors

Barium Cerates and Barium Zirconates

Barium cerate reacts with carbon dioxide to form carbonates at temperature below 1150 °C [240, 241] (Equation 28) and with water vapor to form hydroxides below 400 °C [240] (Equation 29). Figure 19 shows the change of Gibbs free energy for the hydration and carbonation reactions of barium cerate.

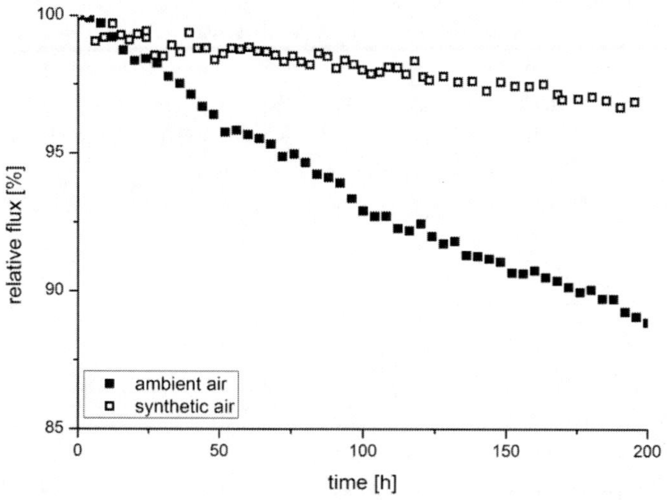

Figure 18. Degradation of the permeation rate of BSCF at 800 °C using ambient (solid squares) and synthetic air (open squares).

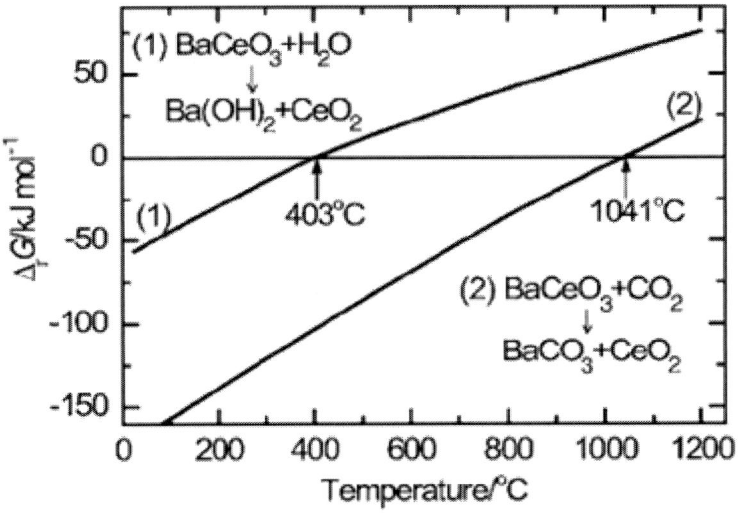

Figure 19. Gibbs free energy of hydration and carbonation reactions of barium cerate. The partial pressure of water vapor and carbon dioxide were assumed to be 10^5 Pa. Reproduced from [240] by permission of the Electrochemical Society.

$$BaCeO_3 + CO_2\,(g) \rightarrow BaCO_3 + CeO_2 \qquad (28)$$

$$BaCeO_3 + H_2O\,(g) \rightarrow Ba(OH)_2 + CeO_2 \qquad (29)$$

A similar instability was observed for 10 mol.% yttrium-substituted barium cerate $BaCe_{0.9}Y_{0.1}O_{3-\delta}$ (BCY10) [242], commonly studied as a proton conductor. Different approaches have been attempted to increase the stability of barium cerate based materials as briefly outlined here.

Ba-deficient compounds: By using an A:B ratio smaller than 1 the reactivity towards carbon dioxide and water vapor is limited. However, this non-stoichiometry is detrimental for protonic conductivity [243-245].

Choice of the substituent: Matstumoto et al. [240] showed a lower stability temperature when using substituents with smaller ionic radii. This resulted also in a lower conductivity (an order of magnitude lower at 600 °C when using In instead of Y).

Double-substitution: For example, simultaneous substitution with Y and Sn [246] resulted in higher chemical stability. However, it also led to decreased conductivity compared to BCY10. Furthermore, $BaCe_{0.7}Nb_{0.1}Gd_{0.2}O_{3-\delta}$ was found to be stable after exposure to 94% N_2 +3% CO_2 +3% H_2O at 700 °C for 20 h [247].

Zr substitution: The higher the difference of electronegativity between the cation and the anion, the more basic the oxygens, resulting in a stronger reactivity towards CO_2 or H_2O [41]. The Allred-Rochow electronegativities of cerium, yttrium and zirconium are 1.07, 1.11 and 1.22 respectively. As a result, substituting part of the cerium cations with zirconium cations makes the compound more stable. When exposed to 100 vol.% CO_2, 30-40 mol.% Zr are necessary to stabilize the solid solution of barium cerate-barium zirconate [65, 67, 248, 249]. Figure 20 [67] is an example of thermogravimetric results recorded in pure CO_2 from room temperature to 1250 °C for 4 different compositions in the $Ba(Ce,Zr,Y)O_3$ system.

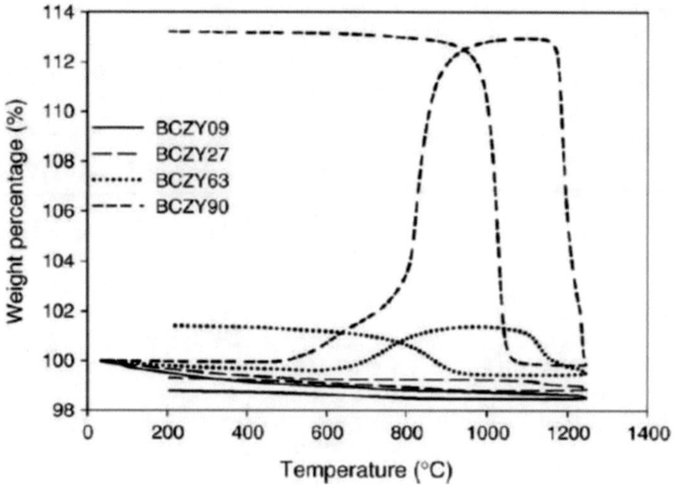

Figure 20. Thermogravimetric measurements in 100% CO_2 atmosphere recorded on four BCZY powders (BCZY09=BZY10:$BaZr_{0.9}Y_{0.1}O_{3-\delta}$, BCZY27:$BaCe_{0.2}Zr_{0.7}Y_{0.1}O_{3-\delta}$, BCZY63:$BaCe_{0.6}Zr_{0.3}Y_{0.1}O_{3-\delta}$ and BCZY90=BCY10:$BaCe_{0.9}Y_{0.1}O_{3-\delta}$). Reproduced from [67] by permission of Elsevier Ltd.

There is a weight increase due to the formation of barium carbonate when increasing the temperature only for compositions with less than 30 mol.% Zr. As shown above, the reactivity of $BaCeO_3$-based compounds is related to the basicity of the oxygen anions. As the proton incorporation is also depending on this basicity (higher incorporation for higher basicity) [50, 250, 251], one has to make a compromise between the stability and the protonic conductivity of the barium cerate-based compounds. The study by Escolástico et al. [191] showed that compounds in the series $BaZr_{1-x-y}Y_xM_yO_{3-\delta}$ (M=Fe, Pr or Mn) are stable after being exposed to a mixture of 10 vol.% CO_2 and 90 vol.% CH_4 in a short term test carried out at 800 °C. As shown in Figure 21, there is no phase degradation of the cubic $BaZr_{1-x-y}Y_xM_yO_{3-\delta}$ compounds after the test, whereas $BaCe_{0.9}Eu_{0.1}O_{3-\delta}$ and $BaZr_{0.9}Y_{0.1}O_{3-\delta}$ suffers from carbonation (see Figure 21B). Brandão et al. [252] published guidelines for improving the stability against CO_2 of ABO_3 perovskites with A = Ca, Sr, Ba and B = Ti, Zr, Ce, Sn. Based on computational thermodynamic analysis the authors represented the reactivity with CO_2 by stability phase diagrams of the relevant AO–BO_2–CO_2 systems. It was shown that the CO_2 tolerance can be strongly affected even with minor compositional changes. Thus, slight A-site deficiency was pointed out as a suitable measure to improve the CO_2 tolerance of mainly (Ba,Sr)TiO_3 and (Ba,Sr)ZrO_3 perovskites. The tolerance to CO_2 was furthermore correlated with to the structural tolerance factor, t (see Equation 1), and it can be furthermore used for predictions when relevant thermodynamic data are not available.

Niobates and Tantalates

For investigation of the stability of doped niobates and tantalates the corresponding powders were exposed to a continuous gas flow of 115 ppm H_2S, 4.43 vol.% CO_2, 2.12 vol.% CO and 92.09 vol.% H_2 at 30 bar and 50 ml·min^{-1} gas flow rate. These conditions were selected to assess the stability in CO_2 and S-containing atmospheres, as those encountered in applications such as water gas shift reactors fed with gasified coal or biomass. The temperature was 500 °C and the samples were held under these conditions for 40 h.

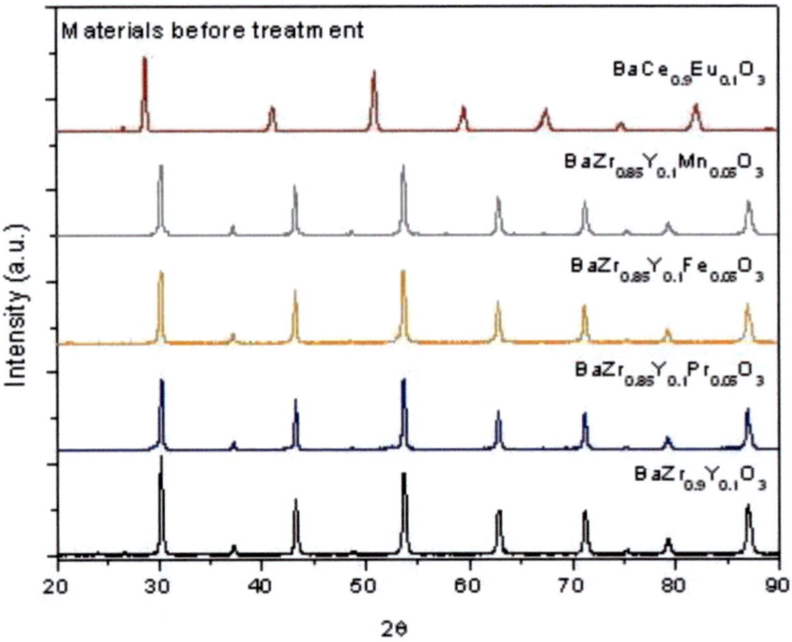

(a)

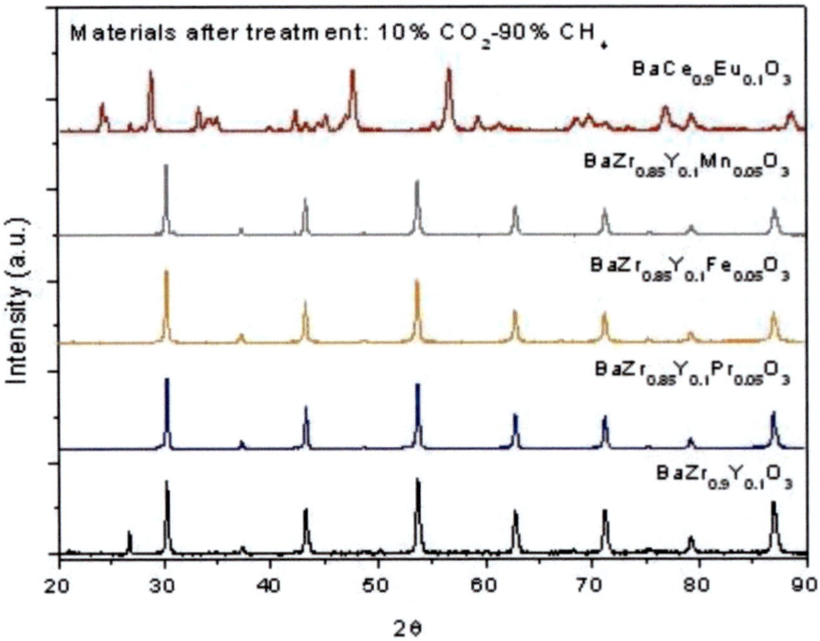

(b)

Figure 21. XRD diffraction patterns of compounds from the $BaZr_{1-x-y}Y_xM_yO_{3-\delta}$-series before and after stability test in a mixture of 10 vol.% CO_2 and 90 vol.% CH_4 carried out at 800 ºC for 72 h. XRD diffraction pattern of $BaCe_{0.9}Eu_{0.1}O_{3-\delta}$ was added for comparison [191]. Reproduced from [191] by permission of the Royal Society of Chemistry.

Figure 22 shows the XRD patterns of the investigated samples before and after exposure to the experimental conditions. Comparing the results before and after the stability test, it can be concluded that niobates and tantalates are stable after this short-term test. Although decreased intensity was observed for the Ga-containing samples after the test, they do not show any new crystalline phase.

Tungstates

As shown in the work by Escolástico et al. [213] $Ln_{6-x}WO_{12-\delta}$ powders were tested at 700 °C and 800 °C for 72 h in dry and humidified (2.5 vol.% H_2O) continuous gas flow of 10 vol.% CO_2 and 90 vol.% CH_4. These conditions were chosen in order to simulate the operating conditions in CO_2/H_2 separation. The experimental temperature was selected to be high enough to initiate carbonation or other reactions, but below the limit of carbonate decomposition. As a result, the whole set of materials remained unchanged after the treatment at both temperatures, regardless of its original sintering temperature and lanthanide ion. Moreover, the addition of water to the CO_2-rich gas stream had not effect on the material stability.

6. Structural, Chemical and Functional Effects of Additives and Substituents in Zeolite

6.1. Modification of the Pore Size, Polarity and Adsorption Behavior

Zeolite NaA (shown in Figure 23) is one of the most technically produced zeolites. With a pore size of 0.41 nm and a Si/Al-ratio of 1 for high hydrophilic character, NaA is commercially used for a number of dewatering processes. The small water molecule (0.26 nm) can enter the pores and is adsorbed by the polar sites inside the cages. In technical adsorption columns the wet gases are streaming through the zeolite bed until the zeolite is nearly saturated with water. For desorption the zeolite bed is heated (temperature swing adsorption) or the pressure is decreased (pressure swing adsorption), while the zeolites are sweeped with a part of the dry gas.

Larger polar gas molecules can enter the cages of zeolite NaA and strongly adsorb inside during the drying of alcohols, amines or H_2S by adsorption technology over long time. Reactive gases like olefins and acetylenic hydrocarbons can be polymerized inside the zeolite pores. By exchanging the extra-framework Na^+ with the larger K^+ the open windows of zeolite A are contracted to 0.38 nm (zeolite 3A) minimizing the risk of pore entering by large molecules.

Zeolite 3A is the first choice for dehydration of polar and reactive gases by adsorption [254]. By replacing two of the extra-framework Na^+ by one smaller Ca^{2+} the window is widened to 0.45 nm. The so-called zeolite 5A allows the separation of normal and iso-paraffins and cyclic hydrocarbons because branched and ring compounds are not able to enter the pores. Exchanging the Na^+ for Ca^{2+} also increases the point charge of the extra-framework cations, changing the adsorption behavior. Therefore zeolite 5A is used for CO_2 removal prior to the adsorptive air separation and for CO and N_2 adsorption in H_2 production from water gas shift (in combination with carbon).

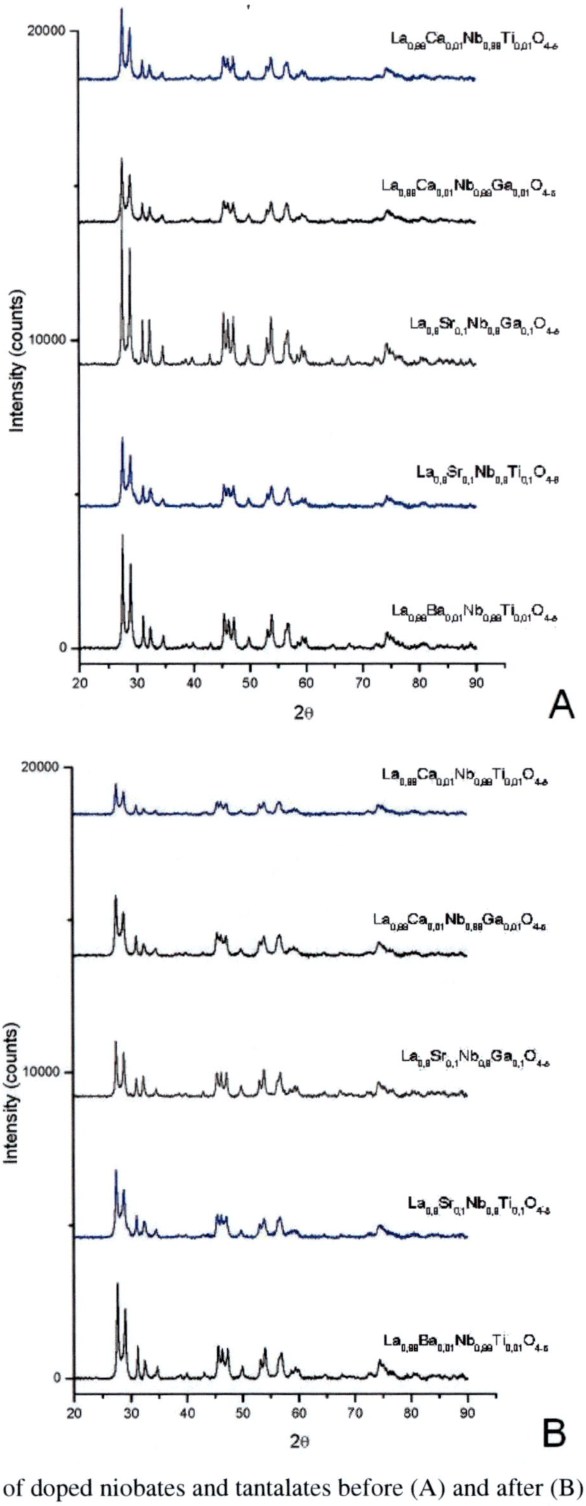

Figure 22. XRD patterns of doped niobates and tantalates before (A) and after (B) treatment in a gas mixture containing 115 ppm H_2S, 4.43 vol.% CO_2, 2.12 vol.% CO and 92.09 vol.% H_2 at 30 bar and 50 ml·min^{-1}. The test was carried out at 500 ºC for 40 h [253].

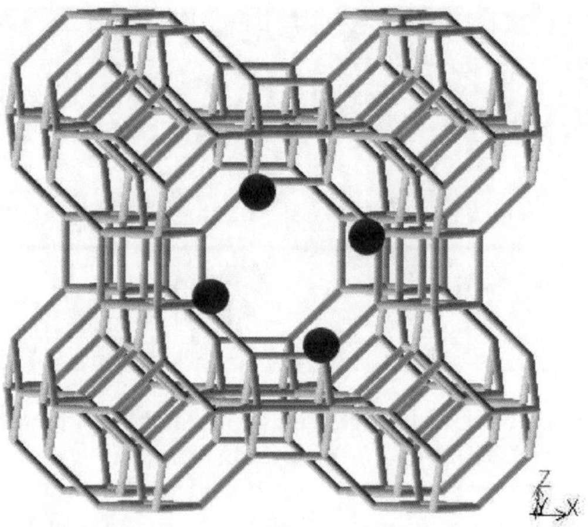

Figure 23. Illustration of extra framework monovalent cation positions in one window of zeolite A [27].

For filtration on atomic scale, water-containing organic liquids or gaseous mixtures are streamed across the membrane surface. The two relevant processes are called pervaporation (PV) and vapor permeation (VP). Driven by a partial pressure difference, only water is permeating through the narrow zeolite pores. The advantages of the membrane processes are the continuous processing, very low product loss and lower energy consumption. A first industrial plant for the dewatering of ethanol by PV was equipped with NaA membranes prepared on the outside of ceramic tubes by the Japanese company Mitsui Engineering and Shipbuilding Co. Ltd. (Mitsui) [255]. A commercial bio-ethanol dewatering by VP with NaA membranes was first preceded by Bussan Nanotech Research Institute Inc. [256]. In Europe, Inocermic GmbH (today Fraunhofer IKTS-Hermsdorf branch) equipped in 2007 equipped a first pilot plant for bio-ethanol dewatering by PV with zeolite membranes prepared inside a four channel structure and in 2009 an industrial scale plant [257] (see Figure 24). Besides ethanol, several other organic solvents like methanol, propanol and higher alcohols, as well as ketones and ethers can be dewatered with NaA membranes. By cation exchange with Ca^{2+}, zeolite 5A membranes can be prepared in order to be permeable also for small organic molecules like methanol.

6.2. Modification of the Thermal and Hydrothermal Stability

The zeolite LTA has narrow pores making the material interesting for gas separating membranes. The high surface charge of the single LTA crystals prevents a defect-free intergrowth. In dewatering processes (PV and VP) remaining narrow mesopores are filled with water.

However, these pores are considered as defects in dry gas separation techniques, where defect-free membrane layers with high degree of intergrowth are required [258]. In the presence of water, the hydrothermal stability of LTA is limited to 200 °C [259] and there is nearly no tolerance against acetic attacks in liquid solutions. Lowering the Al content would extend the potential application of LTA membranes. Dealumination by steam treatment allows the decrease of the Al content of LTA powders but this approach produces mesopores [260].

Figure 24. Industrial pilot plant for dewatering of ethanol with NaA-membranes of Fraunhofer IKTS. Reproduced from [257] by permission of John Wiley and Sons Inc.

Corma *et al.* [261] found a synthesis route for Al-free pure silica LTA. The so-called ITQ-29 was prepared by using a special structure-directing agent (SDA). A very high thermal stability (up to temperature of 800 °C) was observed also in the presence of water (2.5 vol.%). However, the relatively high viscosity of the synthesis gel complicated the membrane manufacturing [262].

Huang et al. [263, 264] used a crown ether to prepare an Al-free LTA with 1/3 Ge^{4+} on the lattice sites of Si^{4+}. Hence, the synthesis conditions and the low surface charge of Al-free LTA enabled the preparation of thin molecular sieving membrane layers on top of porous Al_2O_3-supports.

Zeolite structures can also be constructed from $[AlO_4]^{5-}$- and $[PO_4]^{3-}$-tetrahedrons. No extra-framework cations are necessary to make the AlPO-crystals less polar in contrast to the high Al-containing zeolites. A H_2-selective membrane of AlPO-LTA was prepared by Huang *et al.* [265, 266] using a crown ether-SDA. The hydrothermal stability of this material has not been studied yet.

Sodalite is a zeolite structure built only by the β-cages of zeolite LTA (see Figure 7). The 6-rings are the largest openings with a diameter of 0.28 nm making the material interesting for high selective H_2- and H_2O-separation based on the molecular sieving mechanism. First water and hydrogen selective hydroxo-sodalite (H-SOD, $Na_4(Si_3Al_3O_{12})(OH)$) membranes were prepared by Khajavi et al. [267] and Xu et al. [268].

However, the high Al-content of the material (Si/Al = 1) caused a low hydrothermal stability. Therefore Bibby et al. [269] developed Al-free SOD (schematically shown in Figure 25) by using a SDA-supported synthesis.

Same high hydrothermal stability was found for powder samples as for the Al-free LTA (ITQ-29). Münzer et al. [270] crystallized defect-free membrane layers of Al-free SOD on top of porous Al_2O_3 supports. However, during thermal de-templating to empty the SOD cages the layers always cracked. In conclusion, defect-free hydrothermally stable molecular sieves made of SOD membranes cannot be prepared by a SDA-supported synthesis.

The SDA-free synthesis allowed the manufacturing of crack-free membrane as illustrated in Figure 26.

An additional extra-framework Na^+ is compensated by an extra-framework OH^-, which is localized in the middle of the SOD cage and stabilizes the SOD structure. The small OH^- (0.26 nm) potentially leaves the cage in high thermal and hydrothermal condition causing a collapsing of the SOD structure. The H-SOD synthesis with additional Na_2S causes a partial substitution of the small OH^- by the bigger S^{2-} (0.34 nm) or SO_2^{2-} (0.46 nm).

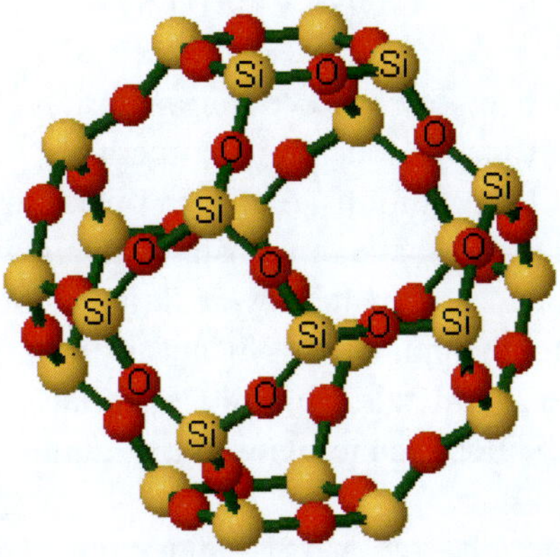

Figure 25. Al-free SOD-structure.

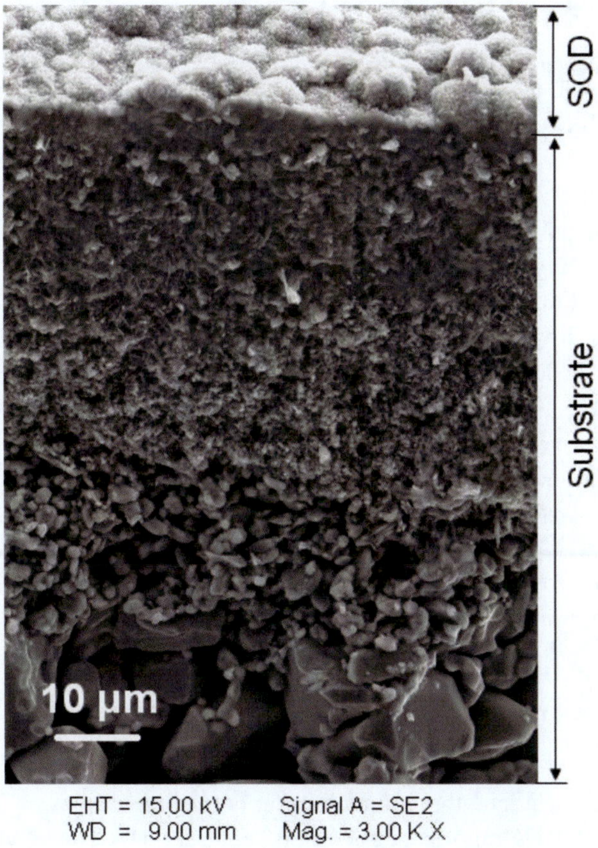

Figure 26. SEM image of a membrane layer made of sulfur-stabilized SOD.

This bigger anions are not able to leave the cage leading to a high thermal and hydrothermal stability identical to that of A-free SOD and LTA (ITQ-29). Molecular sieving behavior was found in H_2-separation tests also at higher temperatures.

CONCLUSION

This chapter gave a brief overview on ceramic materials for energy and environmental applications, and more specifically, on tailored ceramic compositions for membrane applications. The harsh operating conditions, e.g. high operating temperatures and aggressive gas environments that these materials have to withstand, along with their functional properties and mechanical durability, have led to their wide implementation in several applications during the past years. Due to the number of benefits they offer in comparison to other material classes, e.g. polymer or metals, ceramic materials are the choice in component development either for SOFC/SOEC technologies and a number of gas separation tasks using membranes.

The most interesting materials for energy- and environmentally-oriented applications were discussed in terms of their structural characteristics and the transport mechanisms that govern the overall mass transfer through the membranes, either as bulk or microporous solids.

Materials of each structural class were discussed illustrating the effects of doping/substitution on sinterability, chemical and mechanical stability as well as on electrical/electrochemical properties. It was shown how the dopants/substituents directly affect the phase stability, shrinkage during the heat treatments to lowered sintering temperatures, as well as improvement of the ionic/protonic or mixed ionic/ protonic-electronic conductivities with some references to the conductivities of grain interior and boundaries. Certain negative effects, as trapping of charge carriers, were also considered. Strategies based on doping/substitution to enhance the chemical, thermal and hydrothermal stability were also mentioned. It was shown how functional properties of ceramic materials can be modified according to the needs of targeted applications, although this process is rather complex due to the number of specific technological parameters to be taken into account and the whole variety of operational requirements towards materials selection. Several properties can be influenced simultaneously by doping or substitution of a selected material. The optimization of one property often leads to disadvantageous changes of another property resulting in a “best compromise” for the targeted operating conditions.

Modifying the material on an atomistic scale usually leads to a macroscopic response due to the chemical and electrostatic interactions, crystallographic and structural effects, which all are interlinked and sum up to an improved material functionality.

ACKNOWLEDGMENTS

The authors acknowledge the financial support granted from the Initiative and Networking Fund of the Helmholtz Association, contract HA-104 ("MEM-BRAIN"), the Portfolio MEM-BRAIN also funded by the Helmholtz Association, as well as the German Federal Ministry of Education and Research (BMBF) for the Northern European Innovative Energy Research Project, contract 03SF0330 ("N-INNER"), the Danish Council for Strategic Research for the Northern European Innovative Energy Research Project, grant no. 2104-07-0058 ("N-INNER"), and the European Commission for the FP7 NASA-OTM Project (NMP3-SL-2009-228701). The fruitful collaboration with all the partners within these projects is highly appreciated.

REFERENCES

[1] Meulenberg, W. A.; Ivanova, M., van Gestel, T.; Bram, M.; Buchkremer, H-P.; Stöver, D.; Alfaro-Serra, J. M. In: *Hyfdrogen and Fuel Cells: Fundamentals, Theory and Applications*; Stolten, D.; Ed.; Hydrogen Energy ISBN: 978-3-527-32711-9; WILEY-VCH Verlag GmbH and Co. KGaA, Weinheim, DE, 2010; pp 321-349.

[2] Meulenberg, W. A.; Ivanova, M., Roitsch, St.; Serra, J. M. *Advanced membrane science and technology for sustainable energy and environmental applications*; Basile, A.; Pereira Nuñes, S.; Ed.; Woodhead Publishing Series in Energy: Number 25 ISBN ISBN 978-1-84569-969-7; Woodhead Publishing Limited: Cambridge, UK, 2011; pp 541-567.

[3] Yamazoe, N.; Miura, N. *US Patent 4,718,991 to Figaro Engineering Inc.* 1988.

[4] http://www.metafysica.nl/turing/preparation_3dim_3.html

[5] Kharton, V. V.; Yaremchenko, A. A.; Naumovich, E. N.; Marques, F. M. B. *J. Solid State Electrochem.* 2000, 4, pp 243-266.

[6] Minh, N. *J. Amer. Ceram. Soc.* 1993, 76, pp 563-588.

[7] Singhal, S. *Solid State Ionics* 2000, 135, pp 305-313.

[8] Bloom, I. *Fuel Cell Program and Abstracts*, 1992 Fuel Cell Seminar, Tucson, AZ, US.

[9] Bhalla, A. S.; Guo, R.; Roy, R. *Mat. Res. Innovat.* 2000, 4, pp 3–26.

[10] De Souza, E. C. C.; Muccillo, R. *Materials Research* 2010, 13, pp 385-394.

[11] Subramanian, M. A.; Aravamudan, G.; Subba Rao, G. V. *Prog. Solid State Chem.* 1983, 15, pp 55–143.

[12] Tabira, Y.; Withers, R. L. *Philos. Mag. A* 1999, 79, pp 1335–1346.

[13] Minervi, L.; Grimes, R. W.; Sickafus, K. E. *J. Am. Ceram. Soc.* 2000, 83, pp 1873–1878.

[14] Besikiotis, V.; Ricote, S.; Jensen, M. H.; Haugsrud, R.; Norby, T. *to be submitted to Solid State Ionics*, 2012, 229, pp 26-32.

[15] Moon, P. K. *PhD Thesis*, Department of Materials Science and Engineering, Massachusetts Institute of Technology, Cambridge, MA, 1988.

[16] Díaz-Guillén, J. A.; Díaz-Guillén, M. R.; Padmasree, K. P.; Fuentes, A. F.; Santamaría, J.; León, C. *Solid State Ionics* 2008, 179, pp 2160-2164.

[17] Gingerich, K. A.; Bair, H. E. *Advances in X-Ray Analysis* 1964, 7 pp 22-30.

[18] Tsunekawa, S.; Kamiyama, T.; Sasaki, K.; Asano, H.; Fukuda, T. *Acta Crystallographica* 1993, A49, pp 595-600.

[19] Jian, L.; Wayman, C. M.; *Journal of the American Ceramic Society* 1997, 80, pp 803-806.

[20] Aldred, A. T.; *J. of Solid State Chem.* 1985, 59, pp 95-102.

[21] Errandonea, D.; Manjón, F. J.; Somayazulu, M.; Häusermann, D. *J. Solid State Chem.* 2004, 177, pp 1087-1097.

[22] Stubican, V. S. *J. Amer. Ceram. Soc.* 1964, 47, pp 55-58.

[23] David, W. I. F. *Materials Research Bulletin* 1983, 18, pp 749-756.

[24] Vajeeston, N. *Presentation at the 1st Nordic school and seminar on Functional Energy Related Materials*, Gol, NO, 2008.

[25] Cava, R. J.; Roth, R. S. *Journal of Solid State Chemistry* 1981, 36, pp 139-147.

[26] http://dcssi.istm.cnr.it/CORSO%20IPERTESTUALE/StatoSolido/Zeoliti_11/images/Image645.jpg
[27] Baerlocher, Ch.; McCusker, L. B. *Database of Zeolite Structures*: http://www.iza-structure.org/databases/
[28] Sherman, D. *Proc. Natl. Acad. Sci. US* 1999, 96, pp 3471-3478.
[29] Sunarso, J.; Baumann, S.; Serra, J. M.; Meulenberg, W. A.; Liu, S.; Lin, Y. S.; Diniz da Costa, J. C. *Journal of Membrane Science* 2008, 320, pp 13-41.
[30] Bouwmeester, H. J. M.; Burggraaf, A. J. In: *Fundamentals of inorganic membrane science and technology*; Burggraaf, A. J.; Cot, L.; Ed.; Elsevier Science B. V.: Amsterdam, NL, 1996, pp 435-515.
[31] Baumann, S.; Serra, J. M.; Lobera, M. P.; Escolástico, S.; Schulze-Küppers, F.; Meulenberg, W.A. *Journal of Membrane Science* 2011, 377, pp 197-205.
[32] Engels, S.; Beggel, F.; Modigell, M.; Stadler, H. *Journal of Membrane Science* 2010, 359, pp 93–101.
[33] Gozálvez-Zafrilla, J. M.; Santafé-Moros, A.; Escolástico, S.; Serra, J. M. *Journal of Membrane Science* 2011, 378, pp 290–300.
[34] Kreuer, K. D. *Annu. Rev. Mater. Res.* 2003, 33, pp 333-359.
[35] Iwahara, H. *Solid State Ionics* 1996, 86–88, pp 9-15.
[36] Colomban, P. *Ann. Chimi. Sci. Mat.* 1999, 24, 1-18.
[37] Nowick, A. S.; Du, Y. *Solid State Ionics* 1995, 77, pp 137-146.
[38] De Vries, K. J. *Solid State Ionics* 1997, 100, pp 193-200.
[39] Norby, T. *Solid State Ionics* 1999, 125, pp 1-11.
[40] Kosacki, I., Tuller, H. L. *Solid State Ionics* 1995, 80, pp 223-229.
[41] Bagul, R. D. *Master thesis*, University of Florida, US, 2004.
[42] Langguth, J.; Dittmeyer, R.; Hofmann, H.; Tomandl, G.; *Applied Catalysis A: General* 1997, 158, pp 287-305.
[43] Norby, T., In: *Perovskite Oxides for Solid Oxide Fuel Cells*, Ishihara, T.; Ed.; Fuel Cells and Hydrogen Energy, doi: 10.1007/978-0-387-77708-5_11, Springer Science and Business Media, 2009, pp 217-241.
[44] Fontaine, M. L.; Norby, T.; Larring, Y.; Grande, T.; Bredesen, R. In: *Membrane Science and Technology*, ISSN 0927-5193, doi: 10.1016/ S0927-5193(07)13010-2, Elsevier B.V., NL, 2008, 13, pp 401-458.
[45] Krug, F.; Schober, T.; Springer, T. *Solid State Ionics* 1995, 81, pp 111-118.
[46] Schober, T.; Schilling, W.; Wenzl, H. *Solid State Ionics* 1996, 86-88, pp 653-658.
[47] Schober, T.; Bohn, H. G. *Solid State Ionics* 2000, 127, pp 351–360.
[48] Kreuer, K. D. *Solid State Ionics* 2000, 136–137, pp 149–160.
[49] Kreuer, K. D. *Chem. Rev.* 2004, 104, pp 4637-4678.
[50] Kreuer, K. D.; *Solid State Ionics* 1999, 125, 285–302.
[51] Kreuer, K. D. *Chem. Mater.* 1996, 8, pp 610-641.
[52] Agmon, N. *Chemical Physics Letters* 1995, 224, pp 456-462.
[53] Hempelmann, R. *Physica B:Condensed Matter*1996, 226, pp 72-77.
[54] Kreuer, K. D. *Solid State Ionics* 1997, 97, pp 1-15.
[55] Pionke, M.; Mono, T.; Schweika, W.; Springer, T.; Schober, T. *Solid State Ionics* 1997, 97, pp 497–504.
[56] Matzke, T.; Stimming, U.; Karmonik, C.; Soetramo, M.; Hempelmann, R.; Guthoff, F. *Solid State Ionics* 1996, 86–88, pp 621–628.

[57] Munch, W.; Seifert, G.; Kreuer, K. D.; Maier, J. *Solid State Ionics* 1997, 97, pp 39-44.
[58] Shimojo, F.; Hashino, K.; Okazaki, H. *J. Phys. Soc. Jpn.* 1997, 66, pp 8-10.
[59] Weyd, M. *Thesis*: ISBN 978-3-939665-55-7, 2008.
[60] Sladek, K. J.; Gilliland, E. R.; Baddour, R. F. *Ind. Eng. Chem. Fundam.* 1974, 13, 100-105.
[61] Hanebuth, M.; Dittmeyer, R.; Mabande, G. T. P.; Schwieger, W. *Catalysis Today* 2005, 104, pp 352-359.
[62] Kleinlogel, C.; Gauckler, L. J. *Solid State Ionics* 200 135, pp 567-573.
[63] Yan, R.; Chu, F.; Ma, Q.; Liu, X.; Meng, G. *Mater. Lett.,* in press.
[64] Zhang, X.; Decès-Petit, C.; Yick, S.; Robertson, M.; Kesler, O.; Maric, R.; Ghosh, D. *J. Power Sources* 2006, 162, pp 480-485.
[65] Katahira, K.; Kohchi, Y.; Shimura, T.; Iwahara, H. *Solid State Ionics* 2000, 138, pp 91-98.
[66] Duval, S. B. C., Holtappels, P.; Vogt, U. F.; Stimming, U.; Graule, T. *Fuel Cells* 2009, 9, No. 5, pp 613-621.
[67] Ricote, S.; Bonanos, N.; Caboche, G. *Solid State Ionics* 2009, 180, pp 990-997.
[68] Babilo, P.; Uda, T.; Haile, S. M. *Journal Materials Research* 2007, 22, pp 1322-1330.
[69] Cervera, R. B., Oyama, Y.; Miyoshi, S.; Kobayashi, K.; Yagi, T.; Yamaguchi, S. *Solid State Ionics* 2008, 179, pp 236-242.
[70] Guillaume, B.; Boschini, F.; Garcia-Cano, I.; Rulmont, A.; Cloots, R.; Ausloos, M. *J. Eur. Ceram. Soc.* 2005, 25, pp 3593-3604.
[71] Babilo, P.; Haile, S. M. *J. Am. Ceram. Soc.* 2005, 88, pp 2362-2368.
[72] Tao, S.; Irvine, J. T. S. *J. Solid State Chem.* 2007, 180, pp 3493-3503.
[73] Ricote, S.; Bonanos, N. *Solid State Ionics* 2010, 181, pp 694-700.
[74] Azimova, M. A.; McIntosh, S. *Solid State Ionics* 2009, 180, pp 160-167.
[75] Tao, S.; Irvine, J. T. S. *Adv. Mater.* 2006, 18, pp 1581-1584.
[76] Ricote, S.; Bonanos, N.; Wang, H. J.; Haugsrud, R. *Solid State Ionics* 2011, 185, pp 11-17.
[77] Guo, Y.; Ran, R.; Shoa, S. *Int. J. Hydrogen Energy* 2010, 36, pp 5611-5620.
[78] Meulenberg, W. A.; Serra, J. M.; Schober, T. *Solid State Ionics* 2006, 177, pp 2851-2856.
[79] Serra, J. M.; Büchler, O.; Meulenberg, W. A.; Buchkremer, H. P. *Journal of the Electrochemical Society* 2007, 154, pp B334-B340
[80] Serra, J. M.; Meulenberg, W. A. *J. Am. Ceram. Soc*. 2007, 90, pp 2082-2089.
[81] Gibson, St.; Ivanova, M.; Meulenberg, W. A.; Sebold, D.; Etsell, Th., *Proceedings of the 3rd European Fuel Cell Technology and Applications Conference EFC09*, Rome, IT, 2009, pp 135-136.
[82] Badwal, S.; Foger, K. *Ceramics Int.* 1996, 22, pp 257-265.
[83] Ramanarayanan, T.; Singhal, S.; Wachsman, E. *The Electrochemical Soc. Interface* 2001, pp 22-27.
[84] Suzuki, Y. *Solid State Ionics* 1997, 95, pp 227-230.
[85] Atkinson, A. *Lecture at the 1st Real-SOFC IP Summer School on SOFC Technology*, Patras, GR, 2004.
[86] Baumard, J.; Abelard, P. *Advances in Ceramics,* Science and Technol. Zirconia II; Claussen, N.; Rühle, M.; Heuer A.; Ed.; The American Ceramic Society Inc., Columbus, OH, US, 1984, 12, 555-571.

[87] Guo, X.; Maier, J. *J. Electrochem. Soc.* 2001, 148, pp E121-E126.

[88] Van Gestel, T.; Han, F.; Sebold, D.; Buchkremer, H. P.; Stöver, D. *Microsyst. Technol.* 2011, 17, pp 233-242.

[89] Estell, T.; Flengas, S. *Chem. Rev.* 1970, 70, pp 339-376.

[90] Ishizaki, F.; Yoshida, T.; Sakurada, S. *Proceedings of the First International Symposium on SOFCs*, Singhal, S.; Ed.; The Electrochemical Soc., Pennington, NJ, 1989, pp 3-14.

[91] Sahibzada, M.; Steele, B. C. H.; Zheng, K.; Rudkin, R. A., Metcalfe, I. S., *Catal. Today* 1997, 38, pp 459-466.

[92] Serra, J. M.; Vert, V. B.; Betz, M.; Haanappel, V. A. C.; Meulenberg, W. A.; Tietz, F. *J. Electrochem. Soc.* 2008, 155, pp B207-209.

[93] Sun, C.; Stimming, U. *J. Power Sources* 2007, 171, pp 247-260.

[94] Fergus, J. W. *Solid State Ionics* 2006, 177, pp 1529-1541.

[95] Park, H. J.; Choi, G. M. *J. Eur. Ceram. Soc.* 2004, 24, pp 1313-1317.

[96] Fagg, D. P.; Kharton, V. V.; Shaula, A.; Marozau, A. P.; Frade, J. R. *Solid State Ionics* 2005, 176, pp 1723-1730.

[97] Shuk, P.; Greenblatt, M. *Solid State Ionics* 1999, 116, pp 217-223.

[98] Jasinski, P.; Suzuki, T.; Anderson, H. U. *Sensor Actuat B-Chem* 2003, 95, pp 73-77.

[99] Trovarelli, A. *Catal. Rev.* 1996, 38, pp 439-520.

[100] Grirrane, A.; Corma, A.; Garcia H. *Science* 2008, 322, pp 1661-1664.

[101] Esposito, V.; Fonseca, F. C. C.; De Florio, D. Z.; Žunic, M.; Muccillo, R.; Traversa, E. *ECS Trans.* 2007, 7, pp 2093-2101.

[102] Uhlenbruck, S.; Jordan, N.; Serra, J. M.; Buchkremer, H. P.; Stöver, D. *Solid State Ionics* 2010, 181, pp 447-452.

[103] Mori, T.; Drennan, J.; Lee, J. H.; Li, J. G.; Ikegami, T. *Solid State Ionics* 2002, 154-155, pp 461- 466.

[104] Tadokoro, S. K.; Muccillo E. N. S. *J. Eur. Ceram. Soc.* 2007, 27, pp 4261–4264.

[105] Maschio, S.; Aneggi, E.; Trovarelli, A.; Sergo, V. *Ceram. Int.* 2008, 34, pp 1327-1333.

[106] Balaguer, M.; Solis, C.; Serra, J. M. *Chem. Mat.* 2011, 23, pp 2333-2343.

[107] Chatzichristodoulou, C.; Hendriksen, P. V.; Hagen, A.; Grivel, J. C. *ECS Transactions* 2008, 13, 347-359.

[108] Stefanik, T. S.; Tuller, H. L. *J. Eur. Ceram. Soc.* 2001, 21, pp 1967–1970.

[109] Fagg, D. P.; García-Martín, S.; Kharton, V. V.; Frade, J. R. *Chem. Mater.* 2009, 21, pp 381-391.

[110] Blumenthal, R.; Brugner, F.; Garnier, J. *J. Electrochem. Soc.* 1973, 120, pp 1230-1237.

[111] Tuller, H.; Nowick, A. *J. Electrochem. Soc.* 1975, 122, pp 255-259.

[112] Swarr, T.; Karavolis, S.; Maricle, D. *Proceedings of the 25th Intersociety Energy Conversion Engineering Conference*, American Institute of Chemical Engineers, New York, US, 1990, 3, pp 256-261.

[113] Inoue, T.; Setoguchi, T.; Eguchi, K.; Arai, H. *Solid State Ionics* 1989, 35, 285-291.

[114] Cassir, M.; Gourba, E. *Ann. Chim. Sci. Mat.* 2001, 26, pp 49-58.

[115] Takahashi, T.; Iwahara, H.; Nagai, Y. *J. Appl. Electrochem.* 1972, 2, pp 97-104.

[116] Arasteh, S.; Maghsoudipour, A.; Alizadeh, M.; Nemati, A. *Ceram. Int.* 2011, 37, pp 3451-3455.

[117] Miyayama, M.; Nishi, T.; Yanagida, H. *J. Mater. Sci.* 1987, 22, pp 2624-2628.

[118] Jiang, N.; Wachsman, E. D.; Jung, S.-H. *Solid State Ionics* 2002, 150, pp 347-353.

[119] Watanabe, A.; Sekita, M. *Solid State Ionics* 2005, 176, pp 2429-2433.

[120] Capoen, E.; Nowogrocki, G.; Chater, R. J.; Skinner, S. J.; Kilner, J. A.; Malys, M.; Boivin, J. C.; Mairesse, G.; Vannier, R. N. *Solid State Ionics* 2006, 177, pp 489-492.

[121] Hara, N.; Yamanaka, M.; Satou, F.; Hatano, M.; Kushibiki, K.; Fukuzawa, T.; Shibata, I.; Uchivama, M. *US Patent* 6 896 989 B2, 2005.

[122] Doshi, R.; Guan, J.; Minh, N.; Montgomery, K.; Ong, E.; Chung, B.; Lear, G. *US Patent* 6 558 831 B1, 2003.

[123] Hara, N.; Munakata, F.; Isawaki, Y. *US Patent* 6 849 354 B2, 2005.

[124] Sammells, A. *Perovskite solid electrolytes for SOFC, Proceedings of the 4th Annual Fuel Cells Contractors Review Meeting*, US DOE/METC, 1992.

[125] Ishihara, T.; Matsuda, H.; Takita, Y. *J. Am. Chem. Soc.* 1994, 116, pp 3801-3803.

[126] Ishihara, T.; Honda, M.; Shibayama, T.; Furutani, H.; Takita, Y. *Ionics* 1998, 4, pp 395-402.

[127] Huang, K.; Tichy, R.; Goodenough, J. B. *J. Amer. Cer. Soc.* 1998, 81, pp 2565- 2575.

[128] Huang, P.; Petric, A. *J. Electrochem. Soc.* 1996, 143, pp 1644-1647.

[129] Feng, M.; Goodenough, J. B. *Eur. J. Solid State Inorg. Chem.* 1994, 31, pp 663-672.

[130] Goodenough, J. B.; Feng, M. *Solid Oxide Fuel Cells with Gallate Electrolytes*, summary data on research supported by EPRI, 1998.

[131] Kendall, K.; Navas, C.; Thomas, J.; zur Loye, H. *Solid State Ionics* 1995, 82, pp 215-223.

[132] Bloom, I.; Krumpelt, M. *Proceedings of the 4th Annual Fuel Cells Contractors Review Meeting*, US DOE/METC, Morgantown, West Virginia, US, July 14-15, 1992.

[133] Singhal, S. *Proceedings of the 5th International Symposium on Solid Oxide Fuel Cells (SOFC-V)*, Stimming, U.; Singhal, S.C.; Tagawa, H.; Lehnert, W.; Ed.; The Electrochemical Society, Inc., Pennington, NJ, US, 1997, pp 37-50.

[134] Fujita, H.; Inoue, R.; Oshima, T. *US Patent* 6 337 006 B1, 2002.

[135] Akimune, Y.; Shinohara, M.; Munakata, F. *US Patent* 6 447 944 B1, 2002.

[136] Ishihara, T.; Takita, Y. *US Patent* 6 586 127 B1, 2003.

[137] Sugiyama, T.; Matsuo, K.; Munakata, F.; Akimune, Y. *US Patent US* 6 803 140 B2, 2004.

[138] Ishihara, T.; Tsuruta, Y.; Nishiguchi, H.; Takita, Y. *Seminar- Fuel Cell Powering the 21st Century*, Portland, Oregon, US, 2000.

[139] Ishihara, T.; Kilner, J.; Honda, M.; Takita, Y. *J. Am. Chem. Soc.* 1997, 119, pp 2747-2748.

[140] Ishihara, T.; Akbay, T.; Furutani, H.; Takita, Y. *Solid State Ionics* 1998, 113-115, pp 585-591.

[141] Hayashi, H.; Inaba, H.; Matsuyama, M.; Lan, N.; Dokiya, M.; Tagawa, H. *Solid State Ionics* 1999, 122, pp 1-15.

[142] Petric, A.; Huang, P.; Skowron, A. *J. Electrochem. Soc.* 1998, 125, pp 751-760.

[143] Lybye, D.; Poulsen, F.; Mogensen, M. *Solid State Ionics* 2000, 128, pp 91-103.

[144] Nomura, K.; Tanase, S. *Solid State Ionics* 1997, 98, pp 229-236.

[145] Ahmad-Khanlou, A.; Tietz, F.; Stöver, D. *Solid State Ionics* 2000, 135, pp 543-547.

[146] Stevenson, J.; Armstrong, T.; Pederson, L.; Li, J.; Lewinsohn, C.; Baskaran, S. *Solid State Ionics* 1998, 113-115, pp 571-583.

[147] Stevenson, J.; Armstrong, T.; McCready, D.; Pederson, L.; Weber, W. *J. Electrochem. Soc.* 1997, 144, pp 3613-3623.

[148] Kahn, M.; Islam, M. S.; Bates, D. *J. Phys. Chem. B* 1998, 102, pp 3099-3104.
[149] Kharton, V.; Naumovich, E.; Marques, F. *Ionics* 1999, 5, pp 183-193.
[150] Trofimenko, N.; Ullmann, H. *Solid State Ionics* 1999, 124, pp 263-270.
[151] Baker, R.; Gharbage, B.; Marques, F. *J. Electrochem. Soc.* 1997, 144, pp 3130-3134.
[152] Chen, F.; Liu, M. *J. of Solid State Electrochem.* 1998, 3, pp 7-14.
[153] Huang, K.; Tichy, R.; Goodenough, J. B. *J. Amer. Cer. Soc.* 1998, 81, pp 2576-2580.
[154] Huang, K.; Tichy, R.; Goodenough, J. B. *J. Amer. Cer. Soc.* 1998, 81, pp 2581-2585.
[155] Ivanova, M. *PhD Thesis*, UCTM, Sofia, BG, 2008.
[156] Cong, L.; He, T.; Ji, Y.; Guan, P.; Huang, Y.; Su, W. *J. All. Comp.* 2003, 348, pp 325-331.
[157] Strickler, D.; Carlson, W. *J. Am. Ceram. Soc.* 1964, 47, pp 122-127.
[158] Kingery, W.; Pappis, J.; Doty, M.; Hill, D, *J. Am. Ceram. Soc.* 1959, 42, pp 393-398.
[159] Fagg, D. P.; Shaula, A. L.; Kharton, V. V. Frade, J. R. *Journal of Membrane Science* 2007, 299, pp 1-7.
[160] Schmale, K.; Grünebaum, M.; Janssen, M.; Baumann, S.; Schulze-Küppers, F.; Wiemhöfer, H.-D. *Physica Status Solidi B*, 2011, 248, pp 314-322.
[161] van Roosmalen, J. A. M.; Huijsmans, J. P. P.; Plomp, L. *Solid State Ionics* 1993, 66, pp 279-284.
[162] Karim, D. P.; Aldred, A. T. *Physical Review B* 1979, 20, pp 2255-2263.
[163] Teraoka, Y.; Zhang, H.-M.; Furukawa, S.; Yamazoe, N. *Chemistry Letters* 1985, 14, pp 1743-1746.
[164] Kruidhof, H.; Bouwmeester, H. J. M.; Doorn R. H. E. V.; Burggraaf, A. J. *Solid State Ionics* 1993, 63-65, pp 816-822.
[165] Nagai, T.; Ito, W.; Sakon, T. *Solid State Ionics* 2007, 177, pp 3433-3444.
[166] Shao, Z.; Yang, W.; Cong, Y.; Dong, H.; Tong, J.; Xiong, G. *Journal of Membrane Science* 2000, 172, pp 177-188.
[167] Arnold, M.; Xu, Q.; Tichelaar, F. D.; Feldhoff, A. *Chem. Mater.* 2009, 21, pp 635-640.
[168] Vente, J. F.; McIntosh, S.; Haije, W. G.; Bouwmeester, H. J. M. *J. Solid State Electrochem.* 2006, 10, pp 581-588.
[169] Stotz, S.; Wagner, C. *Ber. Bunsenges.Phys.Chem. Bd* 1966, 70, pp 781-788.
[170] Iwahara, H.; Esaka, T.; Uchida, H.; Maeda, N. *Solid State Ionics* 1981, 3/4, pp 359-363.
[171] Bonanos, N. *Solid State Ionics* 2001, 145, pp 265-274.
[172] Fabbri, E.; Pergolesi, D.; Traversa, E. *Chem. Soc. Rev.* 2010, 39, pp 4355-4369.
[173] Malavasi, L.; Fisher, C. A. J.; Islam, M. S. *Chem. Soc. Rev.*, 2010, 39, 4370-4387.
[174] Norby, T.; Kofstad, P. *J. Am. Ceram. Soc.* 1984, 67, pp 786-792.
[175] Norby, T.; Kofstad, P.; *Solid State Ionics* 1986, 20, pp 169-184.
[176] Norby. T.; Dyrlie, O.; Kofstad, P. *Solid State Ionics* 1992, 53-56, pp 446-452.
[177] Norby, T.; Larring, Y. *Curr. Opin. In Solid State and Mater. Sci.* 1997, 2, pp 593-599.
[178] Norby, T. *Solid State Ionics* 1990, 40/41, pp 857-862.
[179] Larring, Y.; Norby, T. *Solid State Ionics* 1995, 77, pp 147-151.
[180] Larring, Y.; Norby, T. *Solid State Ionics* 1997, 97, pp 523-528.
[181] Kreuer, K. D.; Adams, St.; Münch, W.; Fuchs, A.; Klock, U.; Maier, J. *Solid State Ionics* 2001, 145, pp 295-306.
[182] Davies, R. A.; Islam, M. S.; Gale, J. D. *Solid State Ionics* 1999, 126, pp 323-335.
[183] Schober, T.; Friedrich, J.; Condon, J. B. *Solid State Ionics* 1995, 77, pp 175-179.

[184] Iwahara, H.; Yajima, T.; Hibino, T.; Ozaki, K.; Suzuki, H. *Solid State Ionics* 1993, 61, pp 65-69.
[185] Yamanaka, S.; Kurosaki, K.; Maekawa, T.; Matsuda, T.; Kobayashi, S.; Uno, M. *J. Nucl. Mat.* 2005, 344, pp 61-66.
[186] Cheng, S.; Gupta, V.; Lin, Y. S. *Solid State Ionics* 2005, 176, pp 2653-2662.
[187] Lee, W. K.; Nowick, A. S.; Boatner, L. A. *Solid State Ionics* (1986), 18-19, pp 989-993.
[188] Larring, Y.; Norby, T. *Solid State Ionics* 1994, 70/71, pp 305-310.
[189] Oh, T.; Yoon, H.; Li, J.; Wachsman, E. D. *J. Membrane Sci.* 2009, 345, pp 1-4.
[190] Sun, Z.; Fabbri, E.; Bi, L.; Traversa, E. *Phys. Chem. Cchem. Phys.* 2010, 13, pp 7692-7700.
[191] Escolástico, S.; Ivanova, M.; Solis, C.; Roitsch, S.; Meulenberg, W. A.; Serra, J. M. *Royal Society of Chemistry Adv.* 2012, 2, pp. 4932-4943.
[192] Shirsat, A. N.; Kaimal, K. N. G.; Bharadwaj, S. R.; Das, D. *J. Solid State Chem.* 2004, 177, pp 2007-2013.
[193] Scholten, M. J.; Schoonman, J.; van Miltenburg, J. C.; Oonk, H. A. J. *Solid State Ionics* 1993, 61, pp 83-91.
[194] Duval, S. B. C.; Holtappels, P.; Vogt, U. F.; Pomjakushina, E.; Conder, K.; Stimming, U.; Graule, T. *Solid State Ionics* 2007, 178, pp 1437-1441.
[195] Naoki, I.; Matsumoto, H.; Kawasaki, Y.; Okada, S.; Ishihara, T. *Solid State Ionics* 2008, 179, pp 324-329.
[196] Moon, P. K.; Tuller, H. L. *Solid State Ionics* 1988, 28-30, pp 470-474.
[197] Van Dijk, M. P.; Burggraaf, A. J.; Cormack, A. N.; Catlow, C. R. A. *Solid State Ionics* 1985, 17, pp 159-167.
[198] Yamamura, H.; Nishino, H.; Kakinum, K.; Nomura, K. *Solid State Ionics* 2007, 178, pp 233–238.
[199] Shimura, T.; Komori, M.; Iwahara, H. *Solid State Ionics* 1996, 86-88, pp 685-689.
[200] Labrincha, J.; Frade, J.; Marques, F. *Solid State Ionics* (1997), 99, pp 33-40.
[201] Omata, T.; Okuda, K.; Tsugimoto, S.; Otsuka-Matsuo-Yao, S. *Solid State Ionics* 1997, 104, pp 249-258.
[202] Omata, T.; Ikeda, K.; Tokashiki, R.; Otsuka-Yao-Matsuo, S. *Solid State Ionics* 2004, 167, pp 389-397.
[203] Eurenius, K. E. J.; Ahlberg, E.; Ahmed, I.; Eriksson, S. G.; Knee, C. S. *Solid State Ionics* 2010, 181, pp 148-153.
[204] Eurenius, K. E. J.; Ahlberg, E.; Knee, C. S. *Solid State Ionics* 2010, 181, pp 1577-1585.
[205] Eurenius, K. E. J.; Bentzer, H. K.; Bonanos, N.; Ahlberg, E.; Knee, C. S. *J. Solid State Electrochem.* 2011, 15, pp 2571-2579.
[206] Balachandran, U.; Eror, N. *J. Less-Common Met.* 1982, 85, pp 111-120.
[207] Haugsrud, R.; Norby, T. *Proceedings of the 26th Risø International Symposium on Materials Science: Solid State Electrochemistry*, Linderoth, S.; Smith, A.; Bonanos, N.; Hagen, A.; Mikkelsen, L.; Kammer, K.; Lybye, D.; Hendriksen, P. V.; Poulsen, F. W.; Mogensen, M.; Wang, W. G.; Ed.; Risø National Laboratory, Roskilde, DK, 2005
[208] Nowick, A. S.; Vaysleyb, A. V. *Solid State Ionics* 1997, 97, pp 17-26.
[209] Goodenough, J. B. In: *Solid State Electrochemistry*, Bruce, P. G.; Ed.; Cambridge University Press, UK, 1995, pp 54-56.
[210] Shimura, T.; Fujimoto, S.; Iwahara, H.; *Solid State Ionics* 2001, 143, pp 117-123.
[211] Haugsrud, R. *Solid State Ionics* 2007, 178, pp 555-560.

[212] Meulenberg, W. A.; Ivanova, M.; Buchkremer, H. P.; Stöver, D.; Serra, J. M.; Escolástico, S. WO 2012/010386 A1, 26.01.2012

[213] Escolástico, S.; Vert, V. B.; Serra, J. M. *Chem. Mater.* 2009, 21, pp 3079-3089.

[214] Escolástico, S.; Solís, C.; Serra, J. M. *Solid State Ionics* 2012, 216, pp.31-35.

[215] Escolástico, S.; Solís, C.; Serra, J. M. *International Journal of hydrogen energy* 2011, 36, pp 11946-11954.

[216] Seeger, J.; Ivanova, M.; Meulenberg, W. A.; Serra, J. M.; Escolástico, S. *In preparation, 2013.*

[217] Haugsrud, R.; Kjølseth, C. *J. Phys. Chem. Sol.* 2008, 69, pp 1758-1765.

[218] Norby, T.; Haugsrud, R. *Invited lecture at FZJ Workshop*, Jülich, DE, November 16, 2007.

[219] Haugsrud, R.; Norby, T. *Nature Materials* 2006, 5, pp 193-196.

[220] Haugsrud, R.; Norby, T. *Solid State Ionics* 2006, 177, pp 1129-1135.

[221] Mokkelbost, T.; Kaus, I.; Haugsrud, R.; Norby, T.; Grande, T.; Einarsrud, M. A. *J. Am. Ceram. Soc.* 2008, 91, pp 879-886.

[222] Huse, M.; Norby, T.; Haugsrud, R. *Intl. J. Hydrogen Energy* 2001, *in press.*

[223] Brandão, A. D.; Gracio, J.; Mather, G. C.; Kharton, V. V.; Fagg, D. P. *Journal of Solid State Chemistry*, 2011, 184, pp 863–870.

[224] Norby, T.; Haugsrud, R.; Marion, St.; Einarsrud, M. A.; Wiik, K.; Andersen, Oy.; Strom, R. A.; Grande, T. *Patent WO* 2006/066918, 2006.

[225] Ivanova, M.; Ricote, S.; Meulenberg, W. A.; Haugsrud, R.; Ziegner, M. *Solid State Ionics* 2011,213, pp. 45-52.

[226] Fjeld, H.; Kepaptsoglou, D. M.; Haugsrud, R.; Norby, T. *Solid State Ionics* 2010, 181, pp 104-109.

[227] Yamamoto, O. *The International Fuel Cell Conference Proceedings*, NEDO/MITI, Tokyo, JP, 1992.

[228] Singh, J. P.; Bosak, A. L.; Dees, D. W.; McPheeters, C. C. *Fuel Cell Seminar Abstracts*, Courtesy Associates, Washington DC, US, 1988, pp145-148.

[229] Charojrochkula, S.; Wannakittia, S.; Limthongkula, P.; Uedab, J.; Sato, K. *ECS Transactions* 2007, 7, pp 2085-2092

[230] McIntosh, S.; Vente, J. F.; Haije, W. G.; Blank, D. H. A.; Bouwmeester, H. J. M. *Solid State Ionics* 2006, 177, pp 1737-1742.

[231] Niedrig, C.; Taufall, S.; Burriel, M.; Menesklou, W.; Wagner, S. F.; Baumann, S.; Ivers-Tiffée, E. *Solid State Ionics* 2011, 197, pp 25-31.

[232] Švarcová, S.; Wiik, K.; Tolchard, J.; Bouwmeester, H. J. M.; Grande, T. *Solid State Ionics* 2008, 178, pp 1787-1791.

[233] Mueller, D. N.; De Souza, R. A.; Weirich, T. E.; Roehrens, D.; Mayer, J.; Martin, M. *Phys. Chem. Chem. Phys.* 2010, 12, pp 10320-10328.

[234] Yakovlev, S.; Yoo, C.-Y.; Fang, S.; Bouwmeester, H. J. M. *Appl. Phys. Lett.* 2010, 96 , pp 254101-254103.

[235] Lu, H.; Son, S. H.; Kim, J. P.; Park, J. H. *Materials Letters* 2011, 65, pp 702-704.

[236] Efimov, K.; Czuprat, O.; Feldhoff, A. *Journal of Solid State Chemistry* 2011, 184, pp 1085-1089.

[237] Efimov, K.; Halfer, T.; Kuhn, A.; Heitjans, P.; Caro, J.; Feldhoff, J. *Chem. Mater.* 2010, 22, pp 1540-1544.

[238] Huang, B. X.; Malzbender, J.; Steinbrech, R. W.; Singheiser, L. *J. of Membrane Science* 2010, 359, pp 80-85.

[239] Malzbender, J.; Pecanac, G.; Baumann, S.; *Key Engineering Materials* 2012, 488-489, pp 303-306.

[240] Matsumoto, H.; Kawasaki, Y.; Ito, N.; Enoki, M.; Ishihara, T. *Electrochem. Solid-State Lett.* 2007, 10, pp B77-B80.

[241] Scholten, M. J.; Schoonman, J.; Miltenburg, J. C.; Oonk, H. A. *Solid State Ionics* 1993, 61, pp 83-91.

[242] Zakowsky, N.; Williamson, S.; Irvine, J. T. S. *Solid State Ionics* 2005, 176, pp 3019-3026.

[243] Ma, G.; Shimura, T.; Iwahara, H. *Solid State Ionics* 1998, 110, pp 103-110.

[244] Shima, D.; Haile, S. M. *Solid State Ionics* 1997, 97, pp 443-455.

[245] Haile, S. M.; Staneff, G.; Ryu, K. H.; *J. Mater. Sci.* 2001, 36, pp 1149-1160.

[246] Xie, K.; Yan, R.; Chen, X.; Wang, S.; Jiang, Y.; Liu, X.; Meng, G. *J. Alloys Compd.* 2009, 473, 323-329.

[247] Xie, K.; Yan, R.; Xu, X.; Liu, X.; Meng, G. *J. Power Sources* 2009, 187, pp 403-406.

[248] Fabbri, E.; D'Epifanio, A.; Di Bartolomeo, E.; Licoccia, S.; Traversa, E. *Solid State Ionics* 2008, 179, pp 558-564.

[249] Ryu, K. H.; Haile, S. M. *Solid State Ionics* 1999, 125, pp 355-367.

[250] Song, S. J.; Wachsman, E. D.; Dorris, S. E.; Balachandran, U. *Solid State Ionics* 2000, 149, pp 1-10.

[251] Roark, S. E.; Mackay, R.; Sammells, A. F. Hydrogen Separation Membranes for Vision 21 Energy Plants, *28th International Conference on Coal Utilization and Fuel Systems*, Clearwater, FL, March 10-14, 2003.

[252] Brandão, A.; Monteiro, J. F.; Kovalevsky, A. V.; Fagg, D. P.; Kharton, V. V.; Frade, J. R. *Solid State Ionics* 2011, 192, pp 16-20. Guidelines for improving resistance to CO_2 of materials for solid state electrochemical systems

[253] Ivanova, M.; Ricote, S.; Meulenberg, W. A.; Buchkremer, H. P.; Haugsrud, R.; Norby, T.; Stöver, D. Invited lecture at the *NASA-OTM and Mem-Brain Summer School* on Ceramic Material for green chemical production and clean power generation, Valencia, ES, 2010.

[254] Hales, G. E.; *Chem. Eng. Prog.* 1971, 67, pp 49-53.

[255] Morigami, Y.; Y. Kondo, M.; Abe, J.; Kita, H.; Okamoto, K. *Separation/Purification technology* 2001, 25, pp 251-260.

[256] Caro, J.; Noack, M.; *Micropor. Mesopor. Mater.* 2008, 115, pp 215-233.

[257] Weyd, M.; Richter, H.; Kühnert, J.-Th.; Voigt, I.; Tusel, E.; Brüschke, H. *Chemie Ingenieur Technik* 2010, 82, pp 1257-1260.

[258] Noack, M.; Kölsch, P.; Dittmar, A.; Stöhr, M.; Georgi, G.; Schneider, M.; Dingerdissen, U.; Feldhoff, A.; Caro, J. *Micropor. Mesopor. Mater.* 2007, 102, pp 1-20.

[259] Lutz, W.; Fahlke, B.; Lohse, U.; Seidel, R. *Chem. Techn.* 1983, 35, pp 250.

[260] Kokotailo, G. T. *7th Int. Zeolite Conference*, Tokyo, JP, Elsevier, Amsterdam, NL, 1986, p 361.

[261] Corma, A.; Reyl, F.; Rius, J.; Sabater, M. J.; Valencial, S. *Nature* 2004, 431, 287.

[262] Richter, H.; Günther, Ch.; Voigt, I. *Invited lecture at the 12th FZJ Material Symposium*, Jülich, DE, 5-6 October 2011.

[263] Huang, A.; Caro, J. *Micropor. Mesopor. Mater.* 2009, 124, 153-161.

[264] Huang, A.; Weidenthaler, C.; Caro, J. *Micropor. Mesopor. Mater.* 2010, 130, 352-356.
[265] Huang, A.; Caro, J. *Micropor. Mesopor. Mater.* 2010, 129, 90-99.
[266] Huang, A.; Liang, F.; Steinbach, F.; Gesing, T. M.; Caro, J. *J. Am. Chem. Soc.*, in press.
[267] Khajavi, S.; Kapteijn, Fr.; Jansen, J. C. *Journal of Membrane Science* 2007, 299, pp 63-72.
[268] Xu, X.; Bao, Y.; Song, C.; Yang, W.; Liu, J.; Lin, L. *Micropor. Mesopor. Mater.* 2004, 75, pp 173-181.
[269] Bibby, D. M.; Dale, M. P. *Nature* 1985, 317, pp 157-158.
[270] Münzer, S.; Caro, J.; Behrens P. *Micropor. Mesopor. Mater.* 2008, 110, pp 3-10.

In: Doping
Editor: Lixin Yu
ISBN: 978-1-62618-097-0

Chapter 7

EXTERNAL STRAIN: AN EFFECTIVE APPROACH TO TUNE VACANCY FORMATION AND DOPING

Junyi Zhu
National Renewable Energy Laboratory, Golden, CO, US

ABSTRACT

External strain has been proposed to be an effective parameter to control vacancy, interstitial dopants and substitutional dopants in various semiconductor systems. In this paper, some recent progresses in tuning vacancy formation and doping by external strain are reviewed. Vacancies or dopants formation in semiconductors induce volume changes. Vacancies usually induce volume contraction in covalent semiconductors. Interstitial dopants usually induce volume expansion. Substitutional dopants induce volume change based on their atomic size and electronic environment. The strain tuned vacancy or dopant formation is based on the volume change. If the external strain is in the same direction as the volume change, the formation energies of the vacancy or dopant are reduced. Otherwise, the formation energy is increased. This effect can also be used to tune doping sites in a host. The external strain can be a very useful approach to tune the semiconductor defects and dopants and control their device properties.

Keywords: Semiconductor, strain, defects, vacancy, doping

1. INTRODUCTION

External strain is an important parameter in semiconductor device physics. Strain can be applied in many different material systems to change the electronic properties of them. Recently, external strain is found to tune the defects and dopants formation in various semiconductor systems [1-5]. In this paper, the strain tuned vacancy formation, doping incorporation and doping site control will be reviewed.

Strain is a quantity that represents the relative displacement between atoms or particles when a deformation is applied to a material system. Strain has been studied in semiconductor physics since the early days. Deformation potentials of semiconductor systems were measured or calculated when external strains were applied, which can be traced back to early 1960's [6-8].

There are different ways to apply strains to materials. A hydrostatic strain can be applied when the material is merged into liquid and external pressure is applied to the liquid. A uniaxial strain can be applied at a certain orientation of the material mechanically. A biaxial strain can be applied when a material is grown on a lattice mismatched substrate.

When compressive external strain is applied, the material would tend to shrink in volume. When tensile external strain is applied, the material would tend to expand in volume. Volume changes can also be induced when certain defects or dopants are formed in the semiconductor systems [2, 9-12]. For example, a vacancy in covalent semiconductors would often induce a volume shrink [11-12]; an interstitial defect in semiconductors would often induce a volume expansion [9-10]; a foreign dopant would induce volume change that depends on the size and electronic environment difference between the host and the dopant elements [2, 10]. When the direction of the external strain is the same as the defects or dopant induced volume change, the strain would enhance the defect or dopant formation [2, 9-10]. If they are in the opposition direction, the defect or dopant formation would be reduced [2, 9-10]. Therefore, theoretically, using external strain to tune the formation of semiconductor defects or doping is possible.

2. Strain Tuned Vacancy Formation

Vacancies in covalent semiconductors usually induce volume contractions [11-12]. This can be understood by an electron gas model: the removal of an atom would usually lead to the decrease of the electron gas density, which would lead to a contraction of the lattice.

This vacancy induced volume change has been calculated for a few times. One of the most convincing works was done in 1998. It was reported that there are two possible vacancy configurations by density functional calculations in Si at zero pressure [11]. They induce different volume contractions and respond differently following external pressure [11]. In 2005, the intrinsic defect calculation in Si was extended to vacancy, vacancy interstitial pair and at different charge state. Si vacancies are calculated to induce volume contractions, although the values are different in different works [11-12]. Both works calculated the vacancy formation enthalpy as a function of external pressure and found that the vacancy formation enthalpy decrease as the pressure increases [11-12]. These findings indicate that when the defect induced volume change is in the same direction as the external pressure (strain), an enhanced defect formation under the external pressure (strain) should be expected.

In 2010, Bondi et. al studied strain effects on the stability and structure of vacancy clusters in Si using density functional theory [13]. They calculated vacancy formation energy under biaxial and uniaxial strain in Si [13]. Their results also indicate that the formation energy of Si vacancy is reduced when compressive uniaxial and biaxial strain is applied to the lattice [13]. In addition to a single vacancy, they also calculated vacancy clusters in Si under uniaxial and biaxial strain [13]. They found a stable fourfold coordianated structures under strain free condition and a stable partial hexagonal ring structure when certain strain condition is met [13]. Their finding suggests that strain may not only change the vacancy formation energy, but also change their low energy configuration [13].

3. Strain Tuned Substitutional Doping

Doping is very important in semiconductor device. Without dopant, most semiconductors would become useless. Among all types of doping, substitutional doping is the most common and important. Still, for many semiconductors, the doping is limited by the size difference between the host atom and the dopant [14-16].

Substitutional doping occurs when a foreign dopant replaces a host atom in the semiconductor lattice. The difference in size between the dopant and the host atom usually induces a volume change. This volume change makes tuning the substitutional doping with external strains possible [2].

In 2002, Sadigh et. al found that the solubility of B in Si can be enhanced by a compressive biaxial strain due to the small size of B based on their calculations [1]. They concluded that Fermi level change is the major contribution of the enhancement of B solubility [1]. The size difference of B and Si would also lead to a small enhancement [1]. In 2005, Adey et. al suggested that the size difference and the Fermi level change are both important for As solubility in Si [17]. However, they are in opposite sign and cancelled out each other. Both of these works are calculated based on Maxwell Boltzmann statistics [17].

In 2009, Ahn calculated strain energy as strain × stress and obtained energy vs. strain figures for some dopants in Si [4]. They suggested that strain compensation energy is the major contribution of the solubility enhancement [4].

There are also a few experimental observations of strain enhanced doping. Bennett et al. reported that the solubility of donor Sb can reach as high as 10^{21} cm^{-3} under tensile strain [5]. Ikuta shows that Ge incorporation can increase As solubility in Si. SiGe is a complex system [18]. Still, the local strain induced by Ge may play a role in the enhancement of As incorporations [18]. There is no direct experimental observation of strain enhanced doping in III-V systems.

In 2010, Zhu et. al calculated the volume change induced by the substitutional dopant and proposed an external hydrostatic and biaxial strain tuned substitutional doping in GaP [2].

To understand the size effect of foreign dopants, it's important to divide the size difference between the host and dopant into two parts, the intrinsic atomic size difference and the electronic environment difference [2]. For covalent semiconductors, the intrinsic atomic size can be characterized as the covalent radius of the element [2]. The electronic environment can be characterized by the different oxidation states [2]. For example, the covalent radii of Ga and Zn are about the same [19]. When Zn is doped into GaP, it is surrounded by P atoms with -3 nominal ionization states [2]. Its size is not the same as it in ZnO which is surrounded by -2 nominal ionization states [2]. This suggests that Zn in GaP induces a volume change even though Zn and Ga have the same covalent radii [2]. When a p-type dopant replaces a host atom, the electron gas density decreases and a volume contraction should be expected even if the dopant has the same covalent radius as the host atom [2]. When an n-type dopant replaces a host atom, the electron gas density increases and the volume expansion should be expected if the dopant and the host atom have the same covalent radii [2].

To separate the size difference and the electronic environment difference of the dopants and the host atoms, they first calculated the volume change due to the difference of the covalent radii of the dopants and that of the host atoms [2]. They defined it as the volume change due to the size. Then they subtract this value from the total volume change due to the dopant and defined this quantity as the volume change due to the difference of the electronic environment [2].

The calculation was performed using Vienna Ab-initio simulation package [20]. They used generalized gradient approximation and Perdew-Burke-Ernzerhof exchange –correlation [21]. A projected wave method was used to treat the core valence interactions [2]. The plane wave cut off is 400 eV and the Brillouin zone sampling was chosen to be 4x4x4 [2]. They chose 0.01 eV/A as the force cut off to relax the atomic position [2]. Their simulation cell was a 64 atom cell of GaP. The lattice constant was 5.501A, which agrees well with the experimental value [2].

They separated the dopant induced volume change into the intrinsic one and the one based on electronic environment [2]:

$$\Delta V = \Delta V_i + \Delta V_e. \tag{1}$$

here, ΔV_i is the intrinsic volume change, which is due to the difference of covalent radii between the dopant atom and the host atom [2].

$$\Delta V_i = \frac{16}{3\sqrt{3}}[(R_{dopant} + R_p)^3 - (R_{Ga} + R_P)^3] \tag{2}$$

here, R_{dopant} is the covalent radius of the dopant, R_P is the covalent radius of P and R_{Ga} is the covalent radius of Ga [2]. They took the Phillips covalent radii in their calculations [19]. They obtained ΔV from their DFT calculations, ΔV_i from equation 3 and ΔV_e from equation 2 [2]. All the data are listed in Table I [2].

There are a few interesting physical trends observed in their paper. For all the p-type dopants, the ΔV_e are around -1.1 A^3 to -1.5 A^3 [2]. The negative volume changes indicate that the p-type dopants induce a volume contraction based on the difference of the electronic environment [2].

On the contrary, the n-type dopants Ge and Sn induce a ΔV_e of about 3-4 A^3 [2].

The positive volume changes indicate that the n-type dopants induce a volume expansion based on the difference of the electronic environment [2]. The n-type dopants induce a larger absolute value of the volume change than the p-type doapnts do [2]. N-type dopants add one electron to the conduction band, p-type dopants remove one electron from the valence band [2]. Conduction band is relatively more delocalized than valence band [2]. Therefore, n-type dopants induce a large volume change than the p-type dopants [2].

The Phillips covalent radii are the same for Zn, Ge and Ga [19]. Therefore, the intrinsic volume change induced by Zn and Ge in GaP are 0 [2]. However, when Zn or Ge atoms replace a Ga atom in GaP lattice, a noticeable volume change is observed [2]. For Zn, the volume change is -1.46 A^3, for Ge, the volume change is 4. 54 A^3 [2].

Table 1. Calculated dopant induced volume change in GaP. The volume change is divided into ΔV_i due to the intrinsic covalent radii difference and ΔV_e due to the electronic environment

Dopant	$\Delta V(A^3)$	ΔV_i	ΔV_e	p/n	R(A)
Ga	0	0	0	Neutral	1.225
Zn	-1.46	0	-1.46	p	1.225
Al	0.21	0.26	-0.05	Neutral	1.230
Cd	8.66	10.17	-1.51	p	1.405
In	9.54	10.17	-0.63	Neutral	1.405
Ge	4.54	0	4.54	n	1.225
Be	-12.88	-11.78	-1.1	p	0.975
Sn	13.25	10.17	3.08	n	1.405
Zn+Ge	-0.2	0	-0.2	Neutral	1.225

The covalent radii are adopted from Reference [19]. (Reproduced from Reference [2], Copyright 2010 The American Physics Society).

These volume changes are based on the different electronic environment. When Zn and Ge are both doped into the lattice, the total volume change is very small, only 0.2 A^3 [3]. Although p-type and n-type dopants induce very different volume changes, when they are codoped into the lattice, they passivate each other and the overall volume change is very small [2]. This asymmetrical behavior of volume change displays a very unique electronic character of the volume change, which can not be explained by the classic elastic theory, and can only be explained by the quantum theory [2].

The ΔV_e is very small for isovalent dopants [2]. ΔV_e for Al_{Ga} is only -0.05 A^3, and ΔV_e for In_{Ga} is -0.63 A^3 [2]. This could be due to the difference of effective valence between In and Ga. The total volume change for Al_{Ga} is 0.21 A^3, which is also consistent with the experimental fact that the lattice constant of AlP is slightly larger than that of GaP [22-23].

When Ga atom is replaced by Cd and Sn, both of the dopants induce large volume changes [2]. Sn induces a volume expansion of 13.25 A^3 and Cd induces a volume expansion of 8.66 A^3 [2]. Although Cd and Sn have the same covalent radii [19], ΔV_e for Cd and Sn are in opposite directions [2]. The p-type dopant Cd induces a negative ΔV_e and the n-type dopant of Sn induces a positive ΔV_e [2]. Therefore, the large volume difference is due to the large ΔV_e difference [2].

When ΔV_e and ΔV_i are in the same direction, the volume change induced by the dopant is usually the largest [2]. For example, a small p-type dopant induces a large volume shrink; a large n-type dopant induces a large volume expansion [2]. In all the dopants they studied, Be induces the largest volume contraction of 12.88 A^3 because it is the smallest p-type dopant; Sn induces the large volume expansion of 13.25 A^3 because it is the large n-type dopant [2].

Then they calculated the dopant formation energy vs. external hydrostatic strain. The dopant formation energy is defined as following:

$$E_f = E(\text{doped}) - E(\text{host}) + \mu(\text{Ga}) - \mu(\text{dopant}), \quad (3)$$

where E(doped) is the total energy of the doped system, E(host) is the total energy of the undoped system [2]. Both are calculated under the same strain conditions [2]. μ(Ga) and μ(dopant) are the chemical potentials of Ga and the dopant [2]. They also assume that the difference of the chemical potentials is insensitive to the external strain to simplify the calculation [2]. To apply the hydrostatic strain, they changed all three cell dimensions equally [2]. Figure 1 shows the change of the dopant formation energy as a function of external hydrostatic strain [2]. They calculated Be, Al, Zn, Ge, Cd and Sn [2]. They found that dopant formation energy decreases with the external strain when the volume change induced by the dopant is positive (Ge, Cd and Sn) [2]. The formation energy increases with the external strain when the volume change induced by the dopant is negative (Be, Zn) [2]. For Al, since the volume induced by it is very small, the curve is almost flat [2]. Also, the slope of the curve is in the same trend of the volume induced by the dopant [2]. The slope of Sn is larger than that of Cd and Ge [2]. This is the same trend as the volume expansion induced by them.

The slope of Be is larger than Zn, which also agrees well with their volume expansion trend [2]. This volume change factor is an important parameter to determine the dopant formation energy as a function of strain [2].

Also, the change of dopant formation energy with strain is monotonic, which is against a common expectation that the formation energy is the minimum when the deformed lattice has the same size as the dopant [2].

This can be understood by a simple strain model [2]. Figure 2 shows a schematic illustration of total energy of the host lattice and of the doped lattice as a function of the volume [2]. Assume that near equilibrium, the total energy of the host lattice follows: E(host) = $\alpha(V-V_{host})^2$, where α is the elastic constant, V_{host} is the equilibrium volume of the host [2]. The energy of the doped lattice is : E(host+dopant) = $\alpha`(V-V_{host+dopant})^2$, where α' is the elastic constant, V_{host} is the equilibrium volume of the doped lattice [2].

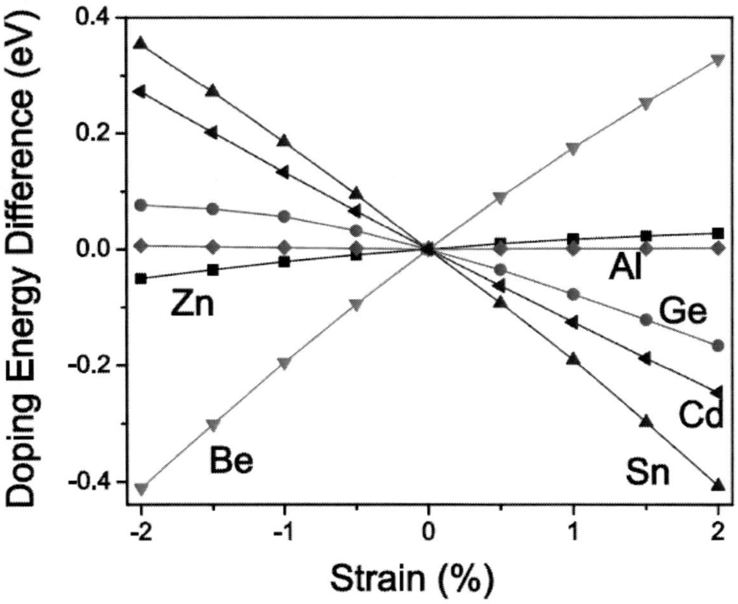

Reproduced from Reference [2], Copyright 2010 The American Physics Society.

Figure 1. the change of the dopant formation energy as a function of external hydrostatic strain.

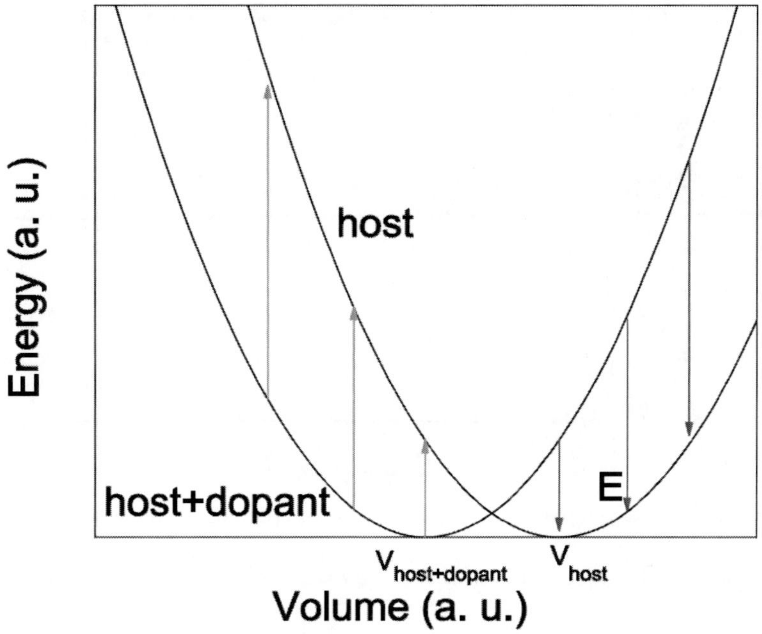

Figure 2. a schematic illustration of total energy of the host lattice and of the doped lattice as a function of the volume.

The change of the dopant formation energy is E(dopant) = E(host+dopant)-E(host) = $(\alpha' - \alpha)V^2 - 2(\alpha' V_{host+dopant} - \alpha V_{host})V + \alpha' V_{host+dopant}^2 - \alpha V_{host}^2$ [2].

Assume the two elastic constants are the same, the change of the dopant formation energy is proportional to $V(V_{host} - V_{host+dopant})$ [2]. This indicates that the formation energy is proportional to V (or strain) and the volume change induced by the dopant [2].

Under an external strain, the size difference between the dopant and the host atoms are about the same and the dopant formation energy is proportional to the product of stress and strain, which is a sub-linear function of the strain [2].

They also calculated the dopant formation energy under the biaxial strain, which may be easy to apply during epitaxial growth [2].

They calculated Be as a dopant because of its large volume contraction and the large formation energy change under the strain [2]. They change the dimension of x and y and allow z axis to relax [2]. Their calculated Poisson ratio is -0.86, which is in good agreement with the experimental value of -0.89 [24]. The change of the formation energy as a function of biaxial strain is shown in Figure 3 [2].

Also, the values as a function of hydrostatic strain are also plotted for comparison [2]. Their calculation indicates that the impurity formation energy is reduced by 80 meV under 1% biaxial compressive strain, which is about 3 times concentration enhancement for a typical OMVPE growth temperature at 900 K for GaP [2].

The trend is about the same for the biaxial case as the hydrostatic one, except the effect for biaxial strain is smaller [2].

In summary, their work demonstrates the origin of the substitutional dopant induced volume change, which can be separated into two parts, the intrinsic size difference and the electronic environment difference [2].

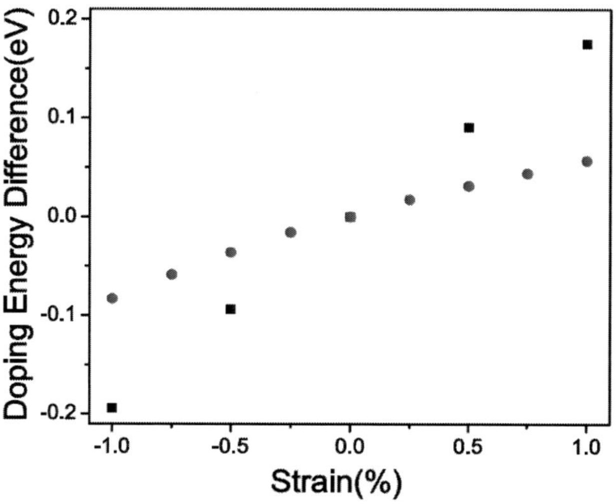

Figure 3. Doping energy difference of Be in GaP as a function of hydrostatic strain (square) and biaxial strain (circle).

The external strain affected doping is directly determined by the dopant induced volume change. They further showed that the dopant formation energy as a function of external strain is monotonic and sub-linear [2].

4. Strain Tuned Interstitial Defects

Interstitial defects are very common defects in semiconductor systems. In many applications, they reduce the device transport properties, and in other applications, they are desired. Therefore a mechanism to tune the doping sites of interstitial and substitutional defects or dopants can be very useful in semiconductor device physics.

ZnO is a very important semiconductor, which has wide applications in light emitting diodes, transparent conducting layer on solar cells and many other devices. It is easy to be doped as n-type and p-type doping is usually considered difficult [25-28].

In 2011, Zhu et al. studied tuning the different doping sites in ZnO. Li is a common p-type dopant for ZnO [29-31]. However, Li atoms are very small and may often occupy interstitial sites that passivate the p-type substitutional sites [31]. The p-type doping condition that moves the Fermi level to the valence band maximum makes this problem more severe [32-33]. Therefore, a mechanism that can enhance the substitutional doping of Li and reduce the interstitial doping of Li in Zno is desired.

In their work, they proposed an approach to apply hydrostatic and biaxial strain to ZnO to enhance the Li substitutional doping and reduce Li interstitial doping [3].

According to the discussion on the substitutional doping, the covalent radius of Li is smaller than that of Zn [19], which leads to a volume contraction when Li replaces a Zn in ZnO. Also, Li is a p-type dopant on Zn substitutional sites, which induces volume contractions. Therefore, substitutional Li would usually induce a volume contraction in ZnO lattice [3]. On the other hand, interstitial Li would increase the electron gas density and tend to expand the lattice [3]. So a compressive strain is expected to enhance the substitutional Li doping and reduce the interstitial Li [3].

They verified this concept by density functional theory calculations. They used a similar calculation set up as they did in the strain tuned substitutional study, except that they chose higher plane wave cut off energy of 600 eV and a different 72 atom ZnO cell.

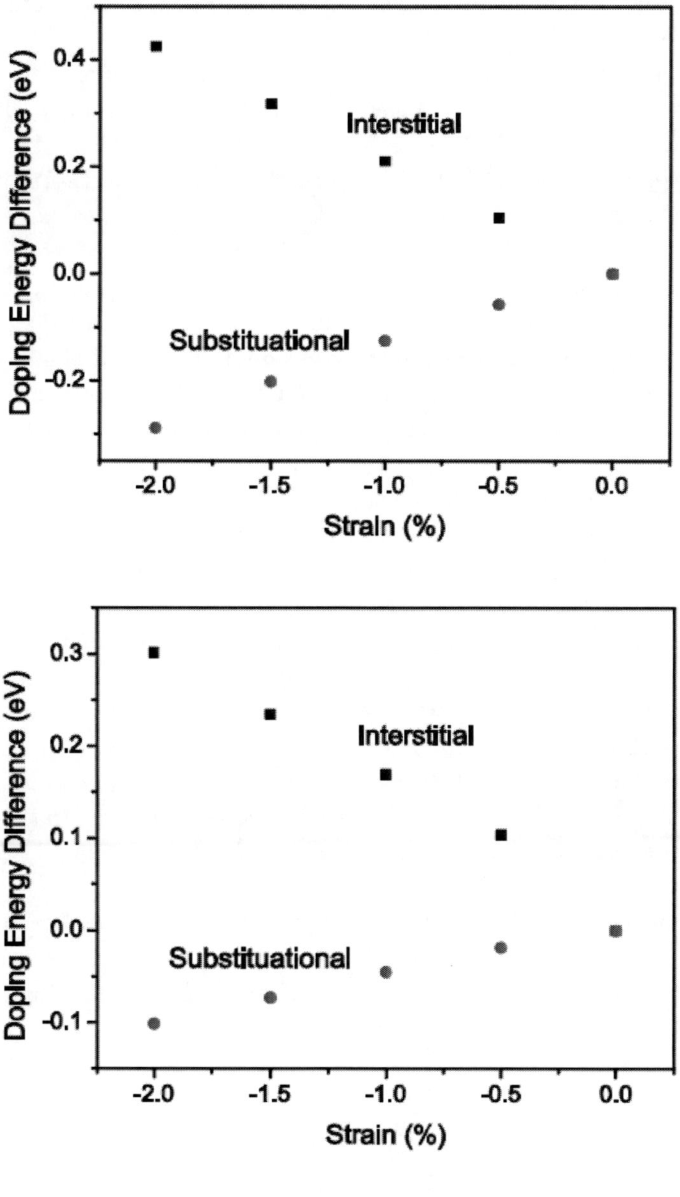

Figure 4. Doping energy difference as a function of (a) hydrostatic strain and (b) biaxial strain. (Square: Li at interstitial sites, dot: Li at substitutional sites.)

Their calculated that ZnO lattice constant a is 3.287 A and the c/a ratio is 1.614, which agrees with experimental value reasonably well [34]. They found that substitutional Li induces a volume contraction of 4.91 A^3 , which is consistent with the experimental observation of a volume reduction when Li occupies substitutional sites in ZnO [31]. They also found that the interstitial Li induces a volume expansion of 8.56 A^3. This can be understood by an electron gas model. The interstitial Li would increase the electron gas density. Therefore, the volume tends to increase to reduce the kinetic energy. This is also in agreement with experiments [31].

Then they calculated the substitutional doping and interstitial doping formation energy of Li as a function of hydrostatic strain and as a function of biaxial strain, as shown in Figure 4 (a) and Figure 4 (b), respectively [3].

Hydrostatic compressive strain indeed reduces the substitutional Li formation energy and increases the interstitial Li formation energy [3]. A 1% compressive strain, lowers the substitutional doping formation energy for about 100 meV, which is about 3-5 times higher concentration enhancement under a typical OMVPE growth temperature at 900K [3]. On the other hand, this amount of strain increases the interstitial doping formation energy for about 200 meV, which would lower the interstitial Li concentration at 900K [3].

The absolute value of the interstitial Li doping energy difference is about twice as large as the substitutional Li, which agrees with the volume change results, because the larger the volume change is, the more effective, the external strain affects the doping [3].

The biaxial strain can also tune the substitutional and the interstitial doping of Li in ZnO [3]. Under biaxial condition, the substitutional Li induces a volume contraction of 2.711 A^3 and the interstitial Li induces a volume expansion of 6.31 A^3, which are both smaller than the hydrostatic condition, because the c direction would relax some strain [3].

When external biaxial strain is applied, the substitutional and interstitial doping formation energies show a similar trend as the study of the hydrostatic one, except that the effect of biaxial strain is smaller [3]. A 1% compressive strain induces a formation energy difference between the substitutional and the interstitial for about 130 meV, which is significant to improve the Li substitutional doping [3].

In summary, the external compressive strain can be an effective strategy to enhance the substitutional doping of a small p-type dopant and reduce the interstitial doping for many semiconductor systems [3].

On the other hand, a tensile strain can enhance the interstitial dopant and reduce the substitutional doping of the small dopant, which can improve device performance in battery applications [35-37]. This strategy should be very useful to control the doping and electronic properties of many semiconductor devices [3].

DISCUSSIONS

The external strain is a very important tool to tune the vacancy formation, substitutional and interstitial doping. In many semiconductor devices, some certain defects are desired and some other defects are unwanted. Therefore, strain engineering can be a useful tool to enhance the wanted defects and reduce the unwanted defects.

A compressive strain would generally reduce interstitial defects and large substitutional dopants, and enhance vacancy formation and small substitutional dopants. Therefore, for semiconductor system with high vacancy formation energy (so compressive strain wouldn't produce a lot of vacancy defects), low interstitial defects formation energy, and desired small substitutional dopants, compressive strain can significantly improve the device quality by reducing the interstitial defects and enhancing substitutional doping. On the other hand, tensile strain would lower the interstitial defects formation energy and the large substitutional dopant formation energy, and increase the vacancy formation energy and small substitutional dopant formation energy. Therefore, if the vacancy formation energy is very low, the interstitial defect formation energy is relatively high (so tensile strain wouldn't produce a lot of interstitial defects) and a large substitutional dopant is desired, tensile strain can be important to improve the device quality by reducing vacancies and enhancing substitutional doping. When the right condition is met, a compressive or tensile strain can effectively tune many different types of defects in semiconductors and serve as a very powerful tool in semiconductor industry.

The author thanks for the contribution of the review materials from Feng Liu, G. B. Stringfellow and Su-Huai Wei.

REFERENCES

[1] B. Sadigh et al., *Appl. Phys. Lett.*, 80, 4738 (2002).
[2] J. Zhu et al., *Phys. Rev. Lett.* 105, 195503 (2010).
[3] J. Zhu and S. –H. Wei, *Sol. State Comm.*, accepted (2011).
[4] C. Ahn et al., *Phys. Rev.* B 79, 073201 (2009).
[5] N. S. Bennett et al., *J. Vac. Sci. Technol.* B 26, 391 (2008).
[6] K. J. Schmidt-Tiedemann, *Phys. Rev. Lett.*, 7, 372 (1961).
[7] S. Riskaer, *Phys. Rev.*, 152, 845 (1966).
[8] K. J. Schmidt-Tiedemann, *Z. Naturforsch.* 16a, 639 (1961).
[9] J. Zhu and S. Wei, *Sol. State Comm.* 151, 1437 (2011).
[10] J. Zhu and S. Wei, *Frontiers of Mat. Sci.*, 5, 335 (2011).
[11] A. Atonelli, E. Kaxiras, D. J. Chadi, *Phys. Rev. Lett.*, 81, 2088 (1998).
[12] S. A. Centoni, et. al. *Phys. Rev. B,* 72, 195206 (2005).
[13] R. J. Bondi, S Lee, and G. S. Hwang, *Phys. Rev. B*, 81, 245206 (2010).
[14] J. Y. Zhu, F. Liu and G. B. Stringfellow, *Phys. Rev. Lett.* 101, 196103 (2008).
[15] G. F. Neumark, *Phys. Rev. Lett.* 62, 1800 (1989).
[16] E. F. Schubert, Doping in III-V Semiconductors, Cambridge Studies in Semiconductor Physics and Microelectronic Engineering (Cambridge University Press, Cambridge, 2005), p. 1.
[17] J. Addey, R. Jones, P. R. Bridon, *Phys. Stat. Sol.* (C), 2, 1953 (2005).
[18] T. Ikuta et al., *Jpn. J. Appl. Phys.* 46, 1916 (2007).
[19] J. A. Van Vechten and J. C. Phillips, *Phys. Rev.* B 2, 2160 (1970).
[20] G. Kresse and J. Hafner, *Phys. Rev.* B 49, 14251 (1994); G. Kresse and J. Furthmuller, *Comput. Mater. Sci.* 6, 15 (1996).

[21] P. E. Blöchl, *Phys. Rev.* B 50, 17953 (1994); G. Kresse and D. Joubert, *Phys. Rev.* B 59, 1758 (1999).
[22] Intrinsic Properties of Group IV Elements and III-V, II-VI, and I-VII Compounds, edited by O. Madelung, M. Schulz, and H. Weiss, *Landolt-Bo¨rnstein,* New Series, Group III, Vol. 22, Pt. a (Springer, Berlin, 1987).
[23] L. I. Berger, *Semiconductor Materials*, (CRC Press, Boca Raton, 1997), p. 134.
[24] R. Weil and W. O. Groves, *J. Appl. Phys.* 39, 4049 (1968).
[25] C. Jagadish, S. Pearton (Eds.), *Zinc Oxide Bulk Thin Films and Nanostructures,* Elsevier, Amsterdam, 2006.
[26] S. B. Zhang, S.-H. Wei, A. Zunger, *J. Appl. Phys.* 83, 3192 (1998).
[27] A. Tsukazaki, A. Ohtomo, T. Onuma, M. Ohtani, T. Makino, M. Sumiya, K. Ohtani, S. F. Chichibu, S. Fuke, Y. Segawa, H. Ohno, H. Koinuma, M. Kawasaki, *Nature Mater.* 4, 42 (2005).
[28] T. Atsushi, K. Masashi, O. Akira, O. Takeyoshi, O. Keita, O. Hideo, C. Shigefusa, K. Masashi, *Jap. J. Appl. Phys.* 44 (21), 643 (2005).
[29] J. R. Duclre, M. Novotny, A. Meaney, R. O. Haire, E. McGlynn, M. O. Henry, J.-P. Mosnier, *Superlattices Microstruct.* 38, 397 (2005).
[30] A. Y. Oral, Z. B. Bahsi, M. H. Aslan, *Appl. Surf. Sci.* 253, 4593 (2007).
[31] S. Majumdara, P. Banerji, Superlattices *Microstruct.* 45 (6) 583 (2009).
[32] C. H. Park, S. B. Zhang, S.-H. Wei, *Phys. Rev. B* 66, 073202 (2002).
[33] S.-H. Wei, *Comput. Mater. Sci.* 30, 337 (2004).
[34] A. Shalimov, W. Paszkowicz, K. Grasza, P. Skupinski, A. Mycielski, J. Bak-Misiuk, *Phys. Stat. Sol.* (b) 244, 1573 (2007). S.-H. Wei, *Comput. Mater. Sci.* 30 (2004) 337.
[35] P. Poizot, S. Laruelle, S. Grugeon, L. Dupont, J.-M. Tarascon, *Nature* 407, 496 (2000).
[36] Y.-S. Hu, Y. G. Guo, W. Sigle, S. Hore, P. Balaya, J. Maier, *Nature Mater.* 5, 714 (2006).
[37] M. V. Koudriachova, N. M. Harrison, S. W. de Leeuw, *Phys. Rev. Lett.* 86 (7), 1275 (2001).

In: Doping
Editor: Lixin Yu
ISBN: 978-1-62618-097-0

Chapter 8

RARE EARTH ELEMENTS DOPING FEATURES IN RENEWABLE ENERGY-RELATED NANO DEVICES: YTTERBIUM CASE

Karem Boubaker*
Département de Physique et de Chimie, École Supérieures des Sciences et des Techniques de Tunis (ESSTT)/ Mahdia, Tunisia

ABSTRACT

This chapter deals with rare earth elements and their doping features. A particular attention is paid to nano-scale effects, mainly Urbach tailing features in the particular case of Ytterbium along with the supply to renewable energy-related nano-devices materials.

I. RARE EARTH ELEMENTS

I.1. Presentation

Rare earth elements (or rare earth metals) are a set of seventeen chemical elements (Figure 1) in the periodic table, specifically the fifteen lanthanides plus scandium and yttrium.

Moreover, scandium (Sc) and yttrium (Y) are considered as rare earth elements (Figure 2) since they tend to occur in the same ore deposits as the lanthanides and exhibit similar chemical properties [1-7].

Rare earth elements became known to the world with the discovery of the black mineral "ytterbite" (renamed to gadolinite in 1800) by Lieutenant Carl Axel Arrhenius in 1787, at a quarry in the village of Ytterby, Sweden (Figure 3).

---

* Corresponding author: Karem Boubaker. Département de Physique et de Chimie, École Supérieures des Sciences et des Techniques de Tunis (ESSTT)/ 63 Rue Sidi Jabeur 5100 Mahdia, Tunisia.

Anders Gustav Ekeberg isolated beryllium from the gadolinite but failed to recognize other elements which the ore contained. After this discovery in 1794, a mineral from a Swedish town, Bastnäs near Riddarhyttan, which was believed to be an iron-tungsten mineral, was re-examined by Jöns Jacob Berzelius and Wilhelm Hisinger. In 1803, there were two known rare earth elements, Yttrium and Cerium, (Figure 4) although it took another 30 years for researchers to determine that other elements were contained in the raw rocks.

Figure 1. Some Rare earth elements.

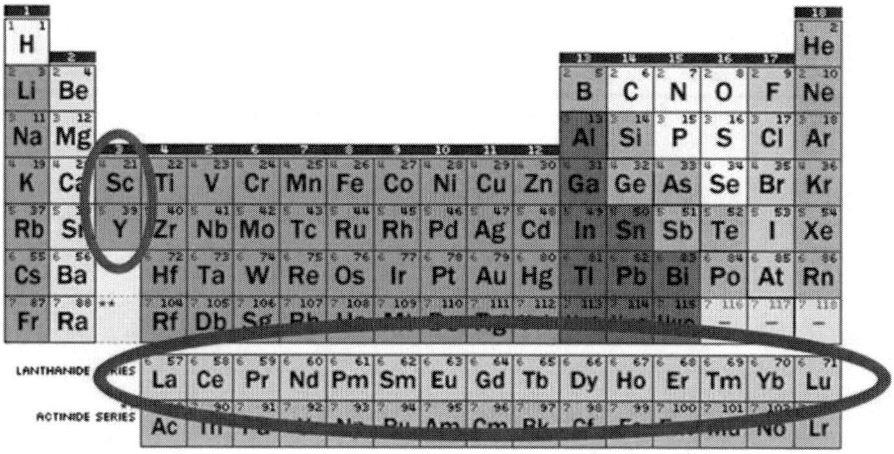

Periodic Table of the Elements

Figure 2. Rare earth elements loci in the periodic table.

Figure 3. A Rare earth element in raw and core form.

I.2. Properties

Rare earth elements are heavier than iron. They hence seem to be produced by supernova nucleo-synthesis or the s-process in asymptotic giant branch stars.

Figure 4. Rare earth elements granular samples.

In nature, spontaneous fission of Uranium-238 produces trace amounts of radioactive promethium, but most promethium is synthetically produced in nuclear reactors. Rare earth elements are not particularly rare. The controversial nomination comes from the fact that long and tedious processes were required to purify the metals from their oxides. Rare earth elements are characterized by 5d and 5f electronic configuration [8-11]. They are divided into two families: the lanthanide series and the actinide series. All lanthanides have an electron configuration resembling Xenon with additional electrons in the 6s, 5d, and 4f orbitals. Actinides are rather radioactive, highly electropositive. They tarnish readily in air and combine directly with most nonmetals. They have numerous allotropes and can react with boiling water or dilute acid to release hydrogen gas. Both series (lanthanides and actinides) share some noticeable properties:

- They have a high luster, but tarnish readily in air.
- Their electrical conductivity is relatively high.
- There have similar solubility and complex formation behaviors.

- They naturally occur together in minerals.
- They have stable valences (mainly + oxidation state)

Rare-earth elements are generally targeted an exploited for their unique orbital arrangement. Shielded 4f shell orbitals are excited during the photoluminescence process and result in sharp-line emission peaks. Since this shell is blocked by outer electron orbitals of 6s and 5d, little or no influence is attributed from the host [12-15]. This is the cause of the spectral transitions and those transitions are responsible for the emission of light which covers the range of ultraviolet to far infrared. Figure 5 presents energy levels diagrams of some Rare-earth elements.

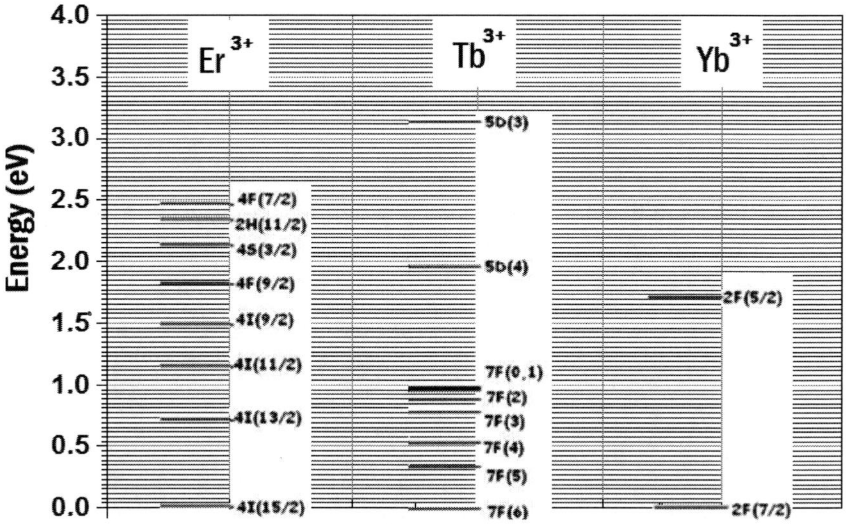

Figure 5. Energy levels diagrams of some Rare-earth elements.

Table 1. Lanthanides

Lanthanide	4f Electrons	ConFigure	Atomic Number	Radius (pm)
Lanthanum	0	[Xe] $5d^1 6s^2$	57	103
Cerium	2	[Xe] $4f^2 6s^2$	58	102
Praseodymium	3	[Xe] $4f^3 6s^2$	59	99
Neodymium	4	[Xe] $4f^4 6s^2$	60	98
Promethium	5	[Xe] $4f^5 6s^2$	61	97
Samarium	6	[Xe] $4f^6 6s^2$	62	95
Europium	7	[Xe] $4f^7 6s^2$	63	94
Gadolinium	7	[Xe] $4f^7 5d^1 6s^2$	64	93
Terbium	9	[Xe] $4f^9 6s^2$	65	92
Dysprosium	10	[Xe] $4f^{10} 6s^2$	66	91
Holmium	11	[Xe] $4f^{11} 6s^2$	67	90
Erbium	12	[Xe] $4f^{12} 6s^2$	68	89
Thulium	13	[Xe] $4f^{13} 6s^2$	69	88
Ytterbium	14	[Xe] $4f^{14} 6s^2$	70	86
Lutetium	14	[Xe]$6s^2 4f^{14} 5d^1$	71	86

Table 2. Actinides

Actinide	4f Electrons	ConFigure	Atomic Number	Radius (pm)
Actinium	0	[Xe] $6d^1 7s^2$	89	203
Thorium	0	[Xe] $6d^2 7s^2$	90	180
Protactinium	1	[Xe] $5f^1 6d^2 7s^2$	91	162
Uranium	4	[Xe] $5f^3 6d^1 7s^2$	92	153
Neptunium	5	[Xe] $5f^4 6d^1 7s^2$	93	150
Plutonium	6	[Xe] $5f^6 7s^2$	94	162
Americium	7	[Xe] $5f^7 7s^2$	95	173
Curium	7	[Xe] $5f^7 6d^1 7s^2$	96	174
Berkelium	9	[Xe] $5f^8 6d^1 7s^2$	97	170
Californium	10	[Xe] $5f^{10} 7s^2$	98	186
Einsteinium	11	[Xe] $5f^{11} 7s^2$	99	186
Fermium	12	[Xe] $5f^{12} 7s^2$	100	176
Mendelevium	13	[Xe] $5f^{13} 7s^2$	101	177
Nobelium	14	[Xe] $5f^{14} 7s^2$	102	175
Lawrencium	14	[Xe] $5f^{14} 7s^2 7p^1$	103	185

I.3. Particular Applications

Thanks to their ability to exchange electrons, Rare Earth elements have become uniquely indispensable in many optical, electronic, magnetic and catalytic applications mainly:

- RE Permanent magnets in hybrids, and electric vehicles : (Nd, Pr, Dy)
- Magnets in wind and hydro-power generators : (Pr, Dy, Tb)
- Energy storage in Ni metal hydride batteries: (La)
- Signal amplification in fiber optics: (Y, Eu, Tb, Er)
- Fluid catalytic cracking (FFC) for gasoline production: (La, Ce)
- Magnets for medical imaging- MRI: (Nd, Pr, Dy, Tb)
- Flat screen displays : (Y, Eu, Tb, Gd, Pr, Ce)

Table 3. Actinides

Physical aggregate	Element		
	Ytterbium	Terbium	Erbium
Atomic Mass (g/mol)	173.04	158.93	167.26
Electronegativity	1.1	1.2	1.24
Melting Point (◦C)	824	1356	1529
Boiling Point (◦C)	1196	3230	2868
Density (g/cm3)	9.321	8.23	9.066
Molar Heat Capacity (J/mol K)	26.74	73.2	28.12

II. Rare Earth Elements Doping

II.1. Doping Applications History

Rare-earth elements were initially produced in small amounts from deposits in granitic pegmatite, the geologic environment in which they were first discovered [16]. During the second half of the 19th century and the first half of the 20th century, Rare-earth elements came mainly from placer deposits. With the exception of the most abundant lanthanide elements (cerium, lanthanum, and neodymium), individual Rare-earth elements were not commercially available until the 1940s. Between 1965 and 1985, most of the available Rare-earth elements came from California. Russia, China and Australia were also a major producer from such sources until the early 1990s. China became the most important Rare-earth elements producer since 1998. (Figure 6).

Rare-Earth (RE) ion doped semiconductors have been the focus of numerous investigations in the last decades because of their unique optical properties and promising applications in optoelectronic devices. It has been discovered that these materials multicolor light could be generated directly from electrical current in monochromatic or colored displays. Growing knowledge of the electronic structure of small semiconductor grains has improved Rare-earth elements use[17-20]. This structure may be very different from that of the bulk counterparts because of the quantum confinement effect, which thus affects the luminescence dynamics of optical activators.

Luminescence of RE ions incorporated in the semiconductor lattice could be sensitized efficiently later by exciton recombination in the host. In this sense, the luminescence of RE ions depends critically on their doping locations in the hosting edifice [21-25]. In this context, Rare-Earth (RE) have been first discovered as having an influence on radiation induced phenomena such as intrinsic defect centre formation, as well as improving the piezoelectric properties of some compounds. Then, they were widely explored for the case of insulating host such as glasses and ceramic materials. A set of Rare-Earth most known applications is gathered in Figure 7.

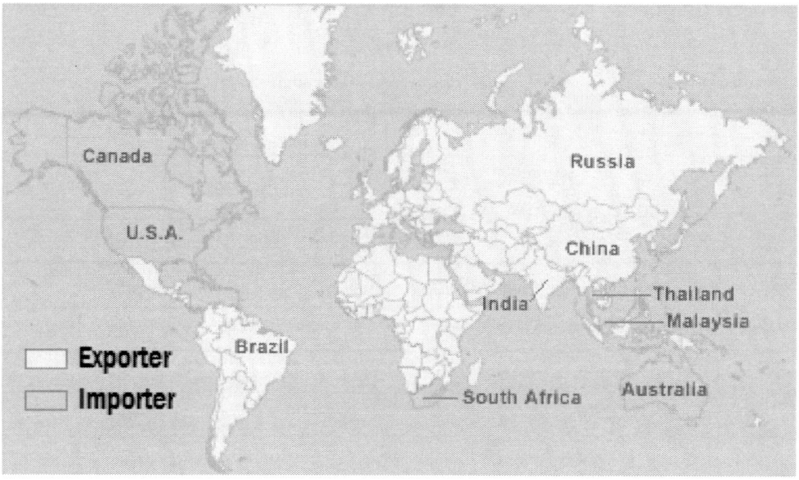

Figure 6. Rare-earth elements throughout the world.

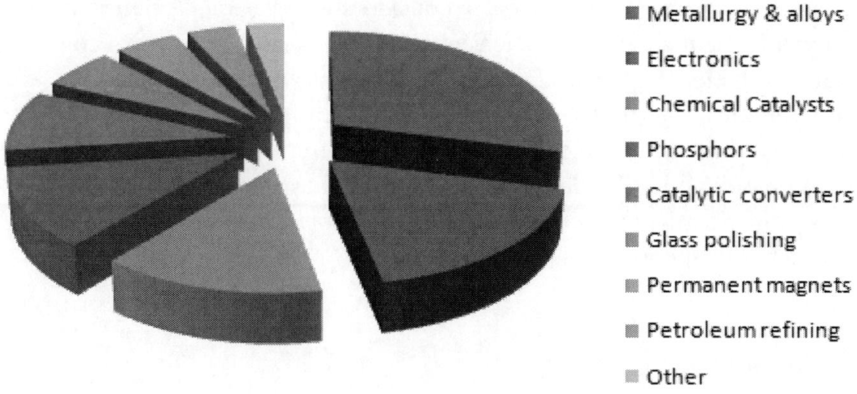

Figure 7. Rare-earth elements most known applications.

II.2. Rare Earth Doping Techniques

There were severl techniques for implementing photo-excitation of doped rare earth ions in a doped materials. For example, resonant excitation by the direct optical absorption or non-resonant excitation by carriers. Direct optical absortion consists of matching the energy of the incident photon with one of the absorption bands of the Rare Erth element. This later absorbs the incident photon energy, and enter an excited state. The other technique is based on carriers intervention: The carriers, either photo-generated or electrically injected, are trapped at RE-related defect states, thus forming an exciton. Recombination energy of these excitons can be transferred to the Rare Erth element ion via an Auger [26].

Since the energy released by exciton recombination in most cases exceeds that necessary for excitation of the Rare Erth elemen, the remaineder is released into the host as phonons. Consequently, photoexcitation of Rare Erth elements in an absorbing host material is nearly always dominated by carrier-mediated excitation, even when the excitation wavelength conincides with one of their absorption bands.

Rare Erth element atoms can also be excited by hot electrons through impact excitation, under a strong electric bias.In phosphors and similar materials, a large amount of work has been carried out during the last decades, and now there are numerous applications, especially in the display industry. Novel material systems and doping techniques are still being developed.

In semiconductors, important contributions have been made to electroluminescent devices based on Er-doped Si crystals or nanocrystals within amorphous matrices. Several techniques have been tried in the last 20 years, and doping Si-based materials and devices with rare earths has emerged as the strongest contender , for visible outputs as well as at mid-infrared wavelengths.

As processing for nano-devices becomes ever more sophisticated, so also does the development of photonic devices based on resonant cavities, waveguides, and Si-related photonic bandgap materials. Progress has been made in understanding luminescence from rare earths sensitized by nano-clusters.

In the last few years, there has been considerable research activity in doping wide-bandgap compound semiconductors, such as GaN, SiC, and ZnO, with Erbium, as well as with other rare-earth elements, in order to obtain efficient visible emission.

New opportunities have been opened up by the recently, evidence of laser emission in Eu doped GaN, operation of visible and infrared light-emitting diodes, room-temperature ferromagnetism based on Gd/Eu doping of GaN as well as the properties of nano-particles of GaN doped with rare earths (garnets, perovskites, etc.) presented new potentials for rare earth element devices multi-functionality [27-29].

II.3. Rare Earth Doping Effects

As mentioned above, rare-earths (RE) ions doping has been confirmed as an efficient mean for inducing optical activity in Si-based materials. Since the electron configuration of RE ions is incomplete 4*f* electronic shell shielded by closed 5*s* and 5*p* shells, 4*f* transition of RE in solid-state shows intrinsic sharp luminescence regardless of the host material and temperature. Moreover, many rare earth ions yield huge luminescence at technologically important wavelengths, like Er^{3+} and Pr^{3+} ions (1.54 μm and 1.33 μm respectively). Room temperature 1.54 μm electroluminescence from Er-doped bulk silicon light emitting diode has been recently reported. In general, this high luminescence can be acquired by optically activating Er^{3+} atoms in the host structure, by both suppressing the non-radiative decay path of excited Er^{3+} ions and increasing the excitation efficiency.

In recent years, a large attention has been noted towards rare earth (RE) doped oxide thin films due to their unique optical properties and also potential applications (photovoltaics, optical fiber telecommunication, photo-catalysis, etc.).

Particularly, the efficient energy transfer from the oxide host matrices to RE ion has been attracting much interest in order to adapt the solar spectrum through photon conversion. Common wide bandgap oxide semiconductors which act as an effective host matrix like ZnO, TiO_2 and SnO_2 have been selected in order to excite the doped rare earth ion efficiently and to yield the high luminescence intensity in the NIR spectral region through energy transfer [30].

In this context, rare earth ions can be divided into two categories: the first one in which ions with radius very close to the Fe ions can enter into the spinel lattice, while the second type of R ions have an ionic radius larger than Fe ions. During the sintering processes , some of these ions will diffuse to the grain boundaries and form an isolating ultra-thin layer around the grains, contributing to the formation of secondary phases

III. The Case of Ytterbium

III.1. Ytterbium Properties

Ytterbium is a rare earth element with the symbol Yb and atomic number 70. A soft silvery metallic element, ytterbium belongs to the lanthanide series and is found in the minerals gadolinite, monazite, and xenotime (Figure 8). Ytterbium is sometimes associated with yttrium or other related elements and is used in certain steels [31].

Natural ytterbium is a mix of seven stable isotopes. Ytterbium-169, an artificially produced isotope, is used as a gamma ray source.

Ytterbium contains unique peak values in its spectrum. Figure 9 represents this spectrum, taken to be within 540 nm to 1560 nm for the green laser and 800 nm to 1560 nm for the infrared laser.

Peak values can be noticed at 966nm for the green laser [32]. This peak is broader and spans the region from 963nm to 969nm. There are No detectable peaks in the infrared. The transition for the peak at 966 nm is determined to be $^2F_{5/2}$ to $^2F_{7/2}$ with an energy corresponding to 1.284 eV.

Figure 8. Ytterbium samples.

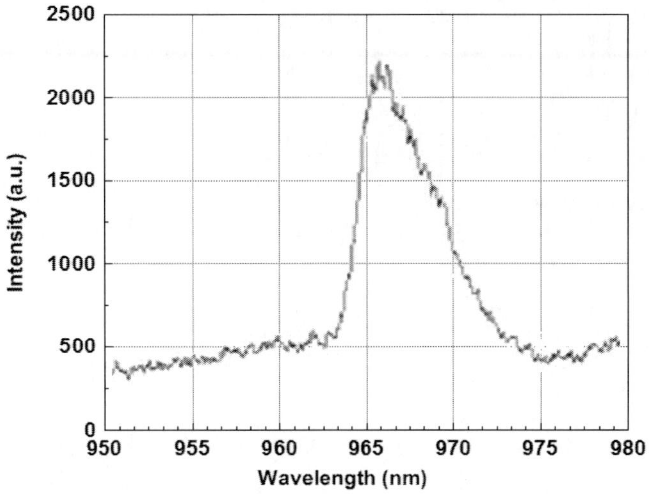

Figure 9. A fragment from Ytterbium spectrum.

III.2. Ytterbium Doped Nano-Applications

Ytterbium is used as a dopant to help improve the grain refinement, strength, and other mechanical properties of stainless steel in nano-devices. ZnO nano-films can be doped with ytterbium (Yb) at different rates [33-37]. Figure 10 shows XRD of ZnO nano-films doped with Ytterbium rates of 100, 200 and 300 ppm. XRD analysis shows a strong (0 0 2) X-ray diffraction line for increasing Yb-doping amounts. This c-axis preferential orientation of ZnO crystallites is naturally required to use this oxide as transparent conductor in optoelectronic applications.

Figure 11 shows AFM images of the same nano-films.

Ytterbium doping is also often used together with erbium doping. Typically, ytterbium ions absorb the pump radiation and transfer the excitation energy to erbium ions.

Even though the erbium ions could directly absorb radiation e.g. at 980 nm, ytterbium co-doping has been proved to be useful because of the higher ytterbium absorption cross sections and the higher possible ytterbium doping density in typical laser glasses, so that a much shorter pump absorption length and a higher gain can be achieved. Ytterbium co-doping is also sometimes used for praseodymium-doped up-conversion fiber lasers.

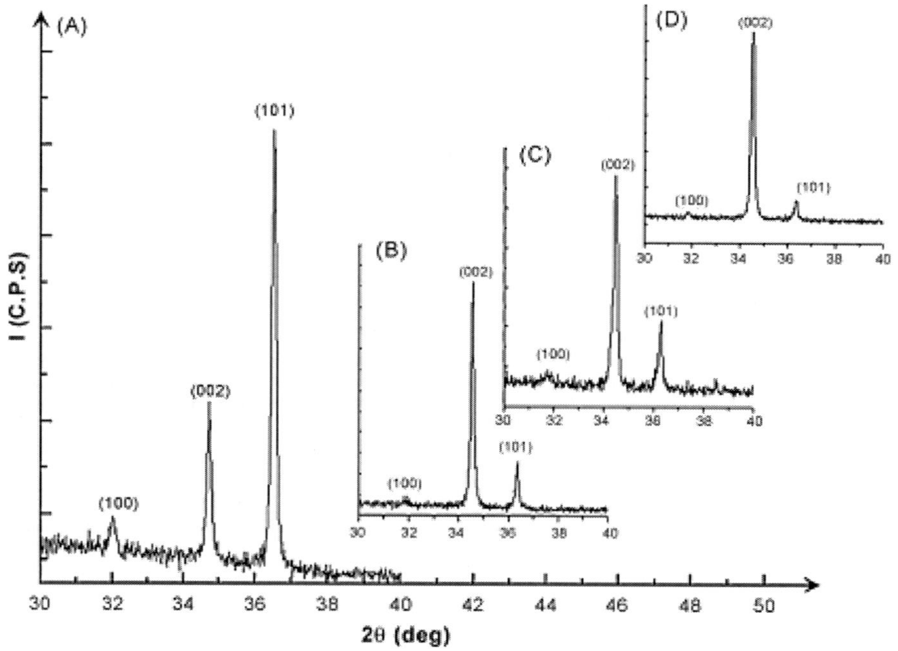

Figure 10. Ytterbium-doped ZnO nano-films XRD.

III.3. Ytterbium Doping Renewable Energy Related Features

Solar energy is one of the most promising renewable energy sources in the world. Recently, it has been proved that the introduction of Ytterbium in the synthesis process of preparation of highly ordered crystallized mesoporous TiO_2 enhaces drastically light conversion near-infrared photocatalytic activity. Ytterbium doping has been also used to realize practical, ultrashort-pulse laser systems. The main advantage of using Yb-based lasers is their simple energy-level structure. The related wavelengths of i. e. Yb:yttrium aluminum garnet (YAG) sources are around 940 and 1030nm. Due to this wavelength difference reduced size, which was also recorded in ZnO Yterbium-doped nano-films (Figure 12), the energy efficiency can be high [38-47].

Ytterbium-doped nano-devices have interesting properties. They have a very simple electronic level structure, with only one excited state manifold ($^2F_{5/2}$) within reach from the ground-state manifold ($^2F_{7/2}$) with near-infrared or visible photons.

Pumping and amplification involve transitions between different sublevels of the ground-state and excited-state manifolds. Quantum defects are usually small. However, complications can arise from the pronounced quasi-three-level behavior as the simple electronic structure excludes excited-state absorption and also a variety of detrimental quenching processes. The upper-state lifetimes are relatively long and gain bandwidth of the laser transitions is large.

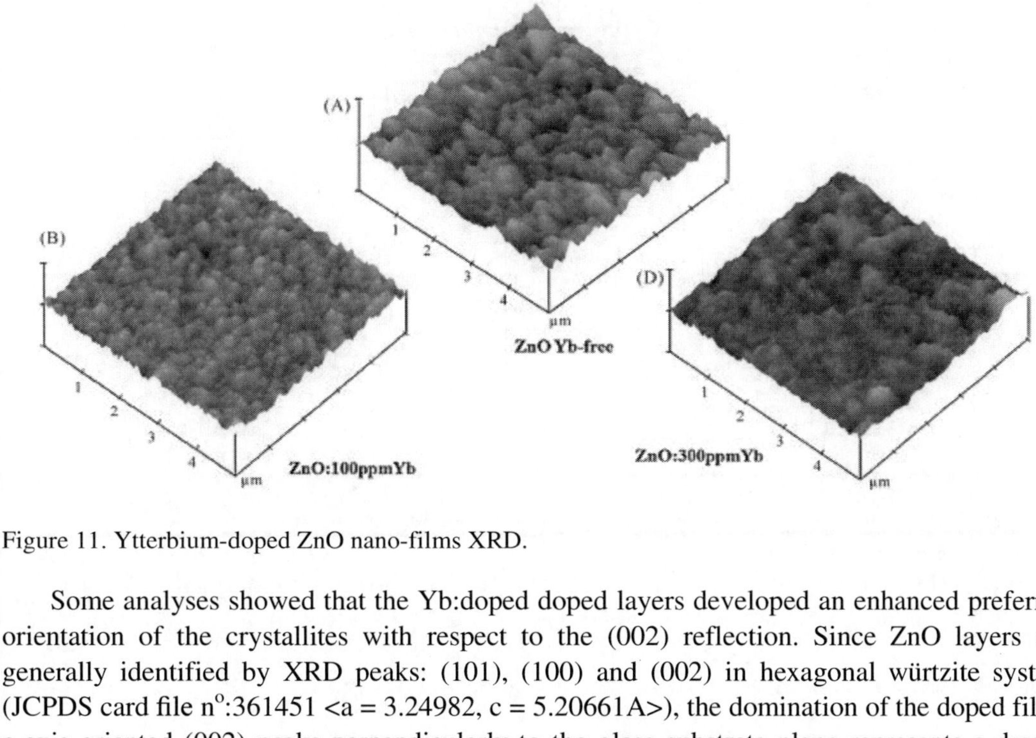

Figure 11. Ytterbium-doped ZnO nano-films XRD.

Some analyses showed that the Yb:doped doped layers developed an enhanced preferred orientation of the crystallites with respect to the (002) reflection. Since ZnO layers are generally identified by XRD peaks: (101), (100) and (002) in hexagonal würtzite system (JCPDS card file n°:361451 <a = 3.24982, c = 5.20661A>), the domination of the doped films c-axis oriented (002) peaks perpendicularly to the glass substrate plane represents a drastic optical enhancement. This feature has been confirmed by mechanical and opto-thermal measurements which recorded an increase of the microhadness of the doped layers along with an obvious decrease of the Amlouk-Boubaker opto-thermal expansivity ψ_{AB}.

The Opto-thermal expansivity ψ_{AB} have been defined as a synthetic physical parameter:

$$\psi_{AB} = D/\hat{\alpha} \qquad (1)$$

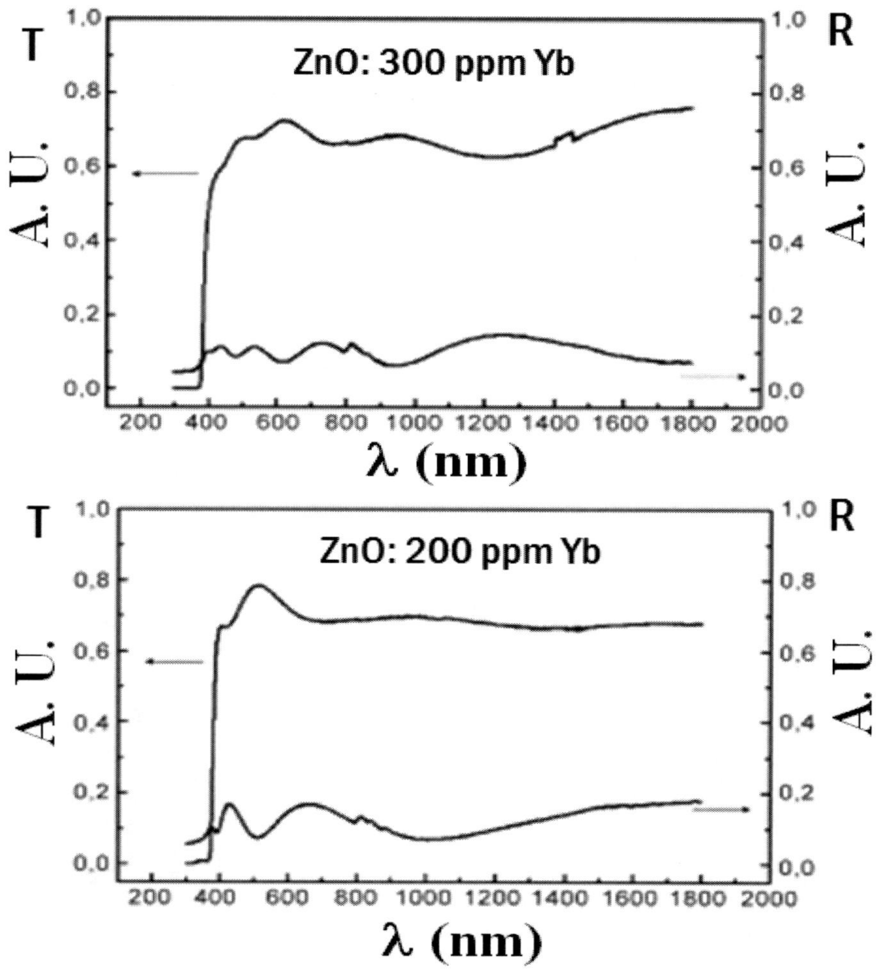

Figure 12. Ytterbium-doped ZnO nano-films Transmission-Reflection spectra.

where D is the thermal diffusivity and $\hat{\alpha}$ is the effective absorptivity, defined [48-51] as the mean normalized absorbance weighted by $I(\tilde{\lambda})_{\mathrm{AM1.5}}$, the solar standard irradiance, with $\tilde{\lambda}$: the normalised wavelength:

$$\hat{\alpha} = \frac{\int_0^1 I(\tilde{\lambda})_{\mathrm{AM1.5}} \times \alpha(\tilde{\lambda}) d\tilde{\lambda}}{\int_0^1 I(\tilde{\lambda})_{\mathrm{AM1.5}} d\tilde{\lambda}} \tag{2}$$

and where: $I(\tilde{\lambda})_{\mathrm{AM1.5}}$ is the Reference Solar Spectral Irradiance, fitted using the Boubaker Polynomials Expansion Scheme BPES.

Since Yb^{3+} performs close Oxygen-related electronegativity, the explanation of concentration-related stark differences of evolution of Urbach energy (Figure 13) and free carriers lifetime had been investigated from the side of the doping element distinctive physical properties [52-66].

In fact, the regular decrease of Urbach energy and disorder (Figure 13) for In-doped layers, for incrementally doped binary compounds, contrasts with the Yb-doped layers behavior [67-71]. A phenomenon of saturation seems to occur for increasing doping amounts inside these latter layers.

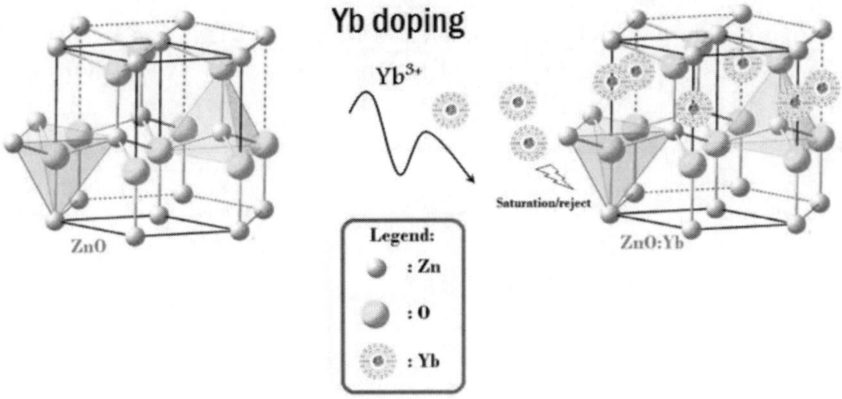

Figure 13. Ytterbium-doped ZnO nano-films Transmission-Reflection spectra.

In this context, nano-scale comparative estimation of atomic and ionic radii revealed an intriguing element: among the prospected entities, Ytterbium represents the highest radius, which is moreover comparable to upper ZnO-O bond dimension inside ZnO würtzite matrix (Figure 12). This analysis is referring to a strong dependence of the Urbach energy E_u versus electron-phonon and exciton-phonon interactions as well as crystalline structural static disorder, radiation inside the crystal and deviation from the ideal stoichiometry, it can consequently be suggested that other rare-earth elements incremental incorporation, doesn't induce any meaningful additional static disorder, due to size adequacy and ion-matrix compatibility, while high-sized atoms (like Ytterbium) cease to be incorporated at a given amount level, and hence disturb the global layer crystalline order and widen Urbach tails [71-85].

Conclusion and Perspectives

Nowadays, the worldwide scientific community should deal thoroughly with rare earth elements doping issues. The future worldwide consumption of rare earth elements is expected to increase, especially in renewable energy industry, automotive pollution catalysts, and nano-devices using permanent magnets. Rare earth applications are expected to need greater amounts of higher purity. Strong demand for ytterbium, erbium, cerium, lanthanum, and neodymium is expected to continue throughout the coming decades as related applications expands with population growth.

Although slight decline is forecast for phosphors used in CRTs and rechargeable nickel-hydride batteries using rare earth elements, as lithium ion batteries gain wider use in portable electronics, tangible growth is forecast for rare earth elements used in lasers, fiber optics, and other medical applications. On the other hand, demand is expected to decline over the next decade for REE, as flat-panel display use increases.

World rare earth elements will be soliciting new searches and investigations on competitiveness and properties of rare earth elements. The long-term outlook is for an increasingly research and technology for new rare earth elements applications which are expected to continue to be discovered and developed.

The intriguing problem of the doping-related Urbach tails distortion is also a long-term research field. Atomistic nano-scale considerations and some explanations have been proposed. Further studies on Vanadium (V) and Molybdenum (Mo) doped samples are actually in progress in order to confirm the suspected effects of ionic size on Urbach tails width and carriers lifetime.

REFERENCES

[1] Ferekids, C., Marinskaya, S., Marinskiy, D., Palekis, V., Morel, D. L. 2nd *world PSEC*, 1998, 1085.

[2] Maqbool, M., H. H. Richardson and M. E. Kordesch, Electron penetration depth in amorphous AlN exploiting the luminescence of AlN:Tm/AlN:Ho bilayers *Current Applied Physics* 9, 417-421 (2009).

[3] Dimitrova, V., P. G. Van Patten, H. Richardson and M. E. Kordesch, Photo-cathodo-, and electroluminescence studies of sputter deposited AlN: Er thin films, *Applied Surface Science* 175-176, 480-483 (2001).

[4] M. L. Caldwell, A. L. Martin, C. M. Spalding, V. I. Dimitrova, P. G. Van Patten, M. E. Kordesch and H. H. Richardson, Visible emission from amorphous AlN thin film phosphors with Cu, Mn, or Cr, *J. Vac. Sci. Technol.* A 19, 1894-1897(2001).

[5] F. S. Liu, Q. L. Liu, J. K. Liang, J. Luo, H. Zhang, Y. Zhang, B. Sun, and G. Rao, Journal of visible and infrared emissions from c-axis oriented AlN:Er films grown by magnetron sputtering, *Applied Physics* 99, 053515 (2006).

[6] M. E. Little and M. E. Kordesch, Band-gap engineering in sputter-deposited $Sc_xGa_{1-x}N$ *Appl. Phys. Lett.* 78, 2891-2893 (2001).

[7] John B. Gruber et al., Spectroscopic properties of Sm^{3+} ($4f^5$) in GaN. *Journal of Applied Physics*, Volume 91, No. 5, 2929-2935 (2002).

[8] H. J. Lozykowski, Kinetics of luminescence of isoelectronic rare-earth ions in III-V Semiconductors. *Physical Review* B 48, 17758 (1993).

[9] Rusu, M., Sadewassar, S., Glatzel, T., Gashin, T., Simashkevich, A., Jorger-waldau, A. *Thin Solid Films*, 2002, 403-404, 344.

[10] Armstrong, S., Datta, P. K., Miles, R. W. *Proceedings of 17th European Photovoltaic Solar Energy Conference*, 22-26 October, 2001, 1184.

[11] Nakanishi, K., Suemune, I., Fuji, Y., Kuroda, Y., Yamanishi, M., *Jpn. J. Appl. Phys.*; 1991, 30, 1399.
[12] Xanthos, M. *Functional Fillers for Plastics*, Wiley-VCH, publications, ISBN 3527604421, 2006, 9783527604425.
[13] Saez-Araoz, R., Abou-Ras, D., Niesen, T. P., Neisser, A., Wilchelmi, K., Lux-Steiner, M. Ch. Ennaoui, A. In situ monitoring the growth of thin film ZnS/Zn(S,O) bilayer on Cu-Chalcopyrite for high performance thin solar cells, *Thin Solid Films*, (article in press 2008).
[14] Kazmerski, L. L., White, F. R., Morgan, G. K. Thin films $CuInSe_2$/CdS heterojunction solar cells. *Appl. Phys. Lett.*, 1976, 29 268-270.
[15] Kazmerski, L. L., Wagner, M. S. *Cu-Ternary chalcopyrite solar cells*, Academic Press, New York, 1985.
[16] Braunger, D., Hariskos, D., Walter, T., Schock, H. W., An 11.4% efficient polycrystalline thin film solar cell based on CuInS2 with a Cd-free buffer layer, Sol. *Energy Materials and Solar Cells*, 1996, 40, 97.
[17] Shay, J., Tell, B., Schiavone, L. M., Kasper, H. M., Tjiel, F. Energy bands of $AgInS_2$ in the chalcopyrite and orthorhombic structures, *Physical Review B*, *9*, 1973, n° 4, 1719.
[18] Ortega-Lopez, M., Vigil-Galan, O., Cruz Gandarilla, F., Solorza-Feria, O. Materials Research Bulletin. *Preparation of $AgInS_2$ chalcopyrite thin films by chemical spray pyrolysis*, 2003, 38, 55.
[19] Ortega-Lopez, M., Morales-Acevedo, A., Solorza-Feria, O. Physical properties of $AgInS_2$ films prepared by chemical spray pyrolysis, *Thin Solid Films*, 2001, 385, 120.
[20] Fiechter, S., Martinez, M., Schmidt, G., Henrion, W., Tomm, Y. Phase relations and optical properties of semiconducting ternary sulfides in the system Cu-Sn-S, *Journal of Physics and Chemistry of Solids*, 2003, 64, 1859.
[21] Bin, Li, Xie, Yi. Jiaxing Huang and Yitai Qian, Synthesis, Characterization, and Properties of Nanocrystalline Cu_2SnS_3, *Journal of Solid State Chemistry*, 2000, 153, 170.
[22] Xiangying Chen, Xiong Wang, Changhua An, Jianwei Liu, Yitai Qian, Preparation and characterization of ternary Cu-Sn-E (E = S, Se) semiconductor nanocrystallites via a solvothermal element reaction route, *Journal of Crystal Growth*, 2003, 256, 368.
[23] Mitsuko Onoda, Xue-an Chen, Akira Sato, Hiroaki Wada, Structure and twinning of monoclinic Cu_2SnS_3, *Materials Research Bulletin Crystal*, 2000, 35, 1563.
[24] Qin, Li, Yi Ding, Xianming Liu, Yitai Qian, Preparation of ternary I-IV-VI nanocrystallines via a mild solution route, *Materials Research Bulletin*, 2001, 36, 2649.
[25] Chamberlin, R., Skarman, H. *J. Electrochem. Soc.*, 1966, 113, 86
[26] Chaouachi, A., Boubaker, K., Bouzouita, H., Amlouk, M. Enhancement of pyrolysis spray disposal performance using thermal time-response to precursor uniform deposition, *European Physical Journal of Physics*: *Applied Physics*, 2007, 37, 105-109.
[27] Gupta, P., Bhattacharyya, D., Chaudhuri, S., Pal, A. K. *Thin Solid Films*, 1992, 221, 154.
[28] Venkata Subbaiah, Y. P., Ramakrishna Reddy, K. T. *Materials Chemistry and Physics*, 2005, 92, 448.
[29] Armstrong, S., Datta, P. K., Miles, R. W. *Proceedings of 17th European Photovoltaic Solar Energy Conference*, 22-26 October 2001, 1184.

[30] Abo-Hassan, K. M. M., Muhamad, M. R., Radhakrishna, S. *Thin Solid Films*, 2005, 491, 117.

[31] Van Vechten, J. A., Bergstresser, T. K. *Phys. Rev. B*, 1970, 1, 3351.

[32] Belgacem, S., Bennaceur, R., *Rev. Phys. Appl.* 1990, 25, 1245.

[33] Kamoun, N., Belgacem, S., Amlouk, M., Bennaceur, R., Bonnet, J., Touhari, F., Nouaoura, M., Lassabatère, L. Structure, surface composition, and electronic properties of β-In_2S_3 and β-$In_{2-x}Al_xS_3$, *Journal of Applied Physics*, 2001, 89, 2766-2771.

[34] Ben Nasrallah, T., Amlouk, M., Bernède, J. C., Belgacem, S. Structure and morphology of sprayed ZnS thin films, *Phys. Stat. Sol. (a)*, 201, No 14 (2004) 3070-3076.

[35] Zouaghi, M. C., Ben Nasrallah, T., Amlouk, M., Belgacem, S. *Proc. 19th European Photovoltaic Solar Energy Conference*, 7 - 11 June, 2004, Paris, France, Vol. II, 1764.

[36] Aissa, Z., Ben Nasrallah, T., Amlouk, M., Bernède, J. C., Belgacem, S. Some physical investigations on $AgInS_2$ sprayed thin films *Solar Energy Materials and Solar cells*, 2006, 90, 1136-1146.

[37] Aissa, Z., Amlouk, M., Ben Nasrallah, T., Bernède, J. C., Belgacem, S. Effect of S/In concentration ratio on the physical properties of $AgInS_2$ spayed thin films, *Solar Energy Materials and Solar cells*, 2007, 91, 489-494.

[38] H. Labiadh, K., Boubaker, A Sturm-Liouville shaped characteristic differential equation as a guide to establish a quasi-polynomial expression to the Boubaker polynomials, *Differential Equations and Control Processes*, 2007, 2, 117-133.

[39] Roger, L., Bagula, Gary Adamson, Triangle of coefficients of Recursive Polynomials for Boubaker polynomials, OEIS *(Encyclopedia of Integer Sequences)*, A137276 (2008).

[40] Neil, J. A. Sloane, Triangle read by rows of coefficients of Boubaker polynomial B_n(x) in order of decreasing exponents, OEIS *(Encyclopedia of Integer Sequences)*, 2008, A138034.

[41] Slama, S., Bessrour, J., Karem, B., Bouhafs, M. Investigation of A3 point maximal front spatial evolution during resistance spot welding using 4q-Boubaker polynomial sequence, *Proceedings of Cotume*, 2008, 79-80,

[42] Labiadh, H., Dada, M., Awojoyogbe, O. B., Ben Mahmoud, K. B., Bannour, A. Establishment of an ordinary generating function and a Christoffel-Darboux type first-order differential equation for the heat equation related Boubaker-Turki polynomials, *Differential Equations and Control Processes*, 2008, 1, 51-66.

[43] The Boubaker-Tuki polynomials (or Modified Boubaker polynomials), Planet-Math Encyclopedia, The Mathematics worldwide Encyclopedia (available also online at: http://planetmath.org/encyclopedia/ *Boubaker TurkiPolynomials.html.*

[44] Abbot, C. G., Fowle, F. E., Aldrich, L. B. *Annals of the Astrophysical Observatory of the Smithsonian Institution*, 1913, 3, 22, Plate 1.

[45] Kimball, H. H. Variations in the total solar radiation intensities measured at the surface of the earth. *Monthly Weather Review*, 1924, 52, 527-29.

[46] La Cour, D., Stagg, J. M. *The International Polar Year*, 1932-1933. Commission Intern. de 1 'Annee Polaire, Lettre Circ. XVII, appendix α.

[47] Nagae, S., Toda, M., Minemoto, T., Takakura, H., Hamakawa, Y. Evaluation of the impact of solar spectrum and temperature variations on output power of silicon-based

photovoltaic modules, *Solar Energy Materials and Solar Cells*, 90, Issue, 2006, 20, 3568-75

[48] Luque, A., Martí, A., Bett, A., Andreev, V. M., Jaussaud, C., van Roosmalen, J. A. M., Alonso, J., Räuber, A., Strobl, G., Stolz, W., Algora, C., Bitnar, B., Gombert, A., Stanley, C., Wahnon, P., Conesa, Wgjhm van Sark, J. C., Meijerink, A., van Klink, G. P. M., Barnham, K., Danz, R., Meyer, T., Luque-Heredia, I., Kenny, R., Christofides, C., Sala, G., Benítez, P., Fullspectrum. a new PV wave making more efficient use of the solar spectrum, Solar Energy *Materials and Solar Cells*, 2005, 87, Issues, 1-4, 467-79

[49] Tinchev, S. S., Nikolova, P. I., Dyulgerska, J. T., Danev, G., Babeva, T., a-CH. absorber layer for solar cells matched to solar spectrum, *Solar Energy Materials and Solar Cells*, 86, Issue, 2005, 3, 421-6.

[50] Gottschalg, R., Infield, D. G., Kearney, M. J. Experimental study of variations of the solar spectrum of relevance to thin film solar cells, *Solar Energy Materials and Solar Cells*, 79, Issue, 2003, 4, 527-37

[51] Moss, T. S. Optical Properties of Semiconductors, *Butterworth, London*, 1959.

[52] Amlouk, M., Dachraoui, M., Belgacem, S., Bennaceur, R. Structural, Optical and Electrical Properties of SnO_2:F and CdS Airless Sprayed Layers, *Solar Energy Materials*, 1987, 15, 453-461.

[53] Belgacem, S., Amlouk, M., et Bennaceur, R. Effet du Rapport Cu/In sur la structure des Couches Minces de $CuInS_2$ Airless Spray. *Application: Conversion Photovoltaique, Revue Phys. Appl.*, 1990, 25, 1213-1223.

[54] Amlouk, M., Belgacem, S., Bouchriha, H., Bnnaceur, R. Effect of ZnO thin layer on the photovoltaic characteristics of CdS/$CuInS_2$ solar cells, *Proc. 12th European Photovoltaic Solar Energy Conference*, 11-15 April 1994, Amsterdam The Netherlands, Vol.2, p. 1519.

[55] Kamoun, N., Belgacem, S., Amlouk, M., Bennaceur, R., Abdelmoula, K., et Bel Haj Amara, A. Caractérisations Structurale et Morphologique des Couches Minces de $CuInS_2$ et d'In-S Airless Spray, *J. Phys. III Fr.*; 1994, 4, 473-491.

[56] Amlouk, M., Brunet, N., Cros, B., Belgacem, S., et Barjon, D. Etude par Microscopie Acoustique de Couches Minces de Ag_2S déposées par Spray, *J. Phys. III Fr.*; 1997, 7, 1741-1753.

[57] Belgacem, S., et Bennaceur, R. Propriétés optiques des couches minces de SnO_2 et $CuInS_2$ airless spray, *Revue Phys. Appl.*; 1990, 25, 1245.

[58] Bhira, L., Amlouk, M., Belgacem, S., Bennaceur, R., Barjon, D. Propriétés Structurales et Optiques des Composés quaternaires de $CuIn_{1-x}Al_xS_2$ Spray, *Physica Status Solidi (a)*, 1998, 165, 141-150.

[59] M. Maqbool, E. Wilson, J. Clark, I. Ahmed and A. Kayani, Luminescence from Cr(+3) doped AlN films deposited on optical fiber silicon substrates for use as waveguides and laser cavities, *Applied Optics* 49, 653 (2010).

[60] N. Managaki, M. Fujii, T. Nakamura, Y. Usui, and S. Hayashi, Enhancement of photoluminescence from Yb and Er co-doped Al2O3 films by an asymmetric metal cavity, *Appl. Phys. Lett.* 88, 042101-042103 (2006).

[61] Y. Mao, J. Y. Huang, R. Ostroumov, K. L. Wang, and J. P. Chang, Synthesis and Luminescence Properties of Erbium Doped Y2O3 Nanotubes, *J. Phys. Chem.* 112, 2278 (2008).

[62] A. J. Steckl and R. Birkhahn, Visible emission from Er-doped GaN grown by solid source molecular beam epitaxy, *Appl. Phys. Lett.* 73, 1700-1702 (1998).

[63] M. A. Ahmed, E. Ateia, S. I. El-Dek, *Fourth International Scientific Conference (F.I.S.C)* (Science, Development and Environment), Al Azhar University, Egypt (2001 March).

[64] Dlala, H., Amlouk, M., Belgacem, S., Girard, P., Barjon, D. Structural and Optical Properties of Ag_2S Thin Films Prepared by Spray Pyrolysis, *European Physical Journal. Appl. Phys.*; 1998, 2, 13-16.

[65] Kamoun, N., Bennaceur, R., Amlouk, M., Belgacem, S., Mliki, N., Frigerio, J. M., Theye, M. L. Optical Properties of InS Layers Deposited Using an Airless Spray Technique, *Physica Status Solidi (a)*, 1998, 169, 97-104.

[66] Michael, E. Levinshtien, S. L. Rumyantsev and M. S. Shur. *Properties of Advances Semiconductor Materials*, John Wiley and Sons, Inc. 2001.

[67] M. L. Caldwell, A. L. Martin, V. I. Dimitrova, P. G. Van Patten, M. E. Kordesch and H. H. Richardson. Emisson properties of amorphous AlN: Cr^{+3} thin film phosphor. *Applied Physics Letters* 78, 1246 (2001).

[68] S. C. Jain, III-nitrides: growth, characterization, and properties. M. Willander, J. Narayan, R. Van Overstraeten, *Journal of Applied Physics* 87, 965 (2000).

[69] H. J. Lozykowski, W. M. Jadwisienczak, I. Brown. Photoluminescence and cathodeluminescence of GaN dope with Pr. *Journal of Applied Physics* 88, 210 (2000).

[70] E. Acar, T. G. Kolda and D. M. Dunlavy. An Optimization Approach for Fitting Canonical Tensor Decompositions. *Technical Report Number SAND2009-0857*, Sandia NationalLaboratories, Albuquerque, NM and Livermore, CA, February 2009.

[71] L. De Lathauwer, B. De Moor, J. Vandewalle. A multilinear singular value decomposition. *SIAM J. Matrix Anal. Appl.,* 21 (2000) 1253-1278.

[72] I. P. Gavrilyuk, W. Hackbusch and B. N. Khoromskij. Hierarchical Tensor-Product Approximation to the Inverse and Related Operators in High-Dimensional Elliptic Problems. *Computing* 74 (2005), 131-157.

[73] I. P. Gavrilyuk, W. Hackbusch and B. N. Khoromskij. Data-sparse approximation of aclass of operator-valued functions. *Math. Comp.* 74 (2005), pp. 681-708.

[74] H.-J. Flad, W. Hackbusch, B. N. Khoromskij, and R. Schneider. Concept of Data-Sparse Tensor-Product Approximation in Many-Particle Modeling. In: *"Matrix Methods: Theory, Algorithms, Applications"*, pp. 313-347, World Scientific publishing, Singapoure,2010.

[75] W. Hackbusch, B. N. Khoromskij and E. E. Tyrtyshnikov. Hierarchical Kronecker tensorproduct approximations. *J. Numer. Math.* 13 (2005), 119-156.

[76] Piper, W. W., DeLuca, J. A., Ham, F. S. Cascade fluorescent decay in Pr 3+

[77] -doped fluorides: Achievement of a quantum yield greater than unity for emission of visible light. *J. Lumin.* 1974, 8 344–348.

[78] Wegh, R. T., Donker, H., Meijerink, A., Lamminmaäki, R. J., Hölsä, J. Vacuum-ultravioletspectroscopy and quantum cutting for Gd3+ in LiYF4. *Phys. Rev.* B. 1997, 56, 13841–13848.

[79] Wegh, R. T., Donker, H., Oskam, K. D., Meijerink, A. Visible quantum cutting in LiGdF4:Eu3+through downconversion. *Science* 1999, 283, 663–666.

[80] B. M., van der Ende, L., Aarts, A. Meijerink. Near-Infrared Quantum Cutting for Photovoltaics. *Adv. Mater.* 2009, 21, 3073–3077.

[81] Vergeer, P., Vlugt, T. J. H., Kox, M. H. F., den Hertog, M. I., van der Eerden, J. P. J. M., Meijerink, A Quantum cutting by cooperative energy transfer in YbxY1-xPO4 : Tb3+. *Phys. Rev*. B 2005, 71014119.

[82] Song, Y., Bin, Z., Jingxin, C., Jin, L., Jian Rong, Q. Infrared quantum cutting in Tb3+,Yb3+ codoped transparent glass ceramics containing CaF2 nanocrystals. *Appl. Phys. Lett*. 2008, 92, 141112.

[83] Zhang, Q. Y., Yang, G. F., Jiang, Z. H. Cooperative downconversion in GdAl3(BO3)4:RE3+, Yb3+(RE = Pr, Tb, and Tm). *Appl. Phys. Lett*. 2007, 91, doi: 10.1063/1.2757595.

[84] Taylor, John R. *An Introduction to Error Analysis: The Study of Uncertainties in Physical Measurements*. University Science Books, 1982.

[85] N. Arora, D. Martins, D. Ruggerio, E. Tousimis, A. J. Swistel, M. P. Osborne, R. M. Simmons, Effectiveness of a noninvasive digital infrared thermal imaging system in the detection of breast cancer, *The American Journal of Surgery* 196, 523-526 (2008).

INDEX

A

B

C

D

E

F

G

H

I

K

L

M

N

O

P

Q

R

S

T